About Island Press

About SCOPE

SCOPE 63

Invasive Alien Species

THE SCIENTIFIC COMMITTEE ON PROBLEMS OF THE ENVIRONMENT (SCOPE)

SCOPE Series

SCOPE 1–59 in the series were published by John Wiley & Sons, Ltd., U.K. Island Press is the publisher for SCOPE 60 and subsequent titles in the series.

SCOPE 60: *Resilience and the Behavior of Large-Scale Systems,* edited by Lance H. Gunderson and Lowell Pritchard Jr.

SCOPE 61: *Interactions of the Major Biogeochemical Cycles: Global Change and Human Impacts,* edited by Jerry M. Melillo, Christopher B. Field, and Bedrich Moldan

SCOPE 62: *The Global Carbon Cycle: Integrating Humans, Climate, and the Natural World,* edited by Christopher B. Field and Michael R. Raupach

SCOPE 63: *Invasive Alien Species: A New Synthesis,* edited by Harold A. Mooney, Richard N. Mack, Jeffrey A. McNeely, Laurie E. Neville, Peter Johan Schei, and Jeffrey K. Waage

SCOPE 64: *Sustaining Biodiversity and Ecosystem Services in Soils and Sediments,* edited by Diana H. Wall

SCOPE 65: *Agriculture and the Nitrogen Cycle: Assessing the Impacts of Fertilizer Use on Food Production and the Environment,* edited by Arvin R. Mosier, J. Keith Syers, and John R. Freney

SCOPE 63

Invasive Alien Species

A New Synthesis

Edited by
Harold A. Mooney, Richard N. Mack,
Jeffrey A. McNeely, Laurie E. Neville,
Peter Johan Schei, and Jeffrey K. Waage

A project of SCOPE, the Scientific Committee on Problems of the Environment on behalf of the Global Invasive Species Program, with partners CAB International (CABI) and the World Conservation Union (IUCN)

Washington • Covelo • London

Library of Congress Cataloging-in-Publication data.

Invasive alien species : a new synthesis / edited by Harold A. Mooney ... [et al.].
p. cm.
Includes bibliographical references and index.
ISBN 1–55963–362–X (cloth : alk. paper) — ISBN 1–55963–363–8 (pbk: alk. paper)
1. Introduced organisms. 2. Biological invasions. I. Mooney, Harold A.
QH353.I5825 2005
578.6'2—dc22

British Cataloguing-in-Publication data available.

Printed on recycled, acid-free paper 

Manufactured in the United States of America
10 9 8 7 6 5 4 3 2

Contents

List of Figures, Tables, and Boxes

Figures

Tables

Boxes

Preface

Between 1982 and 1988 the Scientific Committee on Problems of the Environment (SCOPE) engaged a large number of scientists in an effort to document the nature of the invasive alien species (IAS) phenomenon. The results of this effort appeared in a number of books, including a synthesis titled *Biological Invasions: A Global Perspective* published by John Wiley in 1989. This synthesis clearly established that IAS could have major impacts on ecosystem functioning and that nearly every ecosystem has been affected by IAS, even those protected for conservation purposes. The studies also pointed out that advances in international transportation are enabling humans to establish, often unwittingly, a new biotic order on Earth as biogeographic barriers to migration are overcome. Although the SCOPE program was successful scientifically, it did not offer much practical advice to land managers or nations except to inform them that they were not alone in dealing with IAS problems.

After the 1996 Norway/United Nations Conference on Alien Species held in Trondheim, it was determined that developing a global strategy to address this issue was of paramount importance. In 1997, SCOPE, along with partners from the World Conservation Union (IUCN) and the Centre for Agriculture and Biosciences International (CABI), embarked on a new program on IAS, this time with the explicit objective of providing new tools for understanding and dealing with IAS. This initiative evolved as the Global Invasive Species Program (GISP). The GISP differed substantially from the previous program in that it was designed to engage the many constituencies involved in addressing the problem, including natural and social scientists, educators, lawyers, resource managers, and leaders from both industry and government.

The first phase of GISP was designed with 10 elements, each a building block in a comprehensive approach for dealing with IAS. Each of these contributed to building the comprehensive approach that is needed for dealing with invasive species (Figure P.1). Four elements addressed synthesizing our current knowledge of invasive species. These include an update and analysis of our knowledge of the ecology of invasive species, the current status of invasive species and new methods for assessing changing distributions, how society views and values invasive species, and how global change will influence the success of invaders. The elements were coordinated as follows.

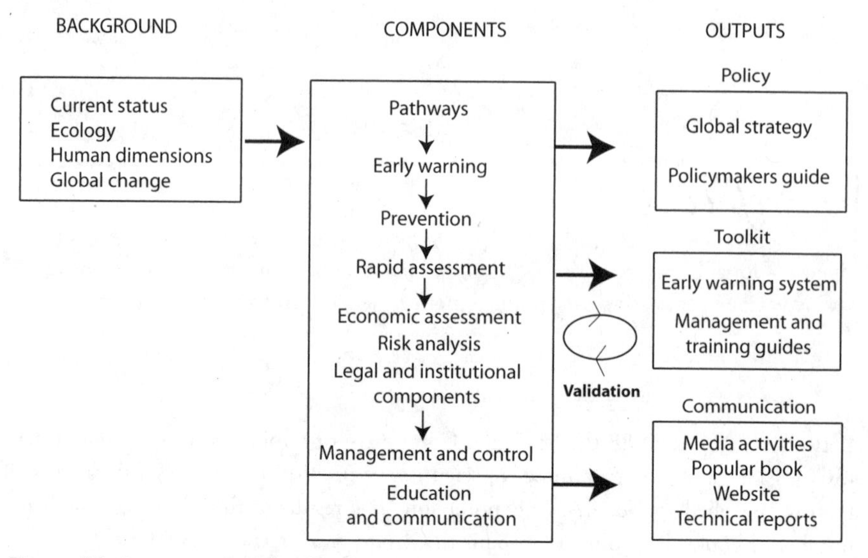

Figure P.1. Structure of GISP Phase I

Biological and Socioeconomic Syntheses

- Update and analysis of our knowledge of the ecology of IAS (Marcel Rejmánek and David Richardson)
- Analysis of the current status of IAS and new methods for assessing their changing distributions and abundance (Mark Lonsdale and Richard Mack)
- Analysis of societal views and values of IAS (Jeff McNeely)
- Analysis of how global change will influence the success of invaders (Richard Hobbs and Harold Mooney)

Policy and Management Syntheses

- Assessment of the best practices for prevention and management of IAS (Jeff Waage)
- Analysis of invasion pathways created by trade and new tools for risk assessment (James Carlton, Gregory Ruiz, and David Andow)
- Assessment of the economic consequences and tools for addressing IAS (Charles Perrings and Mark Williamson)
- Analysis of the legal and institutional frameworks for dealing with IAS (Nattley Williams)
- Development of a pilot database on IAS with early warning capabilities (Mick Clout)
- Development of new approaches for educating the general public on the potential dangers of invasive species (Alan Holt)

Each element was addressed through one or more workshops, resulting in specific products. All of the activities were summarized at a Phase I Synthesis Conference in Cape Town, Republic of South Africa, September 18–22, 2000.

The Phase I Synthesis Conference represented a collective forum convened on behalf of the efforts of the Phase I initiatives. This conference provided an opportunity for experts, governments, nongovernment organizations, and intergovernment organizations to engage. The participants clarified the initial findings, further developed a Global Strategy on Invasive Alien Species, and continued to summarize and refine various prevention and management tools and practices. The group explored future priorities and concerns focusing on a plan for building capacity and fostering international cooperation to address invasive species. This latter process identified areas for further development in management of the problem on a global scale. The final products of GISP Phase I represent an international consensus on the state of knowledge and applications. In addition, a global database has been established, and a popular volume on biological invasions (Baskin 2002) has been published highlighting invasive species issues with special appeal to a broad, nonscientific audience.

GISP has produced other publications for various audiences; these publications provide detailed information and guidance.

Richard N. Mack
Department of Biological Sciences, Washington State University
Pullman, Washington

Laurie E. Neville
Department of Biological Sciences, Stanford University
Stanford, California

John W.B. Stewart
SCOPE Editor-in-Chief
Salt Spring Island, BC, Canada

SCOPE Secretariat
51 Boulevard de Montmorency, 75016 Paris, France
Véronique Plocq Fichelet, Executive Director

References

Baskin, Y. 2002. *A Plague of Rats and Rubbervines: The Growing Threat of Species Invasions.* Shearwater Books/Island Press, Washington, DC.

Lowe, S., M. Browne, S. Boudjelas, and M. De Poorter. 2001. *100 of the World's Worst Invasive Alien Species: A Selection from the Global Invasive Species Database.* IUCN-ISSG, Gland, Switzerland.

McNeely, J. A., ed. 2001. *The Great Reshuffling: Human Dimensions of Invasive Alien Species.* IUCN, Gland, Switzerland.

McNeely, J. A., H. A. Mooney, L. E. Neville, P. Schei, and J. K. Waage, eds. 2001. *Global Strategy on Invasive Alien Species.* IUCN on behalf of the Global Invasive Species Program, Gland, Switzerland.

Perrings, C., M. Williamson, and S. Dalmazzone, eds. 2000. *The Economics of Biological Invasions.* Edward Elgar, Cheltenham, UK.

Ruiz, G., and J. T. Carlton. 2003. *Invasive Species. Vectors and Management Strategies.* Island Press, Washington, DC.

Shine, C., N. Williams, and L. Gundling. 2000. *A Guide to Designing Legal and Institutional Frameworks on Alien Invasive Species.* IUCN, Gland, Switzerland.

Wittenberg, R., and M. J. W. Cock. 2001. *Invasive Alien Species: A Toolkit of Best Prevention and Management Practices.* CAB International on behalf of the Global Invasive Species Program, Wallingford, Oxon, UK.

Acknowledgments

GISP has received support from the Global Environmental Facility, United Nations Environment Programme, with additional support from the United Nations Educational, Scientific and Cultural Organization, the Norwegian government, the government of Brazil, the International Council for Science, the National Aeronautics and Space Administration, La Fondation TOTAL, The David and Lucile Packard Foundation, the U.S. Department of State Office of International Environmental and Scientific Affairs Initiative, the John D. and Catherine T. MacArthur Foundation, the European Commission, the government of Denmark, the government of New Zealand, USAID, U.S. Fish and Wildlife Service, U.S. Bureau of Land Management, U.S. Department of Interior/Office of Insular Affairs, and the Aquatic Nuisance Species Task Force. GISP is a component of DIVERSITAS, an international program on biodiversity science and a project of the Scientific Committee on Problems of the Environment, Cluster I project series titled "Managing Societal and Natural Resources." We are grateful to the participating organizations and to the dedicated groups and individuals who have made substantial in-kind contributions.

—HAROLD A. MOONEY AND LAURIE E. NEVILLE

1

Invasive Alien Species: The Nature of the Problem

Harold A. Mooney

The increasing human population is altering the natural resources on which our societies depend to an ever-greater extent. Many of these changes are purposeful and to the benefit of society. Others, although purposeful, have inadvertent negative impacts on the goods and services that natural resources deliver to society. In order to manage these resources in a sustainable manner we must understand the interactions and trade-offs between resource alteration and the natural, generally renewable processes on which we depend. This book addresses one particular driver of resource alteration: alien species invasions. In aggregate, these invasions are global in extent and are having consequences that are generally unappreciated but quite threatening to many human activities.

The vast numbers of species that populate the earth provide innumerable goods and services that society values. Equally important for society are the services that natural systems provide free of charge (Daily et al. 1997). On the other hand, invasive alien species can represent " ecosystem bads and disservices" (as characterized by Madhav Gadgil, 2000: 16) to systems on which society depends. In this introduction I concentrate not only on how invasive aliens alter ecosystem properties but also more directly on their effects on goods and services valued by society. This information provides a backdrop to the chapters that follow, which focus on what to do about this pervasive problem.

For comprehensive overviews of the problem there are a number of recent summaries (Vitousek et al. 1997; Lonsdale 1999; Parker et al. 1999; Williamson 1999; Mack et al. 2000; D'Antonio et al., 2004), edited volumes (Sandlund, Schei, and Viken 1999; Mooney and Hobbs 2000), and popular books (Bright 1998; Devine 1998).

What Do We Know in General?

The reviews just listed give us some general conclusions about the status and impacts of alien species that can be summarized as follows:

- There has been a massive global mixing of biota.
- This mixing has been both purposeful and accidental.
- There has been both biotic enrichment and impoverishment in any given area (species view).
- A small fraction of alien species have become invasive.
- Invasive alien species come from all taxonomic groups.

We know with less precision the kinds of habitats in which invasive alien species are most successful, the traits of successful invaders, and the mechanisms of habitat degradation caused by invaders. We do know that invasive alien species have altered evolutionary trajectories, can disrupt community and ecosystem processes, are causing large economic losses, and threaten human health and welfare.

Character of the New Biotic World

It is easy to demonstrate that the nature of the biological world is very different from what it was before the age of exploration. The natural ecosystems that evolved in isolation on the various continents and large islands, constrained by biogeographic barriers such as oceans, have become functionally connected through the capacity of humans to transport biological material long distances in a short amount of time. The consequences of this biotic exchange are staggering when you tally up what the biotic world looks like now in comparison with the recent past. In Hawaii, a prime example of the onslaught of alien species, there are 3,500 more species of flowering plants and insects than there were before the age of exploration. Of the flowering plants, there are more alien species than endemics (Eldredge and Miller 1997). In California, more than 1,000 established alien plant species have been added to the approximately 6,300 natives (contrast this to the 64 plant taxa that are threatened or extinct) (Hobbs and Mooney 1998).

Looking across the world, Hawaii is at the extremes of biotic introductions, as are many other islands, and California probably is somewhere in the middle. However, a survey of the number of alien plants in various parts of the world shows that established alien species are everywhere. For plants alone, the following numbers of established alien species have been noted in the large continental areas of the Russian Arctic, 104; Europe, 721; tropical Africa, 536; southern Africa, 824; Canada, 940; continental United States, 2,100; Chile, 678; and Australia, 1,952. For islands, New Zealand has 1,623, the British Isles 945, and the Canary Islands 680. These numbers indicate the extent of the changes in biotic systems that have occurred. Similar numbers, at least proportionately to natural abundance, can be seen for other taxonomic groups (Vitousek et al. 1997).

How Fast Has All of This Happened?

The exchange rate of biological material across biogeographic barriers that have separated continents for millions of years has been extremely low until very recently. Similarly, climate has been fairly constant in recently millennia. However, both climate change, as driven by the changing composition of the atmosphere, and the large-scale intercontinental movement of biological material have greatly accelerated in recent times. To get a sense of the comparative rates of change in atmospheric composition over the past 200 years and in biotic composition caused by biotic introductions, one can examine the detailed records of the changing CO_2 concentration of the atmosphere. In 1800 it was 280 ppm, and in 1990 it had increased to 354 ppm, an increase of 26 percent (Boden et al. 1994). In contrast, to use one well-documented case, the numbers of established alien arthropods in the United States grew from 50 in 1800 to 2,000 in 1990 (Sailer 1983; U.S. Congress 1993: 1263), a forty-fold increase! This is not to say that the potential global environmental consequences of these changes are equivalent, but it does indicate the extent of biotic interchange and the large change in the biotic composition of the earth.

There are indications that the rate of exchange for certain groups of organisms is accelerating. Cohen and Carlton (1998) have shown this for organisms introduced into the San Francisco Bay and Nico and Fuller (1999) for fishes in the United States.

Success of Invasive Alien Species

Only a small fraction of the alien biotic material that lands in new territory actually becomes established, and even a smaller fraction becomes a serious problem (Williamson 1996). However, given the great numbers of alien species imports, the actual numbers of species that become established and do harm can be large. For example, of the 1,500 species of foreign insects that had become part of the insect fauna of the United States 20 years ago, 235 (16 percent) had become pests (Pimentel 1993).

Of course, native species also can harm the biotic systems on which we depend. In an analysis of the numbers of pest species, Pimentel (1993) shows that 73 percent of the 80 major crop weeds of the United States came from other regions, whereas of the 110 pasture weeds 41 percent were from out of the country, and of 70 major forest pests only 23 percent were invaders.

Complex Species Interactions Are Modified by Invasive Alien Species

It has been postulated that one reason for the great success of invasive species is the fact that they escape their natural predators when entering a new biogeographic region. This undoubtedly is the case in many instances, but experimental evidence is not well developed. Some have even postulated that escape from predators allows organisms to devote more of their resources to competitive capacity, making them even more successful (Blossey and Notzold 1995).

These relationships indicate that it is not only the nature of the habitat at the time of invasion but also the relationships that develop subsequently that influence the ultimate success of an invader. It has been proposed that with time an invader will come to a new equilibrium with its environment, and populations will stabilize. In some cases these relationships develop quickly, and others develop over long periods of time. An excellent example of the initial lack of predators on invaders is *Eucalyptus* spp. (native to Australia) in California. *Eucalyptus* species were first introduced into California in 1850. Plantings were so extensive that one could drive for miles and miles and never lose sight of an individual of this genus. Although trees of this genus have had a large impact on their new environment (Robles and Chapin 1995), they actually have not spread much from where they were originally planted. One never saw damage to leaves of these trees by herbivores because evidently none of the local fauna could overcome the abundant natural defense compounds that *Eucalyptus* trees produce, whereas in their native habitat herbivory is extensive.

In 1984, more than 130 years after the introduction of *Eucalyptus* to California, the first herbivore was noted on these trees (a long-horned borer from Australia). Since then on average about one new herbivore has become established per year (all from Australia), all inadvertent introductions. After stem borers came leaf herbivores and then sap suckers; a whole complex of herbivores has become established. However, the food web is still simple: the plant host, the *Eucalyptus,* and a series of herbivores. The complexity of the web is being increased by the purposeful introduction of parasites of these herbivores to protect these trees (Paine et al. 2000), which are of value for some purposes. These parasites are host specific and therefore are also from Australia originally.

This example shows that introducing a new organism into an environment often is just the first step in a complex series of interactions that happen through time, adding to the complexity of predicting consequences of invasions.

Good News about Alien Species

There is a lot of good news about alien species. They serve as the foundation for our food production systems. Only about 20 plant species are major contributors to world food supply, and these are often grown far from their places of origin. They have been purposefully transported and molded through selection, and now through engineering, to fit the local conditions in the most productive manner. Furthermore, alien species grace our gardens and parks. They provide shelter from sun and wind and stabilize our soils.

What Is the Problem: Accidental or Intended Introductions?

Many species that are now invasive to a given region were deliberately introduced. For example, of the woody plants that have been become naturalized in North America, 85 percent were introduced for horticultural purposes (Reichard and Hamilton 1997). Lonsdale (1994) found that of 463 pasture species purposefully introduced into Australia between 1947 and 1985, only 5 percent turned out to be useful in pasture improvement, whereas 13 percent subsequently became weedy.

Purposeful introductions have done much economic damage. The golden apple snail, introduced into Asia from Latin America as a potential source of protein, has escaped and is infesting rice fields. In the Philippines alone, during its initial escape in the 1980s it caused hundreds of millions of dollars of economic damage (Naylor 1996).

Thus, the case can be made that both purposeful and accidental introductions can be equally damaging through time, so developing strategies to deal with the issue is difficult.

How Invasive Alien Species Disrupt Our Lives

Article 8h of the Convention on Biological Diversity addresses only the alien species that do harm. Specifically, it calls for action to "prevent the introduction of, control or eradicate those alien species which threaten ecosystems, habitats or species." What this implies is that for consideration a species must be new to the region (alien) and must threaten the native biota in all of its dimensions. This is generally interpreted as causing ecological harm. This definition is somewhat at odds with a strictly ecological definition of an invasive that relates only to the rate of spread. Furthermore, the negative impact criterion can be ambiguous; Daehler (2000) notes that "whether . . . impacts are great or small, harmful or beneficial, depends on . . . personal perspective." (See the lively discussion of the definition issue in Daehler 2000, Davis and Thompson 2000, and Richardson et al. 2000.) Similarly, the definition of *pest* refers to a destructive or troublesome organism, without reference to origin.

There are examples of ambiguity within a particular society on whether a given alien species with invasive characteristics is doing harm. However, in most cases this is not an issue. An invasive alien species that is extending its range in a new region and is having a large negative impact on the native biota or local economies generally is easily identified and targeted. In the following examples I demonstrate the many ways in which invasive alien species can threaten the goods and services provided by natural systems on which society depends. The following list and the brief examples given illustrate only the extent of the impacts of invasive alien species on systems human societies care about. These examples are intended to show the vast array of human endeavors that are interrupted by the impact of invasive alien species. This list includes species that are

- Fire stimulators and cycle disrupters
- Water depleters
- Animal disease promoters
- Crop decimators
- Forest destroyers
- Fishery disrupters
- Impeders of navigation

- Cloggers of water works
- Destroyers of homes and gardens
- Grazing land destroyers
- Species eliminators
- Noise polluters
- Modifiers of evolution

Fire Stimulators and Cycle Disrupters

Invasive species can have a large effect on the fire regime of an area and thereby completely change the character and dynamics of ecosystems. D'Antonio and Vitousek (1992) describe a pattern that is rather general in many places in the world where woody vegetation is destroyed by land clearing and subsequently invaded by alien grasses. These grasses in turn have a short return time for fires, inhibiting woody vegetation recovery and resulting in a permanent conversion to grasslands. Similarly, alien invasions into native shrub vegetation can change the fire regime to the detriment of the shrubs, resulting in a type conversion, as happened in the Great Basin of the United States with the invasion of cheatgrass (*Bromus tectorum*) (Billings 1990).

Water Depleters

Invasive alien species can significantly alter the water balance of a habitat if they attain greater rooting depths than the native species or if they attain greater biomass. Because water is a limiting resource in many parts of the world, invasives that alter habitat water use are of great concern. In the southwestern United States, more than one-half million hectares of riparian areas, in 25 states, has been invaded by tamarisk (*Tamarix*) species. Zavaleta (2000) recently calculated that these tamarisk species are using excess water with a value of approximately $200 million per year. Similarly, in South Africa invasive alien species are taking over the watersheds in the western Cape Region. They are increasing biomass, resulting in a large loss of water. It is calculated that if they are not removed in time they will result in average losses of 30 percent of the water supply of Cape Town (van Wilgen, Cowling, and Burgers 1996).

Animal Disease Promoters

Some of the most dramatic impacts of invasive alien species have involved disease organisms transmitted by resistant populations to those that have no immunity to the disease. The impact of the so-called Columbian Encounter was devastating. In Mexico, the introduction of human diseases such as smallpox, measles, and typhus from Europe reduced the population from 20 million people to approximately 3 million between 1518 and 1568 and to about 1.6 million in the next 50 years (Dobson and Carper

1996). Although modern medicine has reduced the impact of such diseases, there is still need for concern because of the emerging immunity to antibiotics, the large numbers of people with compromised immune systems, and of course the many more opportunities for rapid transport of disease organisms around the world (Garnett and Holmes 1996). For example, recently it has been shown that cholera bacteria (*Vibrio cholera*) are being transported in ship ballast water around the world (Ruiz et al. 2000).

Invasive diseases also directly affect nonhuman animals, causing large-scale impacts. The introduction of rinderpest into Africa at the end of the nineteenth century devastated not only cattle but also native ungulates. Control of this disease by vaccination of cattle also resulted in a dramatic recovery of native ungulate species populations (McCallum and Dobson 1995). However, in the early days of the outbreak nearly one-quarter of the cattle-dependent Masai pastoralists starved to death (McMichael and Bouma 2000). The introduction of avian pox and malaria into Hawaii from Asia has contributed to the demise of lowland native bird species (Simberloff 1996).

The enormous economic and social costs of the transport of animal disease organisms to vulnerable new localities has been amply demonstrated in both inadvertent cases, such as the recent European mad cow disease outbreak, and the purposeful release of anthrax in the United States.

Crop Decimators

Introduced pests not only have a large economic impact on the crops on which we depend but also have had a major impact on history through the actions of such pests as rusts on wheat, ergot on rye, potato blight, the gypsy moth, and the boll weevil (Horsfall 1983).

Weeds, pest insects, and plant pathogens, most of which are invasive alien species, have a major impact on crop yields. In the United States alone weeds reduce potential crop yields by 12 percent, pest insects by 13 percent, and plant pathogens by a similar amount, resulting in billions of dollars in economic losses (Pimentel 1997; see also Orke et al. 1994).

Large amounts of money are spent to combat the potential threat of invasive pests. For example, the Mediterranean fruit fly, or medfly (*Ceratitis capitata*), which originated in West Africa, is now found throughout Africa and in Europe, the Middle East, Australia, western South America, and Central and North America. It attacks more than 250 types of fruits, vegetables, and nuts. Because of the great damage it does, particularly to stone fruits, importation of products from infected areas is banned. In California it has been calculated that it would cost more than $1 billion if markets were closed because of medfly infestation. Because of this danger, whenever the fruit fly is detected there are extensive campaigns to eradicate it using pesticides and sterile male fly release. Since the first outbreaks occurred in the 1970s there have been almost yearly new occurrences, resulting in annual multi–million-dollar eradication efforts. It is clear that total eradication is quite difficult (Carey 1991) and that continuous large-scale efforts will be needed to keep this pest under control in California. There was a putatively successful medfly eradication in Chile after a 32-year battle capped by an extensive campaign in the mid-1990s costing $13 million. However, the medfly recently reappeared in Chile (USDA, June 2000). Kim (1993) notes that there has been no successful eradication of a pest insect.

Forest Destroyers

The American chestnut (*Castanea dentata*) was a major constituent of the deciduous forests of the eastern United States. This magnificent tree was distributed from Maine to the far south in Alabama and Mississippi and from the Piedmont to the Ohio Valley. It was an important part of the deciduous forest ecosystem, providing food for bear, deer, squirrels, and birds. It was also a tree of high economic value and made up 50 percent of the overall value of the eastern forests. The very large trees (up to 5 feet and even larger in diameter and 100 feet tall) of this species produced lumber of unusual qualities because of its decay resistance and a bark that was widely used in tanning. The fruits provided an important cash crop for rural populations (http://www.acf.org).

In 1904 infected American chestnut trees were discovered in New York City. It appears that chestnut blight (*Cryphonectria* [*Endothia*] *parasitica*) had been accidentally introduced into the United States with nursery stock of Asian chestnut trees. Extensive efforts to combat this disease as it spread throughout the range of *Castanea dentata* were unsuccessful. The devastation caused by this invader resulted in the 1912 passage of the U.S. Plant Quarantine Act (Anagnostakis 1996).

The cinnamon fungus (*Phytophora cinnamomi*) has had a dramatic effect on forest and scrublands of parts of Australia. In contrast to the specificity of chestnut blight, the cinnamon fungus attacks 50–75 percent of the species present in a forest. Because of the impact on so many plant species in a community, dependent wildlife has also been affected (Weste and Marks 1987). There are many other examples of forest devastation by pathogenic invaders (Campbell and Schlarbaum 1994).

Fishery Disrupters

In 1982 the ctenophore *Mnemiopsis leidyi* was first collected in the Black Sea. Its most likely origins were ballast water collected off the coast of Florida that was subsequently discharged into the Black Sea. By the summer of 1988 it extended throughout the Black Sea, reaching biomass levels of about 1.5 kg/m^2. By 1990 biomass reached 10–12 kg/m^2 in several coastal areas and 1.5–3 kg/m^2 in the open sea. After 1990 densities decreased, only to increase to high levels again in 1994. It is thought that the species will continue to have boom-and-bust cycles, as it does in the Americas.

Mnemiopsis is a jellyfish-like organism with a diet of zooplankton, fish eggs, and larvae. As a result of the high density of these organisms and their varied diet, they have had a dramatic effect on local fisheries. The amount and diversity of the fish catch have declined drastically. Before the outbreak the former Soviet states landed 250,000 tons. Recent catches have been about 30,000 tons. Annual losses since the invasion have been calculated at $30–40 million per year. "Fishing vessels are for sale in many countries, and fishermen are abandoning their profession" (GESAMP 1997, Section 5.1.3).

The problem has been so severe that biological control has been proposed for *Mnemiopsis.* Before such action could be approved, however, a comb jelly–specific predator, *Beroe* sp., invaded the Black Sea and is controlling *Mnemiopsis* in some areas. It is too early to see what new dynamics will play out in this dramatic incident.

Impeders of Navigation

Aquatic weeds can interfere with navigation in rivers and lakes. In many developing countries, rivers are the main means of transport of rural people, so infestations by such weeds as water hyacinth (*Eichhornia crassipes*) and *Salvinia molesta* cause serious problems throughout the tropics (Gopal 1990). Millions of dollars have been spent recently on the biocontrol of water hyacinth alone in Lake Victoria. Water hyacinth not only interferes with navigation but also impedes water flow in irrigation channels, causing flooding, and deteriorates fisheries, causes anaerobic conditions, and provides habitat for many disease-causing organisms (Gopal 1987).

Cloggers of Water Works

The zebra mussel (*Dreissena polymorpha*), a native of southern Russia, invaded the United States via ballast water in the late 1980s and has already affected water systems (water intakes to reservoir pumping stations, electric generating plants, and industrial facilities), navigation, boating, and sports fishing, costing $5 billion in the Great Lakes alone (Ludyanskiy, McDonald, and MacNeill 1993). This invasion has resulted in a large industry for producing filtration systems and anti-biofouling screen systems. The remarkable filtration capacity of the very abundant mussels has had large impacts on the aquatic ecosystems of the East Coast of the United States that have yet to be quantified. Concern over the economic and ecological impacts of zebra mussel led to a U.S. congressional study on invasive species (U.S. Congress 1993).

Destroyers of Homes and Gardens

Formosan termites invaded the continental United States in the mid-1960s. They rapidly spread from Texas to the Southeast. They are costing $1 billion per year in property damage, repairs, and control measures, with a third of that in New Orleans alone, where historic buildings and trees have been attacked (Suszkiw 1998).

Grazing Land Destroyers

There are many examples of invasive alien species altering grazing lands, in some cases detrimentally so. These include many thistle species found on most continents. One of these, star thistle (*Centaurea solstitialis*), was introduced into California during the gold rush as a contaminant of alfalfa. By 1960 it had spread to 1–2 million acres. By 1985 it contaminated 8 million acres, and by 1999 it contaminated 14 million acres. Star thistle "plagues agriculture and ranching, poisons horses, lacerates hikers" (California Wild, Dudley 2000) and alters ecological balance. It uses water that would not be tapped otherwise by the vegetation, resulting in habitat water loss (Dudley 2000; Gerlach 2000). In the western United States more than 50 million hectares of rangeland have been infested by five major weeds, three *Centaurea* species, *Bromus tectorum,* and *Euphorbia escula.* These and other rangeland weeds are causing more than $2 billion of damage annually (DiTomaso 2000).

Species Eliminators

On islands the most striking examples of species extinctions are caused by invasive species. On the island of Guam, the invading brown tree snake has driven to local extinction 10 of 13 native bird species, 6 of 12 native lizards, and 2 of 3 bat species, a remarkably grim record (USGS 2000). Other examples of extinctions on islands are given by Case and Bolger (1991) and for lakes by Worthington and Lowe-McConnell (1994).

Noise Polluters

Eleutherodactylus coqui is one of 14 species of this genus of tree frogs found in Puerto Rico. A number of them probably are extinct, and many are threatened. The coqui is an important symbol of Puerto Rico and is used heavily in the design of souvenirs and in art and music. Human interaction with the coqui frog is an acquired taste, however. *E. coqui* has recently invaded some of the Hawaiian Islands and is spreading rapidly, particularly on Maui but also on Hawaii (Kraus et al. 1999). The unusually loud noise that this frog makes at night and the high populations it attains have caused consternation among local residents and hotel owners. These invaders are remarkable for their noise production given their very small size.

Modifiers of Evolution

Mooney and Cleland (2001) recently outlined the large impact that invasive species are having on the trajectory of evolution. Invasives aliens can dominate landscapes, altering the environments of co-occurring organisms. They not only cause extinctions but also modify behavior, compress niche breadths, disrupt mutualisms, and hybridize and introgress with native species. Thus they are causing damage not only in the near term but also over evolutionary time.

The Challenges We Face

Given the impacts of invasive alien species noted in this chapter, we have abundant motivation to alleviate the problem. However, there are many impediments to doing so. These include the basic problem that invasive species are self-replicating. Thus, unlike that of chemical pollution, "cleanup" of invasive species is more difficult. Furthermore, invasive species, particularly short-lived insects and microbes, can quickly evolve mechanisms to overcome control efforts. Once an invasive species integrates into an ecosystem, its control becomes ever more complex because of interactions with both the native biota and other invaders.

One complicating factor is the lag time between when an alien species becomes established and when it truly becomes invasive. Lag times of many decades, or even centuries, have been noted. Lag times complicate risk analyses that are based on traits related to short-term invasive history.

The most fundamental challenge we face relates to the fact that we have little control over the increased movement of biological material around the world, the increasing disruption of natural landscapes, and changes in atmospheric chemistry, all of which favor the success of invasive species.

With the great abundance of invasive species spread throughout many regions of the world and the varied experiences of those who attempt to manage them, one would think that a global database would be available for information sharing to combat these species. Progress is being made toward achieving this goal.

What We Can Do

All of these examples demonstrate the great damage that invasive alien species can do to natural and managed ecosystems and the goods and services they provide.

The Global Invasive Species Program was designed to review our knowledge about certain aspects of invasive species and to develop new approaches and tools for addressing them. The following chapters present the findings of this effort, and the concluding chapter presents a summary of what we learned and what we propose in order to alleviate the impacts of invasive species on society.

References

Anagnostakis, S. L. 1996. "Chestnuts and the introduction of chestnut blight," *Woodworking Times* March–April.

Billings, W. D. 1990. "*Bromus tectorum,* a biotic cause of ecosystem impoverishment in the

Great Basin," in *The Earth in Transition,* edited by G. M. Woodwell, 301–322. Cambridge University Press, New York.

Blossey, B., and R. Notzold. 1995. "Evolution of increased competitive ability in invasive nonindigenous plants: A hypothesis," *Journal of Ecology* 83:887–889.

Boden, T. A., D. P. Kaiser, R. J. Sepanski, and F. W. Stoss, eds. 1994. *Trends '93: Compendium of Data on Global Change.* Carbon Dioxide Information Analysis Center, Oak Ridge, TN.

Bright, C. 1998. *Life out of Bounds.* W.W. Norton, New York.

Campbell, F. T., and S. E. Schlarbaum. 1994. *Fading Forests. North American Trees and the Threat of Exotic Pests.* Natural Resources Defense Council Publications, New York.

Carey, J. R. 1991. "Establishment of the Mediterranean fruit fly in California," *Science* 253:1369–1373.

Case, T. J., and D. T. Bolger. 1991. "The role of introduced species in shaping the distribution and abundance of island reptiles," *Evolutionary Ecology* 5:272–290.

Cohen, A. N., and J. T. Carlton. 1998. "Accelerating invasion rate in a highly invaded estuary," *Science* 279:555–558.

Daehler, C. C. 2000. "Two ways to be an invader, but one is more suitable for ecology," *Bulletin of the Ecological Society of America* 82:101–102.

Daily, G., S. Alexander, P. Ehrlich, L. Goulder, J. Lubchenco, P. Matson, H. Mooney, S. Postel, S. Schneider, D. Tilman, and G. Woodwell. 1997. "Ecosystem services: Benefits supplied to human societies by natural ecosystems," *Issues in Ecology* Second Issue, 1–16.

D'Antonio, C. M., and P. M. Vitousek. 1992. "Biological invasions of exotic grasses, the grass/fire cycle and global change," *Annual Review of Ecology and Systematics* 23:63–87.

D'Antonio, C. M., N. E. Jackson, C. C. Horvitz, and R. Hedberg. 2004. Invasive plants in wildland ecosystems: Merging the study of invasion processes with management needs. *Frontiers in Ecology and the Environment.* Vol. 2, No. 10, pp. 513–521.

Davis, M. A., and K. Thompson. 2000. "Eight ways to be a colonizer; two ways to be an invader: A proposed nomenclature scheme for invasive ecology," *Bulletin of the Ecological Society of America* 81:226–230.

Devine, R. 1998. *Alien Invasion.* National Geographic Society, Washington, DC.

DiTomaso, J. M. 2000. "Invasive weeds in rangelands: Species, impacts, and management," *Weed Science* 48:255–265.

Dobson, A. P., and E. R. Carper. 1996. "Infectious diseases and human population history," *BioScience* 46:115–125.

Dudley, D. R. 2000. "Wicked weed of the West," *California Wild* Fall 2000:32–35.

Eldredge, L. G., and S. E. Miller. 1997. "Numbers of Hawaiian species: Supplement 2, including a review of freshwater invertebrates," *Bishop Museum Occasional Papers* 48:3–22.

Gadgil, M., K. P. Achar, A. Shetty, A. Ganguly. 2000. *Participatory Local Level Assessment of*

Life Support Systems: A methodology manual. Technical Report no. 78, Centre for Ecological Sciences, Indian Institute of Science, Bangalore.

Garnett, G. P., and E. D. Holmes. 1996. "The ecology of emergent infectious disease," *BioScience* 46:127–135.

Gerlach, J. D. 2000. "A model experimental system for predicting the invasion success and ecosystem impacts of non-indigenous summer-flowering annual plants in California's Central Valley grasslands and oak woodlands," PhD thesis, University of California, Davis.

GESAMP (IMO/FAO/UNESCO-IOC/WMO/WHO/IAEA/UN/UNEP Joint Group of Experts on the Scientific Aspects of Marine Environmental Protection). 1997. *Opportunistic settlers and the problem of ctenophore* Mnemiopsis Iedyi *invasion in the Black Sea.* Rep. Stud. GESAMP, (58): 84 pages.

Gopal, B. 1987. *Water Hyacinth.* Elsevier, Amsterdam.

———. 1990. "Aquatic weed problems and management in Asia," in *Aquatic Weeds,* edited by A. H. Pieterese and K. J. Murphy, 318–340. Oxford University Press, Oxford.

Hobbs, R. J., and H. A. Mooney. 1998. "Broadening the extinction debate: Population deletions and additions in California and Western Australia," *Conservation Biology* 12:271–283.

Horsfall, J. G. 1983. "Impact of introduced pests on man," in *Exotic Plant Pests and North American Agriculture,* edited by C. L. Wilson and C. L. Graham, 1–13. Academic Press, New York.

Kim, K. C. 1993. "Insect pests and evolution," in *Evolution of Insect Pests,* edited by K. C. Kim and B. A. McPheron, 3–25. Wiley, New York.

Kraus, F., E. W. Campbell III, A. Allison, and T. Pratt. 1999. "*Eleutherodactylus* frog introductions to Hawaii," *Herpetological Review* 30:21–25.

Lonsdale, W. M. 1994. "Inviting trouble: Introduced pasture species in northern Australia," *Australian Journal of Ecology* 19:345–354.

———. 1999. "Global patterns of plant invasions and the concept of invasibility," *Ecology* 80:1522–1536.

Ludyanskiy, M. L., D. McDonald, and D. MacNeill. 1993. "Impact of the zebra mussel, a bivalve invader," *BioScience* 43:533–544.

Mack, R. N., D. Simberloff, M. W. Londsdale, H. Evans, M. Clout, and F. A. Bazzaz. 2000. "Biotic invasions: Causes, epidemiology, global consequences, and control," *Ecological Applications* 10:689–710.

McCallum, H., and A. Dobson. 1995. "Detecting disease and parasite threats to endangered species and ecosystems," *Trends in Ecology Evolution* 10:190–194.

McMichael, A. J., and M. J. Bouma. 2000. "Global changes, invasive species, and human health," in *Invasive Species in a Changing World,* edited by H. A. Mooney and R. J. Hobbs, 191–210. Island Press, Washington, DC.

Mooney, H. A., and E. E. Cleland. 2001. "The evolutionary impact of invasive species," *Proceedings of the National Academy of Sciences* 98:5446–5451.

Mooney, H. A., and R. J. Hobbs, eds. 2000. *Invasive Species in a Changing World.* Island Press, Washington, DC.

Naylor, R. 1996. "Invasions in agriculture: Assessing the cost of the golden apple snail in Asia," *Ambio* 25:443–448.

Nico, L. G., and P. L. Fuller. 1999. "Spatial and temporal patterns of nonindigenous fish introductions in the United States," *Fisheries* 24:1–27.

Orke, E., H. Dehne, F. Schoenbeck, and A. Weber. 1994. *Crop Production and Crop Protection: Estimated Losses in Major Food and Cash Crops.* Elsevier, New York.

Paine, T. D., D. L. Dahlsten, J. G. Millar, M. S. Hoddle, and L. M. Hanks. 2000. "UC scientists apply IPM techniques to new eucalyptus pests," *California Agriculture* 54:8–13.

Parker, I. M., D. Simberloff, W. M. Lonsdale, K. Goodell, M. Wonham, P. M. Kareiva, M. H. Williamson, B. Von Holle, P. B. Moyle, J. E. Byers, and L. Goldwasser. 1999. "Impact: Toward a framework for understanding the ecological effects of invaders," *Biological Invasions* 1:3–19.

Pimentel, D. 1993. "Habitat factors in new pest invasions," in *Evolution of Insect Pests,* edited by K. C. Kim and B. A. McPheron, 165–181. Wiley, New York.

———. 1997. *Techniques for Reducing Pesticides: Environmental and Economic Benefits.* Wiley, Chichester, UK.

Reichard, S. H., and C. W. Hamilton. 1997. "Predicting invasions of woody plants introduced into North America," *Conservation Biology* 11:193–203.

Richardson, D. M., P. Pysek, M. G. Barbour, F. D. Panetta, and C. J. West. 2000. "Naturalization and invasion of alien plants: Concepts and definitions," *Diversity and Distributions* 6:93–107.

Robles, M., and F. S. Chapin III. 1995. "Comparison of the influence of two exotic species on ecosystem processes in the Berkeley hills," *Madrono* 42:349–357.

Ruiz, G. M., T. K. Rawlings, F. C. Dobbs, L. A. Drake, T. Mullady, A. Huq, and R. R. Colwell. 2000. "Global spread of microorganisms by ships," *Nature* 408:49.

Sailer, R. I. 1983. "History of insect introductions," in *Exotic Plant Pests and North American Agriculture,* edited by C. L. Wilson and C. L. Graham, 15–38. Academic Press, New York.

Sandlund, O. T., P. J. Schei, and A. Viken, eds. 1999. *Invasive Species and Biodiversity Management.* Kluwer, Dordrecht, The Netherlands.

Simberloff, D. 1996. "Impacts of introduced species in the United States," *Consequences* 2:13–22.

Suszkiw, J. 1998. "The Formosan termite a formidable foe!" *Agricultural Research* 46:4–9.

U.S. Congress, Office of Technology Assessment. 1993. *Harmful Non-Indigenous Species in the United States.* U.S. Government Printing Office, Washington, DC.

USDA. 2000. USDA Imposes Treatment Restrictions on Fruit from Two Regions in Chile. Press release. Available at http://www.aphis.usda.gov/lpa/news/2000/05/CHILEMED.htm.

USGS. 2000. *Snake Invader Threatens Island Ecosystems.* U.S. Geological Survey, Biological Resources Division, Reston, VA.

van Wilgen, B. W., R. M. Cowling, and C. J. Burgers. 1996. "Valuation of ecosystem services: A case study from South African fynbos," *BioScience* 46:184–189.

Vitousek, P. M., C. M. D'Antonio, L. L. Loope, M. Rejmanek, and R. Westbrooks. 1997. "Introduced species: A significant component of human-caused global change," *New Zealand Journal of Ecology* 21:1–16.

Weste, G., and G. C. Marks. 1987. "The biology of *Phytophthora cinnamomi* in Australian forests," *Annual Review of Phytopathology* 25:207–229.

Wilcove, D. S., D. Rothstein, J. Dubow, A. Phillips, and E. Losos. 1998. "Quantifying threats to imperiled species in the United States," *BioScience* Vol. 48: 8, pp. 607–615.

Williamson, M. 1996. *Biological Invasions.* Chapman & Hall, London.

———. 1999. "Invasions," *Ecography* 22:5–12.

Worthington, E. B., and R. Lowe-McConnell. 1994. "African lakes reviewed: Creation and destruction of biodiversity," *Environmental Conservation* 21:199–213.

Zavaleta, E. 2000. "The economic value of controlling an invasive shrub," *Ambio* 29:462–467.

2

The Economics of Biological Invasions

Charles Perrings, Silvana Dalmazzone, and Mark Williamson

Identifying the Economic Problem

In this summary of the economic component of the Global Invasive Species Program (GISP) we draw on the introductory and concluding sections of Perrings, Williamson, and Dalmazzone (2000) and extend them. This summary characterizes an economic approach to the problem of invasive alien species (IAS). We review the evidence currently available on the costs and benefits of alternative control options, but we emphasize that the economics of invasions covers much more than calculations of the costs of invasions. Biological invasions are an example of anthropogenic environmental change. Invasions typically are the intended or unintended consequence of economic activities, and hence of the markets, institutions, and property rights that influence economic decisions. An economic approach to the problem inquires into the underlying causes, the consequences, and the options for addressing the problem. These options include preventive measures as well as the more familiar eradication, containment, mitigation, and adaptation. Identifying the cost of invasions is a small part of the problem and is properly embedded in an analysis of the control options. More important is the design of policy instruments and incentives as control measures.

There are two main reasons for current concern over IAS. The first is that the impacts of IAS are already large and are rapidly growing larger. Living organisms have always been transported beyond their original range. However, an increase in the rate of the international movement of people and commodities means that the risk of invasions is also rising. In addition to their impact on output in agriculture, forestry and fisheries, pest control, and health care, invasive species can be one of the main drivers of biodiversity loss (Wilcove et al. 1998).

IAS are a major threat on oceanic islands. The factors that control the growth and spread of IAS in their native range often are not present in such habitats. Moreover, native species may not possess defense mechanisms that allow them to compete successfully for vital resources and therefore may be driven to extinction. Elsewhere, the threat of invasives is variable, but Mack (1997) notes that invasive species have affected whole landscapes in parts of the United States, Central America from Mexico to Panama, Venezuela, Brazil, Argentina, South Africa, Madagascar, India, Myanmar, Australia, and New Zealand.

A second reason for policymakers to be concerned about IAS is contractual. Article 8(h) of the Convention on Biological Diversity (CBD) requires the contracting parties "as far as possible and appropriate to prevent the introduction of, control or eradicate those alien species which threaten ecosystems, habitats or species" (Glowka, Burhenne-Guilmin, and Synge 1994). The CBD was signed by more than 150 governments at the 1992 Earth Summit in Rio de Janeiro, and became effective as international law in December 1993. It is the first international agreement to commit governments to a comprehensive protection of the earth's biological resources. At the time of writing, 188 countries had ratified the agreement (the CBD had not been ratified by Somalia, Macedonia, Timor Leste, Iraq, the United States).

The biological basis for concern over IAS is that they disrupt ecological processes and functions. Specific impacts include predation on or competitive exclusion of other species. Diseases and the addition of new top trophic species (predators, such as cats, or herbivores, such as goats) generally have the most severe biological effects (Williamson 1996). IAS impacts have been measured in a variety of different ecosystems: terrestrial, freshwater and marine, animal, plant, and microbe (Parker et al. 1999). These range from negligible to severe (Williamson 1998). These biological impacts are the source of the economic costs of IAS.

To this point the economic analysis of IAS has been focused on two problems. The first is pests and pathogens in agriculture, forestry, and fisheries; the second is the economics of human disease. So far there has been little systematic economic analysis of other IAS problems. The approach to IAS in agroecosystems provides a reasonable model for the economic analysis of IAS elsewhere. Annual weeds are present in all but the most intensively managed agricultural systems, and many of these species are not native but have been introduced from elsewhere. Ecological models are used to predict the numbers of weeds and the yield losses that result from their presence. They are also used to test the sensitivity of population models in order to target key areas of the life cycle at which control will be most effective (Watkinson, Freckleton, and Dowling 2000). This information is then combined with information on the cost of control options and the value of the yield losses in order to identify the cost of invasions and to calculate the optimum level of control (Perrings, Williamson, and Dalmazzone 2000).

An important feature of such impacts is that they are, ex ante, highly uncertain. This is partly because there seem to be no general laws governing biological invasions (Williamson 1999). Lawton (1999) argues that predicting the population dynamics of a particular species in a particular habitat entails detailed study of that species in that habitat. Law, Weatherby, and Warren (2000) make a similar point, that understanding invasions depends as much on detailed knowledge of idiosyncratic biological interactions as on general properties of community structure. Kareiva, Parker, and Pascual (1996) argue that models and short-term experiments are inadequate predictors of invasions, so new situations necessitate extensive monitoring. Up to now the best predictors of invasions have been a history of invasions by the same species and propagule pressure (Williamson 1999).

A second reason for uncertainty is the potential irreversibility of many of the consequences of IAS. If a species is driven to extinction, people forgo the option to make use of that species in the future—its option value. But this is clearly difficult to calculate if the range of possible uses is unknown. The value of the genetic information lost through species extinction is called a quasi–option value. It is the value of the future information protected by preserving a resource (Arrow and Fisher 1974; Fisher and Hanemann 1983). There are no reliable estimates of the magnitude of these costs. What is certain is that the maximum potential costs are not trivial (Pearce and Moran 1994; Heywood 1995).

To illustrate this, take the example of introduced cultivated crop varieties in agroecosystems. Most cultivated crop varieties and some livestock strains contain genetic material from related wild or weedy species or from primitive stocks still used and maintained by traditional agricultural peoples. Such traditional varieties are the result of millennia of selection by farmers. It has been estimated that at least half of the increase in agricultural productivity achieved in the last hundred years is attributable to artificial selection, recombination, and intraspecific gene transfer procedures (Heywood 1995). The introduction of high-yielding varieties and supporting pesticide regimes has displaced many land races and traditional varieties, a number of which are now extinct.

Although this is managed competitive exclusion, its effect is similar to the loss of species in any other ecosystem. The loss of traditional varieties imposes costs in terms of both the information value of the lost species (their quasi–option value) and the functionality of the system. Such systems become more susceptible to exogenous shocks or changes in environmental conditions (Conway 1993). The adoption of crops with a narrow genetic base has increased average yields, but it has also increased risks (the variance in yields) and the degree to which risks are correlated across agroecosystems. That is, the systems become less resilient with respect to variation in climatic conditions or the relative abundance of pathogens. This is an irreversible and continuing cost of the loss of traditional varieties. If the introduced high-yielding variety were withdrawn, the costs would remain. Another example concerns the fishhook waterflea (*Cercopagis pengoi*), introduced into the Great Lakes via the ballast water of seagoing vessels. Even if the ballast water pathway were closed—vessels were prohibited from discharging ballast water in the lakes—the flow of damages would not stop (MacIsaac et al. 1999; Ricciardi and MacIsaac 2000).

A third source of uncertainty is the fact that the risks of invasions are partly endogenous. They depend on human responses to the threat of invasions. Shogren (2000) points out that decision makers protect themselves against the risks of invasive species in two related ways: by mitigation and by adaptation. Mitigation includes the conventional categories of eradication and containment and has the effect of reducing the likelihood that a species will establish or spread. Adaptation, on the other hand, implies some change in behavior in order to reduce the impact of the establishment and spread of a species. That is, it works on the value of the effect rather than the likelihood of the effect. Nonetheless, mitigation and adaptation jointly determine the risks and the costs of invasions. If mitigation and adaptation are linked in this way, there is another layer of uncertainty (Shogren and Crocker 1991).

Aside from the direct impacts of IAS, ecosystems are indirectly affected in a variety of ways. The loss of functionality of agroecosystems is an example of an indirect ecological effect. The introduction of IAS in other ecosystems similarly leads to changes in ecological services that are locally important by disturbing the operation of the hydrological cycle, including flood control and water supply, waste assimilation, nutrient recycling, soil conservation and regeneration, and crop pollination. These services have both current and option value. For example, *Acacia, Hakea,* and *Pinus* species have established and spread in the South African fynbos with significant implications for hydrological flows. This has reduced water supplies to the whole community (Turpie and Heydenrych 2000). One aim of the papers in Perrings, Williamson, and Dalmazzone (2000) is to clarify how the direct and indirect costs of invasions may be assessed.

The underlying causes of invasions are quite varied. From an economic perspective the most important are the economic factors that predispose a country to IAS and the set of incentives offered by the structure of property rights to the resources. This, in turn, reflects the public good nature of both control activities and the ecosystem services most affected by IAS. Although the destruction of crops and harvests caused directly or indirectly by IAS is reflected in the market prices of agricultural, fishery, or forestry products, these costs are not borne by those responsible for the introduction or spread of the IAS. They are externalities: costs that a given activity unintentionally imposes on another, without the latter being able to exact compensation for the damage received.

Where IAS disrupt ecosystem services such as the regulation of specific biogeochemical cycles over a range of climatic conditions or the protection of crop yields over a range of pests and pathogens, they disrupt the supply of a public good. Ecological services in this category include watershed protection, flood and drought mitigation, waste assimilation, detoxification and decomposition, microclimatic stabilization, air and water purification, soil generation and renewal, pollination, agricultural pest control, seed dispersal, and nutrient transport (Daily 1997). We discuss this in more detail later.

Predisposing Economic Conditions

To identify the institutional and policy conditions that predispose countries to biological invasion, Dalmazzone (2000) considers the relationship between the establishment of alien plant species in 29 countries; trade flows and their composition; arable, pastoral, and forested land; and other socioeconomic variables. She finds that economic activities can increase the inherent susceptibility of ecosystems. Specifically, the variables affecting the recipient environment (land tenure, gross domestic product, population density) are responsible for explaining a high proportion of the variation in the share of alien species in different countries. All ecological communities are vulnerable to invasion to some degree. The degree of human activity (human disturbance) appears to increase that vulnerability further.

Trade has a weaker but still significant impact on the share of alien species hosted by any given country. Preliminary results also add some insight into the problem of biological invasions on islands. Island ecosystems are generally considered highly susceptible to invasions because of a particularly vulnerable native biodiversity. Also, island states typically are small, open economies, often geared to the production of primary products. The average percentage of merchandise imports as a share of the gross domestic product in the sample considered is about 43 percent for island countries and 26.8 percent for continental countries. Island states are both ecologically vulnerable and open in terms of the movement of goods and services across ecological boundaries.

Some interactions between human behavior and biological invasions are intuitive. For example, the probability of establishment of intentionally introduced species is higher than that of unintentionally introduced species. One reason is that intentionally introduced species have been selected for their ability to survive in the environment where they are introduced (Smith, Lonsdale, and Fortune 1999). Another is the link between intentional and repeated introduction. Exotic species that are marketed over a period of time have a greater probability of establishment than those that are marketed once (Enserink 1999). The probability of both establishment and spread also depends on the way in which the environment is altered by human behavior. The introduction of specific disease- or pest-resistant crops, for example, selects in favor of other pests and predators in a way that is well understood (Heywood 1995).

Most invasions are not intended; they are the unintended consequence of a market transaction. This is due partly to the lack of markets for many of the effects of invasions (the problem of incomplete property rights) and partly to the effects of policy. In the South African fynbos case mentioned earlier, individual landowners have not historically had the legal right to resist the spread of *Pinus, Hakea,* or *Acacia* species from neighboring farms. This has made it difficult to prevent the spread of those species. Elsewhere, fiscal, price, and income policies have all promoted management regimes that have increased the susceptibility of agroecosystems to invasion. For example, subsidies designed to promote cash cropping as a means of increasing export revenue have encouraged the use of farm inputs that open agroecosystems to invasion.

Economics of IAS Control

The classic method for evaluating control options is benefit–cost analysis. Despite difficulties associated with the nonmarketed and public good nature of many environmental resources, the consensus is that this is still the most appropriate method for evaluating environmental projects (Arrow et al. 1996). The benefit–cost test requires that the expected present value of the benefits of the control program (the net costs avoided by the control program) be no less than the expected present value of the costs of control (the foregone benefits of the program). For a screening program of the sort applied in Australia, for example, this implies a test of the following form:

$$E\sum_{t=0}^{T} r^{t}\left(\frac{B_{it} - C_{it}}{B_{nt} - C_{nt}}\right) \geq 1$$

where E indicates expected value, $B_{it} - C_{it}$ indicates the net benefits of the exclusion of potentially invasive species at time t, $B_{nt} - C_{nt}$ indicates the net costs of excluding noninvasive species, and (ρ is a discount factor. In other words, the test requires the capacity to estimate the stream of net benefits associated with the action or investment, taking all relevant effects over the whole time horizon into account.

We are unaware of any cases in which estimates of the net benefits of IAS control options would support such a test. Some studies summarize the direct costs of control measures for particular invasive species. Most estimate an annual damage cost for an invasive species at some moment in time. There have been only two attempts to aggregate the damage costs of invasions for some region, and those vary widely. Both refer to the United States, and both seek to identify a damage cost (OTA 1993; Pimentel et al. 2000). The numbers generated by these exercises do not mean very much by themselves, but they give a general indication of both the scale of the problem and the degree of uncertainty attached to the process. In 1993, the U.S. Office of Technology Assessment estimated damage costs of $97 billion from 79 particularly harmful species over the preceding 85 years. In 2000, however, Pimentel and others estimated damage costs of $137 billion per year from all species.

Most studies offer estimates of damage costs of individual IAS (Table 2.1). The literature includes estimates of the potential control costs of the screwworm fly (*Chrysomya bezziana*) in Australia (Anaman et al. 1994); the benefits of clearing alien species from fynbos ecosystems in South Africa (Higgins et al. 1997); the impact of knapweed and leafy spurge on the economy of several U.S. states (Bangsund, Leistritz, and Leitch 1999); the damages to North American and European industrial plants from the zebra mussel and other invaders (Khalanski 1997); the damage cost of the green crab (*Carcinus maenas*) on the North Pacific Ocean fisheries (Cohen, Carlton, and Fountain 1995); control costs of weed species in Australian agroecosystems (Watkinson, Freckleton, and Dowling 2000); and damage costs of *Pinus, Hakea,* and *Acacia* species in the South African fynbos (Turpie and Heydenrych 2000), plant and fish species in the African lakes (Kasulo 2000), rabbits in Australia (White and Newton-Cross 2000), and the tree *Maeopsis eminii* in the eastern arc montane forests of Tanzania (Lovett 2000). Sharov and Liebhold (1998) analyze the costs of eradicating, stopping, or slowing the spread of invasive species in North American ecosystems, using the case of the gypsy moth (*Lymantria dispar*) as an example.

Table 2.1. Estimates of selected economic impacts of invasive species

Species	*Impact (US$ million)*[1]	*Location and impact*	*Source*
Pinus, Hakea, and *Acacia* species	160	South Africa (estimated restoration costs in the fynbos, Cape Province)	Higgins et al. 1997; Turpie and Heydenrych 2000
Leafy spurge and knapweed	129.5/year	USA (losses to rangelands, three states)	Bangsund, Leistritz, and Leitch 1999; Hirsch and Leitch 1996
Zebra mussel	3,000–5,000	USA, Europe (cumulative damage to industrial plants)	Khalanski 1997
Green crab	44/year	USA (impact on North Pacific Ocean fisheries)	Cohen, Carlton, and Fountain 1995
Various weeds	80/year	Australian (annual control costs of six weeds)	Watkinson, Freckleton, and Dowling 2000
Water hyacinth in the African lakes	71.4/year	East Africa (impact on fisheries in the African Lakes)	Kasulo 2000
Tamarix	7,000–16,000	USA (cumulative ecological impacts over a 55-year period)	Zavaleta 2000
Rabbits	280/year	Australia (damage and control cost)	White and Newton-Cross 2000
8 spp. "ancient" invasive plants, 4 spp. "modern" invasive plants	145/year (ancient) 170/year (modern)	UK (annual cost of herbicide control)	Williamson 1998

[1]The base year is not always stated but is between 1995 and 2000.

Hill and Greathead (2000) assess the evidence on the net benefits of biological control as a tool to counteract invasions. Approximately 10–15 percent of some 5,000 classic biological control introductions against arthropods have proved completely successful. Against weeds, about 30–40 percent of some 900 introductions have achieved their objective. Although most attempts at classic biological control are failures or have adverse side effects, a review of 27 analyses of successful programs shows some to have been extremely profitable, so much so that Hill and Greathead conjecture that the successes may cover the costs of the failures. However, this is an area where very little work has been done. Much of what is available consists of either highly aggregated estimates of the cost of invasions in particular sectors or cost–benefit analyses of successful control strategies. This is equivalent to evaluating the net present value of the purchase of a winning lottery ticket. There is a clear need to improve the basis on which control strategies are evaluated and on which the potential impacts of species introductions are valued.

It is hard to avoid the conclusion that valuation of the costs and benefits of invasion control options is still quite primitive. It focuses on partial damage estimates or direct control costs. It does not always distinguish capital from income. It pays little attention to the interactive effects between invasive and native species. Some of the current tests of the effectiveness of controls on IAS have at least elements of a standard test for the efficiency of investments. In most cases, however, the current tests include only a partial assessment of the expected benefits and costs. They often ignore the opportunity cost of control programs and the cost of controls that prove to be unnecessary. Three problems must be addressed if we are to offer realistic evaluations of control alternatives.

The first is to identify the full set of IAS impacts. In lakes or wetlands, for example, this entails an understanding of their effect on a wide array of economically valuable functions including storm and pollution buffering, flood alleviation, and recreation. There is also a profile to the stream of such costs. Kasulo (2000) analyzes the ecological and socioeconomic impact of invasive species in African lakes. His focus is on introduced fish species and water weeds—the Nile perch (*Lates niloticus*), the Tanganyika sardine (*Limnothrissa miodon*), and water hyacinth (*Eichhornia crassipes*)—into lakes Victoria, Kyoga, Nabugabo, Kariba, Kivu, Itezhi-tezhi, and Malawi. The introduction of Nile perch has increased profits from commercial fishing and contributed to the generation of foreign exchange. However, the Nile perch is believed to have caused the extinction of numerous endemic species. The introduction of the Tanganyika sardine also resulted in an increase in productivity, with less dramatic impact on the ecosystems of the lakes to which it was introduced. The water hyacinth, introduced in Africa as an ornamental plant, has proliferated explosively in most African lakes, obstructing water passages and displacing native aquatic plants, fish, and invertebrates by cutting out light and depleting dissolved oxygen. The weed is also believed to harbor disease-carrying organisms and has little potential for economic use.

In all cases, the impacts of the invasive species involve other interacting species. Those interacting species may themselves be directly harvested, as are the haplochromine cichlids in Lake Victoria, or may themselves interact with harvested species. Therefore, the time profile of an IAS involves periods of exponential growth of the invasive species itself, along with a set of lagged and often conflicting changes in the relative abundance of interacting species. For this reason the time profile of the damage costs of the introduced species may be quite complicated.

Our other example, the fynbos of the South African Cape Floral Kingdom, was evaluated at the point where approximately 66 percent of the fynbos area in the Western Cape was affected. The damage costs estimated there included both a reduction in biodiversity and scenic beauty, with associated implications for the flower trade, and a change in ecosystem functioning. Turpie and Heydenrych (2000) argue that fynbos mountain catchments are extremely valuable in terms of their water yield and that water yields have been significantly reduced by alien invasions. There has been some previous research on the fynbos's contribution to the hydrological cycle, but this study also provides estimates of the value it yields in the form of consumptive use benefits as well as nonconsumptive use values.

In both cases the time path for the spread of invasives depends on the control strategy. The time path of an invasive species subject to biological control reflects the population dynamics of the control agent because the effectiveness of the control effort depends on the interaction between the control and the invasive species. Mechanical or chemical control options reflect evolution of the control effort. Any evaluation of benefits of particular control strategies must take account of these sorts of interactions.

A second and related problem is the interdependence between human behavior and the dynamics of control. Estimation of the benefits of a control strategy depends on an understanding of the way in which human behavior and invasive species interact. This in turn reflects the institutional and policy environment within which people make their decisions. The importance of human behavior in the control strategy is easiest to see in the case of invasive species with direct effects on human health, such as the human immunodeficiency virus (HIV). Indeed, such species control depends on the ability to influence the behavior of infected and susceptible people. Delfino and Simmons (2000) look at how infectious diseases interact with the economy in which they occur. They show that control entails changing the behavior that affects the introduction and spread of such diseases.

The dynamic interactions between pathogens, human behavior, and economic development turn are quite complex. Whereas the probability of infection influences decisions in a way that is reasonably well understood, the interactions between the virulence of a disease, infected and susceptible populations, the pattern of settlement, and the level of development are complex. The relationship between the level of economic development, the epidemiology of an invasive pathogen, and settlement and migration is particularly interesting. In this case the spread of the invasive species depends directly on the mobility of infected and susceptible people and is highly sensitive to parameter values.

The same is true of other IAS. Like the spread of disease, the spread of other potentially invasive species depends on people's behavior. This includes the use people make of invasive species and their predators and competitors, demographic patterns, transport networks, and the like. Knowler and Barbier's (2000) study of *Mnemiopsis leidyi* shows how land–sea interactions and fisher strategies have combined to favor the spread of a species that has severely affected the Black Sea anchovy fishery. In this case both the cost and benefits of a control strategy will be highly sensitive to human responses.

A third problem is the treatment of risk and uncertainty. Because IAS often are quite different from existing species, their establishment, naturalization, and spread can induce novel responses. Evaluation of control options is sensitive to the way in which invasion control risks are treated. In general, risk may be allowed for by adjusting the discount rate upward, using a risk premium, or finding the certainty equivalent of the expected environmental damage (the certain damage that would yield the same disutility as the expected damage). We have already noted that it is difficult to estimate the probabilities attaching to different outcomes on the basis of the characteristics of species or their habitat. This is partly because the risks are not wholly exogenous. The probability of establishment of intentionally introduced species is higher than that of unintentionally introduced species. But the probability of establishment and spread of any species depends on how the environment is altered by human behavior. For example, the introduction of specific disease- or pest-resistant crops selects in favor of other pests and predators.

Societies reduce the risks posed by invaders through eradication, containment, mitigation, and adaptation. The first three reduce the odds against invasions; adaptation reduces the consequences when an invasion does occur. Shogren (2000) shows how a society can mix mitigation and adaptation strategies to reduce the risk from exotic invaders, taking into account the interaction of biological and economic factors to assess risk, the value of risk reduction, and the impact of additional risk of damages.

However, a particular difficulty with invasive species is that many biological invasions fall into the category of low-probability events with a high potential cost (Williamson 1992). The probability that any one introduced species will establish and become a pest or pathogen is very low, but the costs to society if it does can be very high. Smith, Lonsdale, and Fortune (1999) suggest 2 percent for the probability of plant introductions into Australia becoming pests, somewhat higher than the 0.01–1.6 percent, mostly based on British examples, indicated by the tens rule of Williamson and Fitter (1996). Of course such figures are markedly affected by the definitions of *introduced* and *pest* (Williamson 1996). At the same time, the control and damage costs of species that do become significant pests or pathogens can be extremely high, as in the case of HIV.

Such invasive pests or pathogens provoke dread, a phenomenon that increases subjective risk assessments connected with certain types of health risk (McDaniels, Kamlet, and Fischer 1992). It has been shown that risk rating has a higher effect on willingness to pay for risk reduction measures in infrequent high-dread situations than in frequent low-dread situations (cf. Loomis and du Vair 1993). In fact, the power of the expected utility hypothesis declines as the probabilities of outcomes tend to unity or to zero. In the former case the probability of an "almost sure" event tends to be approximated by certainty. In the latter case, people facing a "very unlikely" event tend either to overestimate the probability or to identify it with zero. For very low probabilities the weighting function often is not defined. In the liability insurance markets it has long been established that where the probability is very low but the potential loss is very high, insurers demand a risk premium that exceeds the expected losses, whereas the insured are willing to pay less than predicted by expected utility calculations (Katzman 1988). The dread effect is likely to bias upward private estimates of the costs if a potentially invasive species becomes a pest, whereas the low probability of outcomes is likely to induce a distortion in the perception of risk.

Institutions and Incentives for Controlling Biological Invasions

What this says is that even with the best information available we may not know enough to estimate the net present value of control options with much confidence. Where does this leave the economic control of biological invasions? First, notice that even if ecological systems are neither observable nor controllable, they may still be stabilizable (Perrings 1991). What this means is that policy may be used to restrict pressure on the ecological system so that it can continue to function over the expected range of environmental conditions. Generic controls may be devised that protect the resilience of the ecological system. Such controls comprise ecologically sensitive safe minimum standards (Bishop and Ready 1991) whether or not supported by appropriate economic incentives.

Instruments that safeguard the range of future options by protecting thresholds of resilience are generally defined as sustainability constraints. The stabilization of ecological–economic systems through the application of sustainability constraints implies a precautionary approach. The precautionary principle holds that where the effects of some activity are uncertain but are potentially both costly and irreversible, society should take action to limit those effects before the uncertainty is resolved. The rationale for the principle is generally that the conjectured costs of not taking action are much greater than the known costs of preventative or anticipatory action (Taylor 1991). This is partly just the notion that an implicit benefit–cost analysis of activities with highly uncertain environmental effects should err on the side of caution. But there is another side to the precautionary approach. Costanza et al. (1998) argue that where an activity is potentially damaging, the burden of proof should lie with those whose activities are the source of damage. That is, it is also a principle about who should bear the burden of proof.

Because the general problem of biological invasions, like the more specific problem of the spread of communicable diseases, depends on the independent decisions of millions of individuals, the control of that problem requires instruments and institutions that alter the incentives they face. These should reflect differences in the level of uncertainty and the potential cost of the establishment and naturalization of invasive species. Some introductions are entirely accidental. For example, the spread of communicable diseases is seldom deliberate. Most introductions involve deliberate imports to support agriculture, horticulture, forestry, and fisheries. Yet market prices for seeds, foods, fibers, pesticides, and fertilizers generally do not reflect the ecological risks associated with their use. It follows that farmers have little incentive to take account of costs such as the depletion of indigenous species through predation, browsing, or competition; genetic alteration of indigenous species through hybridization; or the alteration of biogeochemical, hydrological, and nutrient cycles, soil erosion, and other geomorphological processes.

What we need are instruments that protect key thresholds where the costs of crossing those thresholds are uncertain but are conjectured to be high or irreversible and that confront people with the full cost of their behavior wherever the risks are known. This implies a regulatory regime to protect key species, habitats, and ecological services by controlling the introduction of potential invaders; an appropriate set of property rights in natural resources (along with their supporting institutions); a compensation mechanism; and a supporting structure of incentives and disincentives to induce the desired response. The instruments typically will differ for the prior control of species introductions and the posterior control of introduced species that have become invasive.

An additional problem is that there is a strong public good element in the control of invasive species. If control is left to the market, it will be undersupplied. More importantly, the public good involved in the control of infectious diseases and many other invasive species is of the weakest link variety. That is, the benefits from control to a whole society depend on the level of control exercised by the least effective member (Sandler 1997). If control over a communicable disease involves eradication campaigns in all nations, for example, that control will be only as good as the campaign run by the least effective nation. Rich nations typically have more effective public health programs than poor nations, and the public health programs in the poorest of the poor are almost nonexistent.

This points to two things. The first is intervention to change the incentives to those whose behavior determines the spread of an invasive species. The second is public investment in the control of IAS either nationally (where the potential range of the invader lies within national boundaries) or internationally (where the potential range of the invader crosses national boundaries).

Because of the public good nature of strategies to protect ecological services, environmental authorities might be expected to insist on higher levels of environmental protection than would be provided by the market. But because the introduction of exotic species may also yield substantial benefits or have no adverse effect, environmental authorities would not be expected to insist on total protection. Although the potential irreversibility of the costs of invasions and uncertainty over damages both indicate a precautionary approach to the control of introductions, this does not imply the exclusion of all nonindigenous species.

Maintaining biodiversity in agroecosystems by precluding the introduction of crop species that displace traditional crops certainly would reduce the risks of biological invasions. It might also reduce the risks of crop failure. Higher levels of genetic diversity in cultivated crops and wild relatives generally mean that yields can be maintained over a wider range of environmental conditions. Indeed, the genetic simplification of agriculture and forestry has already reduced the resilience of agroecosystems. More than 90 percent of world food supply derives from a small number of grasses (wheat, rice, and corn), nightshades (tomato and potato), mammals (cattle, sheep, and pigs), and birds (chickens and ducks) (Heywood 1995). The narrow genetic base of the food supply means that it is highly susceptible to disease and pest epidemics.

However, preventing new introductions would also entail substantial losses in welfare, especially if forgone development benefits turn out to be high and the importance of existing biodiversity turns out to be low. Enhanced disease resistance avoids enormous damage costs. To take just one example, the use of Ethiopian barley to protect California barley from dwarf yellow virus was valued at $160 million a year in 1995. These are very substantial benefits to forgo. The exclusion of all exotics would impose great costs on society.

Precautionary Instruments

What we should be looking for is a regime that allows the social benefits of new introductions while protecting society from the associated risks. The difficulty with new introductions is that the associated risks generally are uninsurable commercially for the reason that they are fundamentally uncertain and potentially very large. It is impossible to compute such risks actuarially. At present the risks of new introductions typically are borne by the state in the receiving country. They are limited only by the quantity and effectiveness of resources committed to screening and the exclusion policy adopted. The effectiveness of the screening process in turn depends on the resources committed to research into the consequences of the establishment and naturalization of introduced aliens. The effectiveness of the exclusion policy depends partly on the nature of the constraints (black list or white list), partly on the resources committed to the detection and prosecution of noncompliance, and partly on the incentive effects of the penalty regime. Moreover, just as at the international level, the effectiveness of the national system is only as great as the effectiveness of the weakest link in that system.

However, it is possible to make more constructive use of incentives in this area. There are precautionary instruments that can be used to protect society by altering the incentives to importers. The environmental assurance bond is an instrument developed to address the fact that, without market incentives, experimental research conducted by agents proposing innovative activities typically will not include all relevant potential future costs (Perrings 1989; Costanza and Perrings 1990). At present, environmental bonds are used by the state of New South Wales to protect against the IAS risks of aquaculture and there are similar schemes elsewhere (Shine, Williams, and Gündling 2000).

Because innovative activities are historically unique, there is no basis on which to establish ex ante markets in all potential future effects. Sequentially determined environmental bonds offer incentives to research the socially interesting outcomes of innovative activities. In the case of potentially invasive species they would work in the following way. Importers of new species or those undertaking high-risk activities would be required to post a bond equivalent to the conjectured damage if the species established, naturalized, and became a pest. This information might derive from the national screening service. But it might also derive from a central world or regional data and information source that could operate for invasive species generally in the same way as the U.S. Centers for Disease Control operates for communicable diseases.

The value of the bond would then be reassessed as additional data on the environmental risks emerged. The bond would be refunded if it could be shown that there was no risk or used to fund an eradication or control campaign if the risks were realized. An environmental assurance bond has the advantage that it protects society and, by shifting the burden of proof to those responsible for the introductions, provides an incentive to research the ecological consequences of introductions. The use of an equivalent to the U.S. Centers for Disease Control to provide data on the risks of invasive species has the additional advantage that it partly solves the problem of the weakest link.

The most effective control regimes still will not eliminate the risk of invasions. As in the case of screening and the control of species imports, the effectiveness of eradication, control, or mitigation measures for introduced species that do become invasive depends both on the quantity and quality of public resources committed to these measures and on the structure of incentives. The latter depends on an understanding of the way in which human behavior and invasive species interact. We have already made the point that it is difficult to estimate the probabilities of different outcomes of some control strategy solely on the basis of the characteristics of species or their habitat. The control of invasive species depends on human behavior, and the key element in any control strategy is likely to be the regulation of human behavior. This may well imply the use of penalties to deter behavior that increases the risks of invasions, but it may also imply the use of positive incentives to encourage behavior that reduces those risks.

In this case, however, even if the incentives are right, the weakest link problem remains. The protection provided to all is only as good as the protection provided by the weakest. This is a very practical problem with no ready solution. Although a central source of information on invasive species may provide the data and technical advice to support eradication, control, or mitigation, it would not have the resources to mount campaigns against particular invaders, nor, at present, would the United Nations Environment Programme in isolation. In the absence of a World Environment Organization commanding sufficient resources to fill this role, there seems to be little alternative to the Global Environment Facility sponsors, the United Nations Environment Programme, the United Nations Development Programme, and the World Bank. They should be urged to consider the establishment of a resource with the capability of protecting both global and regional interests from the threat of biological invasions by strengthening the weakest links in the chain.

Conclusion

Two processes above all else are the proximate causes of global biodiversity loss. The first is the destruction and fragmentation of habitat associated with the expansion of mining, forestry, and agriculture. Habitat fragmentation and loss in areas of high endemism are considered to be the major cause of species extinction worldwide. In most terrestrial systems a high proportion of original habitats have been converted to some specific economic use, and much more land is indirectly affected by economic activity. Indeed, it is no longer useful to describe any of the world's habitats as undisturbed. The main proximate cause of species loss therefore is better described as habitat disturbance, with the effect of disturbance depending on its intensity, nature, and location.

The second is the introduction of species—the problem of invasions. Most introductions consist of controlled imports to support agriculture and fisheries. However, an important subset of introductions are entirely uncontrolled. These include a range of pests and pathogens that affect the health of human and nonhuman species alike. Only a small proportion of introduced species establish themselves and spread, and not all invasive species are undesirable. However, the Global Biodiversity Assessment (Heywood 1995) concluded that invasives generally have negative effects on both species and genetic diversity at local and global levels. These effects include the depletion of indigenous species through predation, browsing, or competition; the genetic alteration of indigenous species through hybridization; and the alteration of ecosystem structure and function including biogeochemical, hydrological, and nutrient cycles, soil erosion, and other geomorphological processes.

Behind these proximate causes is a set of institutional and market conditions that increases the susceptibility of countries to invasive species and encourages resource users to ignore the consequences of their actions. Markets fail to accommodate the risks posed by invasive species. Although many private benefits of species introductions (or land use that increases the susceptibility of ecosystems to species introductions) are captured in market prices, many of its social costs are not. Markets for seeds, foods, fibers, pesticides, and fertilizers drive specialization in agriculture but do not signal its social costs. Any policy for biodiversity conservation accordingly has four key elements:

- A regulatory regime to protect key species, habitats, and ecological services and to control the introduction of invasive species
- An appropriate set of property rights in natural resources (along with their supporting institutions)
- A compensation mechanism
- A supporting structure of incentives and disincentives to induce the desired response

In the case of biological invasions, however, there is an extra dimension. The most common justification for action to conserve biodiversity lies in the fact that the genetic information it contains is a global public good. This is the rationale both for international effort to conserve hotspots and for the incremental cost approach adopted by the Global Environment Facility. We have argued here that the main costs and benefits of actions to control biological invasions are local. However, biological invasions almost always involve two or more countries, with the actions of one affecting the welfare of another. They involve a transboundary externality.

Similarly, where the costs of failure to control invasives affect more than one country, the solution entails international cooperation. The fact that centers of endemism or range size rarity can be in different locations from centers of species richness is important in the development of control strategies. Species richness typically is correlated with habitat heterogeneity, particularly in mountains, but may be associated with variation in soils and landscape or interannual variation in rainfall. These need not coincide with centers of endemism where the vegetation is composed of rare species with small geographic ranges. Species of restricted distribution in centers of endemism would not generally be expected to be resilient to anthropogenic stress or shocks. In centers of unrestricted range species richness, however, it may be possible to alter environmental conditions without unduly adverse effects. Control strategies and international cooperation to control invasives should be sensitive to this possibility.

Because of the public good nature of strategies to protect environmental health, all will entail public investment in control. The particular difficulty in strategies to control invasive species is the uncertainty associated with different control options. We have suggested that despite the particular difficulties with the base load problem in the case of invasive species, it is still reasonable to evaluate public investment in control options using a benefit–cost framework. However, as in other areas of environmental management where there are high levels of uncertainty and where the costs of errors are potentially very high, it is important that the control protects the capacity of the system to absorb the stresses and shocks of biological invasions. Although this does not favor the blanket eradication of all invaders, it does call for caution in the designation of the release area for screened species.

The potential irreversibility of the costs of invasions and the uncertainty of the damages they may cause both indicate a conservative approach to their management. But this has to be tempered by a realistic appraisal of the costs and benefits of the options. Environmental assurance bonds are an example of the sort of mechanism needed to address the uncertainty inherent in estimating the costs of biological invasions. Instruments of this sort exploit the fact that private research into the consequences of species introductions is profit driven. Just as research into the side effects of new drugs is motivated by the threat of legal action, so research into the ecological consequences of new introductions may be motivated by the liability built into assurance bonds. In other words, the problem is both to make sure that the potential costs of species introductions or species adoption are properly researched and to use the information generated to make rational decisions about the appropriate form of control.

References

Anaman, K. A., M. G. Atzeni, D. G. Mayer, and J. C. Walthall. 1994. "Economic assessment of preparedness strategies to prevent the introduction or the permanent establishment of screwworm fly in Australia," *Preventive Veterinary Medicine* 20(1–2):99–111.

Arrow, K. J., M. L. Cropper, G. C. Eads, R. W. Hahn, L. B. Lave, R. G. Noll, P. R. Portney, and M. Russell. 1996. "Is there a role for cost–benefits analysis in environmental, health and safety regulation?" *Science* 272:221–222.

Arrow, K. J., and A. C. Fisher. 1974. "Environmental preservation, uncertainty, and irreversibility," *Quarterly Journal of Economics* 88(2):312–319.

Bangsund, D. A., F. L. Leistritz, and J. A. Leitch. 1999. "Assessing economic impacts of biological control of weeds: The case of leafy spurge in the northern Great Plains of the United States," *Journal of Environmental Management* 56:35–43.

Bishop, R. C., and R. C. Ready. 1991. "Endangered species and the safe minimum standard," *American Journal of Agricultural Economics* 73:309–312.

Cohen, A. N., J. T. Carlton, and M. C. Fountain. 1995. "Introduction, dispersal and potential impacts of the green crab *Carcinus maenas* in San Francisco Bay, California," *Marine Biology* 122(2):225–237.

Conway, G. R. 1993. "Sustainable agriculture: The trade-offs with productivity, stability and equitability," in *Economics and Ecology: New Frontiers and Sustainable Development,* edited by E. B. Barbier. Chapman & Hall, London.

Costanza, R., F. Andrade, P. Antunes, M. van den Belt, D. Boersma, D. Boesch, F. Catarino, S. Hanna, K. Linburg, B. Low, M. Molitor, J. G. Pereira, S. Rayner, R. Santos, J. Wilson, and M. Young. 1998. "Principles for sustainable governance of the oceans," *Science* 281:198–199.

Costanza, R., and C. Perrings. 1990. "A flexible assurance bonding system for improved environmental management," *Ecological Economics* 2:57–76.

Daily, G., ed. 1997. *Nature's Services: Societal Dependence on Natural Systems.* Island Press, Washington, DC.

Dalmazzone, S. 2000. "Economic factors affecting vulnerability to biological invasions," in *The Economics of Biological Invasions,* edited by C. Perrings, M. Williamson, and S. Dalmazzone, 17–30. Edward Elgar, Cheltenham, UK.

Delfino, D., and P. Simmons. 2000. "Infectious diseases as invasives in human populations," in *The Economics of Biological Invasions,* edited by C. Perrings, M. Williamson, and S. Dalmazzone, 31–55. Edward Elgar, Cheltenham, UK.

Enserink, M. 1999. "Biological invaders sweep in," *Science* 285:1834–1836.

Fisher, A. C., and W. M. Hanemann. 1983. *Option value and the extinction of species,* Working Paper 269, Giannini Foundation of Agricultural Economics, University of California, Berkeley.

Glowka, L., F. Burhenne-Guilmin, and H. Synge. 1994. *A Guide to the Convention on Biological Diversity.* IUCN, Gland, Switzerland.

Heywood, V., ed. 1995. *Global Biodiversity Assessment.* Cambridge University Press, Cambridge, UK.

Higgins, S. I., E. J. Azorin, R. M. Cowling, and M. J. Morris. 1997. "A dynamic ecological–economic model as a tool for conflict resolution in an invasive alien-plant, biological control and native-plant scenario," *Ecological Economics* 22(2):141–154.

Hill, G., and D. Greathead. 2000. "Economic evaluation in classical biological control," in *The Economics of Biological Invasions,* edited by C. Perrings, M. Williamson, and S. Dalmazzone, 208–223. Edward Elgar, Cheltenham, UK.

Hirsch, S. A., and J. A. Leitch. 1996. *The Impact of Knapweed on Montana's Economy.* Department of Agricultural Economics, North Dakota State University, Fargo, North Dakota, Agricultural Economics Report 355.

Kareiva, P., I. M. Parker, and M. Pascual. 1996. "Can we use experiments and models in predicting the invasiveness of genetically engineered organisms?" *Ecology* 77:1670–1675.

Kasulo, V. 2000. "The impact of invasive species in African lakes," in *The Economics of Biological Invasions,* edited by C. Perrings, M. Williamson, and S. Dalmazzone, 183–207. Edward Elgar, Cheltenham, UK.

Katzman, M. T. 1988. "Pollution liability insurance and catastrophic environmental risk," *Journal of Risk and Insurance* 55:75–100.

Khalanski, M. 1997. "Industrial and ecological consequences of the introduction of new species in continental aquatic ecosystems: The zebra mussel and other invasive species," *Bulletin Français de la Peche et de la Pisciculture* 344–345:385–404.

Knowler, D., and E. B. Barbier. 2000. "The economics of invading species: A theoretical model and case study application," in *The Economics of Biological Invasions,* edited by C. Perrings, M. Williamson, and S. Dalmazzone, 70–93. Edward Elgar, Cheltenham, UK.

Law, R., A. J. Weatherby, and P. H. Warren. 2000. "On the invasibility of persistent protist communities," *Oikos* 88:319–326.

Lawton, J. 1999. "Are there general laws in ecology?" *Oikos* 84:177–192.

Loomis, J., and P. du Vair. 1993. "Evaluating the effect of alternative risk communication devices," *Land Economics* 69(3):287–298.

Lovett, J. 2000. "The economics of invading species: A theoretical model and case study application," in *The Economics of Biological Invasions,* edited by C. Perrings, M. Williamson, and S. Dalmazzone, 70–93. Edward Elgar, Cheltenham, UK.

MacIsaac, H. J., I. A. Grigorovich, J. A. Hoyle, N. D. Yan, and V. E. Panov. 1999. "Invasion of Lake Ontario by the Ponto-Caspian predatory cladoceron *Cercopagis pengoi,*" *Canadian Journal of Fisheries and Aquatic Sciences* 56:1–5.

Mack, R. N. 1997. Plant invasions: Early and continuing expressions of global change. Pages

205–216 in Huntley, B., W. Cramer, A. V. Morgan, H. C. Prentice, and J. R. M. Allen, eds., *Past and Future Rapid Environmental Changes: The Spatial and Evolutionary Responses of Terrestrial Biota.* NATO ASI Series. Series 2: Global Environmental Change, Vol. 47. Berlin: Springer-Verlag.

McDaniels, T., M. Kamlet, and G. Fischer. 1992. "Risk perception and the value of safety," *Risk Analysis* 12(4):495–503.

OTA. 1993. *Harmful Non-Indigenous Species in the United States.* U.S. Office of Technology Assessment, United States Congress, Washington, DC.

Parker, I. M., D. Simberloff, W. M. Lonsdale, K. Goodell, M. Wonham, P. Kareiva, M. Williamson, B. Von Holle, P. B. Moyle, J. E. Byers, and L. Goldwasser. 1999. "Impact: Toward a framework for understanding the ecological effects of invaders," *Biological Invasions* 1:3–19.

Pearce, D. W., and D. Moran. 1994. *The Economic Value of Biodiversity.* Earthscan, London.

Perrings, C. 1989. "Environmental bonds and environmental research in innovative activities," *Ecological Economics* 1:95–115.

———. 1991. "Ecological sustainability and environmental control," *Structural Change and Economic Dynamics* 2:275–295.

Perrings, C., M. Williamson, and S. Dalmazzone, eds. 2000. *The Economics of Biological Invasions.* Edward Elgar, Cheltenham, UK.

Pimentel, D., L. Lach, R. Zuniga, and D. Morrison. 2000. "Environmental and economic costs associated with non-indigenous species in the United States," *BioScience* 50:53–65.

Ricciardi, A., and H. J. MacIsaac. 2000. "Recent mass invasion of the North American Great Lakes by Ponto-Caspian species," *Trends in Ecology and Evolution* 15:62–65.

Sandler, T. 1997. *Global Challenges.* Cambridge University Press, Cambridge, UK.

Sharov, A. A., and A. M. Liebhold. 1998. "Bioeconomics of managing the spread of exotic pest species with barrier zones," *Ecological Applications* 8(3):833–845.

Sharov, A. A., A. M. Liebhold, and E. A. Roberts. 1998. "Optimizing the use of barrier zones to slow the spread of gypsy moth (*Lepidoptera: Lymantriidae*) in North America," *Journal of Economic Entomology* 91(1):165–174.

Shine, C., N. Williams, and L. Gündling. 2000. *A Guide to Designing Legal and Institutional Frameworks on Alien Invasive Species.* IUCN, Gland, Switzerland.

Shogren, J. 2000. "Risk reduction strategies against the 'explosive invader,'" in *The Economics of Biological Invasions,* edited by C. Perrings, M. Williamson, and S. Dalmazzone, 56–69. Edward Elgar, Cheltenham, UK.

Shogren, J., and T. Crocker. 1991. "Risk, self-protection, and ex-ante economic value," *Journal of Environmental Economics and Management* 20(1):1–15.

Smith, C. S., W. M. Lonsdale, and J. Fortune. 1999. "When to ignore advice: Invasion predictions and decision theory," *Biological Invasions* 1:89–96.

Taylor, P. 1991. "The precautionary principle and the prevention of pollution," *ECOS* 124:41–46.

Turpie, J., and B. Heydenrych. 2000. "Economic consequences of alien infestation of the Cape Floral Kingdom's fynbos vegetation," in *The Economics of Biological Invasions,* edited by C. Perrings, M. Williamson, and S. Dalmazzone, 152–182. Edward Elgar, Cheltenham, UK.

Watkinson, A. R., R. P. Freckleton, and P. M. Dowling. 2000. "Weed invasion of Australian farming systems: From ecology to economics," in *The Economics of Biological Invasions,* edited by C. Perrings, M. Williamson, and S. Dalmazzone, 94–114. Edward Elgar, Cheltenham, UK.

White, P. C. L., and G. Newton-Cross. 2000. "An introduced disease in an invasive host: The ecology and economics of rabbit calicivirus disease (RCD) in rabbits in Australia," in *The Economics of Biological Invasions,* edited by C. Perrings, M. Williamson, and S. Dalmazzone, 117–137. Edward Elgar, Cheltenham, UK.

Wilcove, D. S., D. Rothstein, J. Dubow, A. Phillips, and E. Losos. 1998. "Quantifying threats to imperiled species in the United States," *BioScience* 48:607–615.

Williamson, M. 1992. "Environmental risks from the release of genetically modified organisms (GMOs): The need for molecular ecology," *Molecular Ecology* 1:3–8.

———. 1996. *Biological Invasions.* Chapman & Hall, London.

———. 1998. "Measuring the impact of plant invaders in Britain," in *Plant Invasions. Ecological Mechanisms and Human Responses,* edited by S. Starfinger, K. Edwards, I. Kovarik, and M. Williamson, 57–70. Backhuys, Leiden, The Netherlands.

———. 1999. "Invasions," *Ecography* 22:5–12.

Williamson, M., and A. Fitter. 1996. "The varying success of invaders," *Ecology* 77:1661–1666.

Zavaleta, E. 2000. "Valuing ecosystem services lost to *Tamarix* invasion in the United States," in *Invasive Species in a Changing World,* edited by H. A. Mooney and R. J. Hobbs. Island Press, Washington, DC.

3

Vector Science and Integrated Vector Management in Bioinvasion Ecology: Conceptual Frameworks

James T. Carlton and Gregory M. Ruiz

> The seedsman inadvertently brushes against the purchaser's coat. . . . His hand touches the tiny seeds, or his sleeves; dusting himself with his handkerchief, they are transferred from pocket to pocket; they hide themselves beneath his finger-nails, they fall into his shoes; departing, he carries away seed lifted unbeknown in a dozen ways. Reaching the station, they are shed on floors and carpets, they are swept out dry in dust; they are carried abroad glued to wet boots; they adhere to gaiters and saddle-gear. . . . Had Romeo to such matters seriously inclined . . . I am confident that, after his interview with Juliet, her nightie and hair must have been plastered with seeds.
>
> —GUTHRIE-SMITH (1921) on the spread by humans of the seed of yarrow, *Achillea millefolium* (the genus named after Achilles)

The unintentional transfer of species by human activities is the primary driver of biological invasions, which in turn have become a major force of ecological and economic change operating on a global scale. Tens of thousands of species of terrestrial, freshwater, and marine organisms are moved accidentally around the world every day by myriad human-mediated vectors. With steadily increasing global trade and with a steadily increasing number of people moving around the world, the number and rate of such accidental movements show no sign of decline. Without due attention to the means by which species gain access to new regions, the number of new successful invasions of exotic species (i.e., established nonnative populations) will continue to increase around the world on land and in lakes, rivers, and the ocean.

Clearly the easiest means to prevent new invasions is vector interception or disruption, whereby the capacity of the vector to move species is greatly constrained. Such vector-level management has many distinct advantages, including the ability to affect simultaneously the delivery of many species or indeed of even entire communities and no requirement for a priori knowledge, prediction, or risk assessment of possible impacts for particular species (Ruiz and Crooks 2001). Thus, recognition of the vector as the most vulnerable and directly manageable portion of the invasion sequence should lead to the measurable prevention and reduction of species invasions.

However, fundamental to vector management is an understanding of vector operation and the development of a conceptual framework through which vectors can be characterized and quantified adequately to guide effective interception strategies. We set out that framework here. We also set forth an expanded concept of integrated vector management (IVM) as a necessary component of invasion management theory and science.

Historical Analyses of Vectors

The mechanisms by which human activities have led to the dispersal of animals, plants, and other organisms have been the subject of scholarly discussion since at least the nineteenth century. General works include Darwin (1859); Guthrie-Smith (1921); Clark (1949); Runnstrom and Vivier (1953); Elton (1958); Purseglove (1965); IUCN (1967); Crosby (1972, 1986); Mann (1979); Wilson and Graham (1983); Courtenay and Stauffer (1984); Shotts and Gratzek (1984); MacKenzie et al. (1985); Wells et al. (1986); Kruger, Richardson, and van Wilgen (1986); Mooney and Drake (1986); Welcomme (1988, 1992); Drake et al. (1989); Carlton (1992); Rosenfield and Mann (1992); McKnight (1993); OTA (1993); Munro, Utting, and Wallentinus (1999); Sandlund, Schei, and Viken (1999); Claudi and Leach (2000); Ruiz et al. (2000); and Wittenberg and Cock (2001).

Numerous additional monographs and individual studies have focused on specific regions, routes, vectors, corridors, or the transport of selected taxa. Hehn (1885) argues extensively for early historical episodes of anthropogenic dispersal, particularly of Asian plants to Europe. A rarely cited monograph is that of Kew (1893), who reviewed many early means of human-mediated (synanthropic) dispersal of terrestrial and freshwater mollusks. For example, he noted how visits by French vessels led to the introduction of the European land snail (*Helix aspersa*), carried aboard as food, to such places as Cape Town (South Africa) and the Loyalty Islands (South Pacific). Guthrie-Smith's (1921) magnificent treatise on the alteration of the flora of New Zealand grouped introduced species by their transport purpose or vector, with separate chapters for "stowaways," "garden escapes," "children of the Church," "pedestrians," and other more obscure categories, such as "burdens of sin" (an allegory to the shedding of species by other species, i.e., species transported accidentally with other taxa). As early as the 1920s, motor vehicles were being implicated as vectors (Guthrie-Smith 1921: 240–241; Pereira 1926; Rice 1930).

Mead (1961) reviews late-nineteenth- and early-twentieth-century examples of human-mediated vectors for terrestrial invertebrates, especially snails, including by people on their clothing, with agricultural products (such as bunches of bananas, which provide a crevicolous habitat), soil, nursery plant stock, horticultural exhibitions, stored and abandoned war equipment, and railroads. Mead further notes a case in which attempts at control and eradication of the giant African snail (*Achatina fulica*) led to further dispersal when attempts were made to drown the snails in a stream.

Desmond (1995) provides a detailed overview of the methods by which plants were transported for centuries around the world to and from the Kew Gardens in England, citing such early works as Ellis's (1770) *Directions for Bringing Over Seeds and Plants from the East Indies and Other Distant Countries.* The most famous of the devices for facilitating the global transport of live plants was a sealed box invented in 1835 by Nathaniel Ward. The box became universally known as the Wardian case (Spongberg 1990; Desmond 1995). Desmond quotes Sir William Hooker as congratulating Ward in 1851 that his cases "have been the means in the last fifteen years of introducing more new and valuable plants to our gardens than were imported during the preceding century," to which we add that the Wardian case must have also spread around the world an incalculable number of associated terrestrial organisms, including protists, nematodes, flatworms, nemerteans, annelids, mollusks, crustaceans, insects, and other arthropods.

The long and often bizarre histories of "acclimatization societies," dedicated to the deliberate movements of plants and animals around the world, have been treated in part by Guthrie-Smith (1921), Lever (1992), McDowall (1994), and Baskin (2002). A classic treatise on the role of sea level canals in invasions is that of Por (1978). The role of shipping vectors in transporting marine life was reviewed by Woods Hole Oceanographic Institution (1952); Carlton (1985); Carlton and Geller (1993); Carlton and Hodder (1995); and Carlton, Reid, and van Leeuwen (1995). Fuller, Nico, and Williams (1999) provide an overview of fish introductions by a wide range of vectors, including both intentional stocking and unintentional transfers, in fresh and brackish waters of the United States. A rich literature exists on terrestrial biocontrol and the hundreds of species released since the nineteenth century to attempt to manage prior invasions (Howarth 1991; Hokkanen and Lynch 1995; Jervis and Kidds 1996; Andow, Ragsdale, and Nyvall 1997).

This is by no means an exhaustive list of the literature on vectors, underscoring the pervasive and complex nature of species transfer mechanisms that intersect with so many human enterprises. A workshop held in 1999 (Ruiz and Carlton 2003) further highlighted the operation of vectors across various taxonomic groups and ecosystems. The results of this workshop were sobering: the rate of known invasions is increasing over time and results from a diverse array of human activities. However, the increased rate of invasion within each ecosystem (or taxonomic group) often was driven by one or few dominant vectors (e.g., in marine systems; Ruiz et al. 2000). This suggested that a rigorous evaluation of vector operation (or vector analysis) can be used to target these dominant vectors and can greatly reduce the rate of new invasions. Our intent in this chapter is to outline such an approach, developing key elements for vector analysis.

Framework to Characterize Vectors

We present here a framework (Boxes 3.1 and 3.2) to characterize the human-mediated movement of living organisms, that is, a framework for vector science (named after one component—the dispersal mechanism—of the entire framework). There are six primary elements in this framework: cause, route, vector, vector tempo, vector biota, and vector strength. Terminology has varied widely (and thus created ambiguity in meaning) in invasion science in general and in vector science in particular. For example, *pathway* often is used with very different meanings; it is used in the literature for four of the phenomena in our framework (cause, route, vector, and corridor). We avoid the use of this term here. For clarity and precision, we suggest using separate, more specific terms for each of these different elements. Boxes 3.1 and 3.2 indicate common synonyms for the terms that we apply to components of vector science and management. We briefly elaborate each of these six elements.

Cause

The cause is the reason why a species is transported, which may be accidental (unintentional or inadvertent) or deliberate (intentional or planned). Purpose is the reason why a species is deliberately introduced. For many introductions, especially of taxa such as fish and terrestrial plants, whether the release was accidental or deliberate may not be known or may be obfuscated for political, legal, or other purposes. For a number of taxa, both accidental and deliberate introductions have been part of their dispersal his-

Box 3.1. A Framework for Vector Science: Cause, Route, and Vector

CAUSE **Why a species is transported, that is, whether accidentally or deliberately (if the latter, see *Purpose*)**

Examples: The European shore crab *Carcinus maenas* was accidentally introduced through shipping operations to eastern North America and elsewhere.

Synonyms: Pathway, enterprise, activity, trade, endeavor, commerce, motive, rationale, incentive, reason

of accidental: Unintentional, inadvertent, escape, chance

of deliberate: Intentional, planned, purposeful, premeditated, planted, direct

Purpose: Why a species is deliberately introduced

Examples: Food resource, ornamentation (aesthetics), biocontrol, pets, medicine

Synonyms: As above

(*continues*)

Box 3.1. Continued

ROUTE **The geographic path over which a species is transported from the origin (donor area) to the destination (target area)**

Examples: A route may be from Rio de Janeiro (Brazil) to Le Havre (France).

Synonyms: Pathway, path, passageway, course, corridor

Corridor: The physical conduit over or through which the vector moves (i.e., within the route)

Examples: Footpaths, roads, highways, canals, shipping lanes, trails, railroad beds

Synonyms: Pathway, conduit, path

VECTOR How a species is transported, that is, the physical means or agent

Examples: Ballast, ships' hulls, movement of commercial oysters, clothing, animal feed, vehicles

Synonyms: Pathway, mode, dispersal mechanism, transport mechanism, manner, carrier, bearer, method

Cryptovectic and Polyvectic Species: Species for which the vector of introduction is not known are cryptovectic, a term introduced here (from the Greek *crypt-*, "secret," and *vect-*, Latin, *vectus,* past participle of *vehere,* "to carry")

Species having two or more means of being transported are polyvectic, a term introduced by Cohen (1997); from the Greek *polys,* "much, many"; *vect-,* Latin (as above). Cohen defined *polyvectic* as "having many vectors or means of being transported.

Box 3.2. A Framework for Vector Science: Vector Tempo, Vector Biota, and Vector Strength

VECTOR TEMPO **How a given vector operates through time, in terms of size and rate, speed, and timing**

Size and Rate: The frequency with which the vector operates to deliver propagules to the target, measured as the quantity of the vector (in units appropriate to the vector) expressed per unit time

Examples: Gallons of ballast water per hour, number of container boxes per hour, number of logs per day

Duration: The length of time it takes for the vector to move species from the donor area to the target area

(*continues*)

VECTOR TEMPO (CONTINUED)	*Timing:* The period (such as time of day, season, or other intervals) when the vector is active and delivers propagules to the target area
VECTOR BIOTA	**Description of the biota (the propagules) transferred by a given vector, in terms of diversity, density, and condition** *Synonyms:* Propagule pressure, inoculant *Diversity:* The species richness, or number of different organism types, associated with the vector *Density:* The concentration or abundance of organisms, often expressed per taxon *Condition:* The physiological condition or quality of propagules upon delivery to target area
VECTOR STRENGTH	**The relative number or rate of established invasions that result within a specified time period from a given vector in a particular geographic region** *Synonyms:* Magnitude, importance

tory. Deliberate introductions may be authorized (legal) or unauthorized (which may be illegal or legal).

Some workers (P. Moyle, pers. comm. 2000) have noted that referring to certain types of releases as accidental is problematic if those involved in the releases are informed of their actions but continue thereafter to release species (e.g., the release of plankton-rich ballast water by ships), even if their actions are done for other reasons (e.g., the need to discharge the water aboard ship in order to load cargo).

Route

For any route, the geographic path over which a species is transported from one area to another, the contributing region is the origin or donor area; the destination region is the receiving or target area. The area from which a particular introduction originates may not be the area to which the species is native, of course. The physical conduit over or through which the vector moves within a route is the corridor. Examples of corridors are roads, paths, highways, canals (both lock and sea level), shipping lanes, and railroad tracks and beds (the rock substrate of which is called ballast). Changes in corridors may be important in vector history, and details of such often are lacking in vector descriptions. On mountainous islands, for example, plant invasions may be restricted to one part of the island until roads and paths are constructed. Railroads have played a similar major function as corridors in plant invasions, which were previously restricted by physical barriers. Intracontinental canals have long played roles in the interbasin and intrabasin homogenization of aquatic floras and faunas (Crossman and Cudmore 2000; Mills, Chrisman, and Holeck 2000).

It is often difficult to determine the exact origin of a new invasion because many species that now occur widely may be entrained on many routes. In this case, genetic comparisons of the new population with an array of other populations (both in the native region of the species and elsewhere) may prove of value in determining an origin and thus a route (Davies and Roderick 2000).

Vector

The vector is the physical means or agent by which a species is transported. On land, such agents are vehicles (automobiles, jeeps, trucks, buses, wagons, trains) and carts (wheelbarrows, handcarts). In the air, agents include airplanes, helicopters, balloons, and other airborne devices. In the water, agents are vessels (ships, boats) of many types and purposes.

It is often important to define the physical subcomponents of each agent in order to understand its full potential and to fully define control and management strategies. Thus, a land vehicle may carry species internally (e.g., engine block, passenger spaces, under and in floor mats, or trunk) and externally (e.g., tire wells, window wells, hood and trunk grooves, or window washer grooves). A ship may carry species internally (e.g., in ballast tanks, cargo holds, piping systems, bilge spaces, or anchor chain lockers) and externally (e.g., on the hull, rudder, propeller, or anchor).

Finer-grain sampling may be needed: for example, in a ballast tank of a ship there may be organisms in the water column, in sediment on the tank bottom, as a biofilm on tank walls, or as macroorganisms in a fouling community. Sampling only one or two of these components will lead to an underestimate of biotic diversity and density. Full sampling of these finer components of a vector may entail considerable investment of personnel, time, money, and other resources. Coarser assessments undoubtedly have led to many invasions through the failure to recognize the full complement of vector subcomponents.

Living organisms may be associated with the conveyance itself or with goods, articles, and products the conveyance is transporting or conveying. These goods may be organic animal or plant products (e.g., edible fruit, edible vegetables, flowers, tobacco, or wood) or inorganic (e.g., ore, soil, tires, machinery, clothing, or household goods). The condition of the goods may influence the associated biota. Transported goods may have associated protection (cushioning), packaging, or weighting materials. Material used to protect cargo is called dunnage; such materials include other cargo, sawdust, wood, and sacks. Packing materials include pallets, cartons, boxes, crates, and other containers. Weighting materials usually are called ballast. Over the course of millennia, materials used for ballast have ranged from every conceivable dry or solid object that would provide weight (e.g., rocks, sand, debris, bricks, and other building materials) to wet resources (freshwater or saltwater). Thousands of insect and plant species have been transported globally in such cushioning, packaging, and ballast.

Polyvectic versus Cryptovectic Species

The history of the transport of many species is cryptic. Species for which a vector is truly not known are rare; these are cryptovectic species (see Box 3.1). An example of a cryptovectic species was the discovery of an epibenthic Western Pacific (Japanese) star snail (*Guildfordia yoka*) at 34 meters' depth in 1912 in the Eastern Pacific, in the open ocean about 1 kilometer west of San Francisco, California (Carlton 1979). No vector is known that would explain the one-time collection of this species in North America.

Many species have been transported by multiple means, either historically over time as vectors have changed (e.g., the replacement of ballast rocks and sand by ballast water) or simultaneously. Cohen (1997) has introduced the term *polyvectic* for taxa transported by multiple mechanisms (see Box 3.1). Polyvectic species may have both verified and suspected vectors of transport. In turn, the vector for many introduced species often is indicated in the literature as "unknown," when *polyvectic* is meant. That is, possible or probable vectors are known, but which exact vector is responsible for transport of a given introduction event (i.e., established population) is not known.

Polyvectism is a significant management challenge. We use the history of global transport of the European shore (green) crab (*Carcinus maenas*) as an example. *Carcinus maenas* is native to western and northern Europe. Since the 1700s it has been transported to eastern North America (at the turn of the eighteenth century), Australia (at the turn of the nineteenth century), and western North America, South Africa, and Japan (in the twentieth century). This dispersal history was accompanied by an ever-increasing number of intercontinental or transoceanic vectors or vector subcomponents (Fig. 3.1). In 1800, only two transport mechanisms existed to transport *Carcinus*

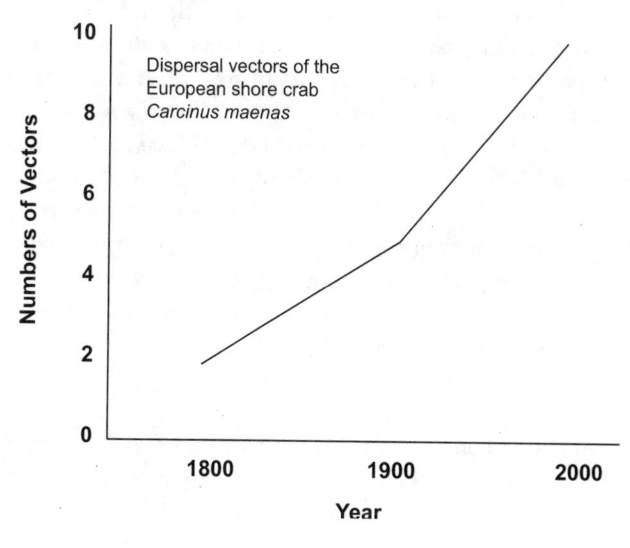

Figure 3.1
Dispersal vector example, *Carcinus maenas*

out of Europe: ships' solid (rock) ballast and ships' hull fouling. By 1900, the number of mechanisms had more than doubled: solid ballast and hull fouling were still in existence, but water ballast, ships' seawater piping systems (for both ballast and washroom use), and the movement of commercial oysters were also available as means by which *Carcinus* could be transported. By 2000, the number of mechanisms had doubled again: *Carcinus* could now be moved by all the previous mechanisms (except for solid ballast) as well as for fish bait, in algal dunnage for bait worms, with other shellfish (such as lobsters), in the aquarium trade, on semisubmersible exploratory petroleum platforms, and through biological supply houses. Given this plethora of transport mechanisms, what strategies should be instituted to prevent the further invasion of *Carcinus* to new regions? Critical to addressing this question is knowledge of the diversity of vectors, and their subcomponents, that impinge on the area of concern. Unfortunately, such knowledge often is either nonexistent or incomplete.

Vector Tempo

Vector tempo is the means by which a vector operates through time. There are three principal components of tempo: frequency, duration, and timing.

Frequency: Vector Size as a Rate Function

A major component of the description of a vector's scale of activity is the frequency with which the vector operates in terms of quantity per unit time (e.g., container boxes or gallons of ballast water per hour, per day, or per week). The quantity is a measure of the vector itself rather than the number of propagules delivered (which falls under the biotic characterization of the vector); examples are shown in Box 3.2.

Obtaining accurate historical data on the frequency of a given vector, especially over long periods of time (e.g., decades or even centuries) often proves difficult. Such data often are critical factors in understanding changing patterns of invasions. Being aware of predicted changes in trade patterns and relating them to potential changes in invasions is rarely addressed because most biological scientists do not follow the literature of global trade economics. For example, Skjerve and Wasteson (1999) observed that "an increased volume of international food trade after the establishment of the World Trade Organization (WTO) will also speed up the transfer of pathogenic agents and microbial genes around the world. . . . So far, only minor effects of the WTO agreements have appeared, due to the high toll barriers replacing the previous legal barriers. Within a few years the full effect of WTO will emerge, due to a reduction in toll barriers, and a major increase in the trade of agricultural products is expected."

Duration

Vector duration—the length of time in hours, days, weeks, or sometimes months—that it takes to deliver propagules from one region to another also is a critical descriptor of vector tempo. Over historical time, vector duration has decreased. Automobiles, trains, airplanes, ships, and other agents have become faster, thus reducing voyage duration. Decreased voyage duration may lead to increased survival of the propagules in terms of all aspects of biotic characterization discussed in this chapter (species diversity, density, and condition).

Timing

The period during which a vector delivers propagules may have a major influence on multiple aspects of associated biota. For example, the seasonal timing of vector operation may influence density and diversity of entrained biota, survival during the transfer process, and survival upon arrival to a recipient environment. For the latter, if the vector is transiting through a mild climatic regime and delivers propagules to a seasonably hospitable donor area, initial survival of the propagules may be high. Time of day may be locally or regionally important in terms of whether the vector is more active at night or during the day. Trucks transiting at night during a hot summer may retain a cooler internal environment and thereby increase survival of certain entrained species.

Vector Biota

The propagules being carried by a given vector may be described in terms of diversity, density, and condition. Importantly, the biotic attributes described here may vary as a function of the components of vector tempo and source region. Rarely are such data sets available for most vectors.

Diversity

A fundamental component of vector management is the diversity of species being transported at any given time and their origins. With the possible exception of agricultural entomology, increasing limitations on taxonomic knowledge for most taxa (NRC 1995) will make assessments of vector biodiversity progressively more challenging. Potential and nonexclusive solutions include using genetic tools, such as molecular probes, to identify some taxa or taxonomic groups and considering the number of different organism types (in terms of supraspecific-level phylogenetic categories, feeding guilds, or other categorizations) associated with the vector as a surrogate means of assessing taxonomic or ecological diversity. Although this method may have value in certain contexts, it will not support risk assessment or decision models that seek to target specific potentially invading taxa.

Density

Assessing the abundance of a given taxon (in terms of number per unit space, either by square or by volume, depending on the vector) may assist in long-term vector assessment, particularly if quarantine, interception, or other management scenarios are in place that seek to reduce the abundance of life associated with a given vector. Although the abundance of organisms transported by a vector is one potential measure of vector strength, we suggest here that organismal density be used to understand the patterns of propagule delivery (propagule pressure) over time rather than as a direct measure of vector potency or importance.

A goal of invasion theory is to attempt to correlate the abundance of propagule inoculation with the probability of a species becoming established. Data on initial inoculum density (or multiple, sequential events, known as metainvasions; Davies and Roderick 2000) for accidental introductions are largely nonexistent. Therefore, this remains an important area for research.

Condition

The potential for postrelease survival of species as a result of short- or long-distance transport is a key but often missing component of the description of species flow and inoculation along a given vector. The initial performance of an organism is influenced by both its condition upon arrival and the environmental (including biological) attributes of the recipient region (which is external to the framework outlined here).

Condition of the species during and at the end of the transport event can include both life stage and physiological state upon arrival. Assessment of the physiological state of arriving propagules is especially rare and entails interception and sampling of the vector's transported biota, followed by studies of properly preserved material or living organisms. For example, preserved specimens may be analyzed for their lipid content, stress protein expression, or other biochemical responses, or for their histopathology, if comparative baseline physiological data are available.

In turn, living organisms recovered from the vector may be cultured under the physicochemical regime of the vector itself or under the physicochemical regime of the target environment to determine their capability (as larvae, juveniles, or adults) to survive, grow, and reproduce, as a reflection of their condition upon arrival. Thus metabolic aspects (e.g., oxygen consumption, locomotion, and feeding), growth aspects (e.g., development rate and the ability to undergo gametogenesis), and other characteristics of arriving organisms can be measured directly.

Temporal Characterization of Diversity, Density, and Condition along the Invasion Route

In addition to a static picture for each of the biotic characteristics associated with a vector at a specific time and place, it may be particularly valuable to understand the biological dynamics, especially if the goal is to consider appropriate targets for management action.

A sequence of four key points, or steps, along a transport route may permit direct biotic sampling for changes in diversity, density, and condition of propagules if access to the vector can be gained during and along the route.

The importance of understanding the impact of methods (sampling techniques and sample design) on the perception of diversity and density cannot be overemphasized. One of the major banes of vector science is the many different methods, applied with a wide range of quality control, that are used to sample the same vector, resulting in large volumes of noncomparative data.

Origin: The diversity, density, and condition of the biota associated with a given vector at the point of origin and at the time of departure.

Measurements of the biota at the departure place and time provide baseline data for all elements of the ensuing transport events. Important on specific regional bases for risk assessment purposes may be the determination that a given vector during a given transport event is actively transporting noxious, pestiferous, or other well-known harmful organisms. Such determinations could result in quarantine at the port of departure or severe limitations on subsequent ports of call relative to the release of contaminated cargo, ballast water, or other involved vectors. This said, a significant challenge is that departure (source) port biotas may be under a constant state of change, with some preexisting organisms becoming more abundant and thus more available for transport (and other species perhaps becoming less abundant) and with the arrival of new invasions. Much effort is needed to maintain an up-to-date assessment of the biota of a given region.

En route: The diversity, density, and condition of the biota while the species are being transported by a given vector along the route.

Measurements of the associated biota at one or more times along or during the transport process provide data for two critical fields: changes in diversity, density, and condition over time and documentation of organisms that are acquired along the route, after departure from the point of origin. The timing and location of en route measurements should reflect natural bottlenecks (e.g., major changes in temperature) and management "nodes" (e.g., vector management requirements or quarantine inspections).

Destination: The diversity, density, and condition of the biota associated with a given vector at the point of arrival but before the vector is exposed to the environment.

Measurements of the associated biota at the terminus of the transport event provide measures of survivorship (such empirical data may aid in calculations of survival probabilities) and baseline data on the diversity and condition of organisms released alive into the new environment. The species that are released into the open environment thus have been locally inoculated; they are not introduced species until they have established naturally sustaining and reproducing populations.

Postexposure or postrelease: The diversity, density, and condition of the biota remaining associated with the vector at the point of arrival, as measured after some fixed time interval after the vector has been exposed to the new area or the associated organisms released into the environment.

Measurements of the residual biota remaining with the vector after any associated organisms have been free to enter the new environment (e.g., through ballast water release, opening of shipping containers, or laying out of wooden pallets previously held in such containers) provide a critical data set on the proportions of taxa actually released. This residual biota remaining with the vector may then continue along the transport route, gaining additional species along the way. That is, certain organisms may not depart or leave the vector before it is in motion again.

Marine examples of measuring differential survival over some or all parts of these transport sequences include work on biota associated with shipping. Samples of entrained biota associated with ships have been taken at the beginning of, during, and at the completion of voyages. For ship ballast water, studies include Carlton (1985); Lavoie, Smith, and Ruiz (1999); Gollasch et al. (2000); Olenin et al. (2000); Wonham et al. (2001). For ship hull fouling, Carlton and Hodder (1995) assessed all four components (origin, en route, destination, and postrelease) for the voyage of a vessel along the American Pacific coast. Carlton and Hodder demonstrated that a vessel's hull fouling community may accumulate species at each port visit; a vessel arriving in San Francisco Bay had species on its hull that were recognizable as having fouled the vessel in three other ports before arrival in San Francisco Bay (where additional species were added to the assemblage).

Ship ballast studies have found that not all taxa survive the entire voyage. As Wonham et al. (2001) emphasize, determining mortality patterns for individual taxa (and by extension broader taxonomic or ecological guilds) along the invasion route can be a key element in predicting invasion success. They note that "invasion success of a particular taxon may be predicted both by high density at the end of a voyage (which is comparatively easy to measure) and by low mortality during a voyage (which may indicate good body condition, but is harder to measure)." Density and condition may not be independent predictors, depending on other factors such as voyage length. Of course, invasion success is subsequently modified by the nature of the receiving environment.

For management purposes, it is critical to stress that the number of species being imported often is underestimated, as is the number of resulting invasions. It is difficult to overstate the number of species in motion at any given time. This is a result of the physical and economic inability to intercept and examine the biotic diversity associated with all inbound shipments of species, whether they are the animals and plants in the millions of pieces of luggage crossing national borders daily or the tens of millions of gallons of ballast water released worldwide in ports hourly. A working assumption must be that vector management operations serve as a filter and not as a wall to exotic species invasions.

Vector Strength

We define *vector strength* as the number of established invasions that result from a given vector within a specified time period and location. It is thus an ultimate measure of invasions. Ideally, these data could be converted to rate of invasion measures if accurate timing data of sufficient quality exist for enough individual species (e.g., Ruiz et al. 2000). Furthermore, with adequate description of propagule pressure, a rate of establishment (relative to the inoculum pool) could also be estimated. Although the latter may be possible for some intentional introductions (e.g., Simberloff 1986), sufficient data rarely exist to estimate such rates for accidental introductions. In both cases, the number of invasions in an area should be quantified when possible, using natural biogeographic regions—or some other construct with biological or ecological meaning—rather than political boundaries. We note that later reducing such data to be political regimes may permit assessment of regional (and potentially different) management programs.

Alternative data sets often are used as proxies for the strength of a given vector; these include vector tempo and various measures of vector biota (including diversity and density). However, the relationship between vector tempo or propagule supply and invasion success remains largely untested in the field. For this reason, we suggest that the long-term signal of vector strength is best expressed as the number of invasions that result from the vector's activity (i.e., the ultimate outcome). It is important to emphasize that the vector simply delivers viable propagules and that invasion success is the result of an array of complex interactions between the invader and the physical, chemical, and biological nature of the recipient environment.

Vector Analysis and Vector Management

We discuss here several components of vector science and management that emerge from the aforementioned considerations: approaches to vector analysis, including polyvectism as a management challenge; the broad application of the concept of IVM; and tracking changes in vector operations.

Approaches to Vector Analysis

The most appropriate dependent variable to use as a measure of vector strength is the number of invasions (i.e., established introduced species) that result from the activity of a given vector. Thus the strongest vectors are those that have brought the most invaders to a given region. These vectors, if still active, would then be targeted for management action. Although we place highest priority on the use of vector strength to precipitate management action because of our confidence in it to identify vectors of known potency, we recognize the limitations of this approach. These include polyvectism (whereby one species may be transported by multiple means) and the fact that a lower-strength vector may lead to a major invasion. For example, in San Francisco Bay, California, a region impinged upon by numerous vectors, the European shore crab (*Carcinus maenas*) appears to have gained successful access to the area by being transported as juvenile crabs in algae (seaweed) used as packing material for bait worms (polychaete annelids) shipped from Maine to California. Fishers then discard the seaweed into the water when fishing. What appears to be a lower-strength vector (as measured by the criteria listed earlier) has resulted in what is widely considered to be a major invasion of the northeastern Pacific Ocean (Grosholz et al. 2000). Therefore, ranking and categorization of vectors do not mean that lower-strength vectors should be entirely ignored but rather that priorities may be driven first by the relative contribution of a vector to known invasions and second by the additional quantitative descriptors of a vector as noted earlier.

The scale of delivery (size and rate from a repeating donor area) combined with the biotic characteristics of the vector is a reasonable first (i.e., temporary) proxy or alert that should precipitate serious scrutiny and management action. Data that suggest that the environment into which the propagules are being released is hospitable, given the timing of such release matched with the seasonal climatic regime, may further support management action. (Note that taxa from a source region that experiences wide chemical and physical fluctuations may have a range of physiological plasticity that equals or exceeds any seasonal climatic ranges in the target area). In making such decisions, one should conceive the assessment of relative strength of a given set of vectors as a continuum rather than a set of discrete categories. For example, ranking a given vector globally or regionally as either minor or major may incorrectly reduce a complex state of affairs to an overly simplistic dichotomy and thus lead to an inadequate management strategy.

Despite the incomplete nature of existing data and the resulting limitations, we suggest that the best strategies are to target priority vectors and then launch IVM approaches (described in the next section).

Integrated Vector Management

As a fundamental management strategy and tool, the concept of IVM should be expanded to encompass the overall field of vector science, in all its elements in the framework described here. IVM was independently proposed by us as a strategy to accompany integrated pest management (IPM) at a meeting of the Global Invasive Species Program in Cape Town, South Africa, in September 2000. However, it has existed as a concept within the public health field of malarial epidemiology and mosquito control since at least the 1980s (Vasuki 1990; World Wildlife Fund 1998). For example, IVM has been described as being "nested within integrated disease management" (World Wildlife Fund 1998). Rather, we apply IVM to all vectors and to all vector management strategies.

We define IVM as a program that brings to bear available strategies and technologies at multiple stages of a vector's period of active life, from origin to destination, with the goal of reducing or preventing the transmission and release of living organisms. Ideally, such targeted action would focus on the sequences in the vector or life stages in which the greatest impact can be achieved and the approaches that are the most technically feasible and cost-effective, reducing impediments to full implementation.

As an example of the application of IVM, the uptake, transport, and potential release of living organisms in ships' ballast water can be broken into temporal components (Box 3.3). At each stage—origin, en route, and arrival—a series of integrated

Box 3.3. Integrated Vector Management: Ballast Water–Mediated Transport as an Example

At Origin

Before or during ballasting in departure port

- Assess shipside presence of
    - –Harmful algal blooms (toxic phytoplankton blooms)
    - –Sewage effluent
    - –Known noxious, pestiferous, and harmful species
    - –High sediment loads
- Assess timing of ballasting relative to
    - –Presence of nocturnal vertical migrating benthic and planktonic species
    - –Depth of water and potential to ballast up benthic organisms and tychoplankton due to turbulence
- Undertake ballast water treatment measures
- Educate ship's personnel about issues involved and rationale for the need for management

En Route

While vessel is at sea

- Exchange port or coastal water for open ocean water
- Undertake ballast water treatment measures

At Destination

When vessel arrives at destination port

- Undertake ballast water treatment measures
- Do not release or minimize ballast water release
- Transfer water to onshore facility or lighter vessels
- Return to open ocean for ballast water exchange

Source: Events and management strategies selected from Carlton, Reid, and van Leeuwen (1995).

strategies can be applied, the total effect of which should be to reduce the biotic diversity successfully transported. Limiting management strategies to any one stage of the process would be less effective than an integrated approach.

Tracking Vectors through Time and Space

Perhaps the central challenge in predicting the roles of the many elements of vector science and identifying the resulting appropriate management decisions is the changing nature of vectors over time and space (OTA 1993; Carlton 1996, updated in part in Carlton 1999). OTA (1993) discussed examples of factors affecting pathways and rates in terms of technological innovations and changes in social and political factors.

Carlton (1996, 1999) discusses the broad suite of factors that influence the timing of invasions in ecosystems, many of which are directly linked to changes in vectors. Changes in vector tempo can result in increases in species diversity, increases in abundance, and improvements in propagule condition (i.e., increases in the ability of post-transport individuals to survive and reproduce; Carlton 1996, 1999). Fundamental to appreciating such changes is an awareness of their existence, either now or as planned innovations. OTA (1993) thus noted several specific examples of changes in vectors and the potential to alter the biotic diversity, abundance, or condition of the transported biota. The change from dry ballast (soil, sand, and debris) to wet ballast (water) commencing in the mid- to late nineteenth century led to a change in biota transported from terrestrial arthropods and plants (and other taxa) to aquatic organisms. The increased speed resulting from the advent of steamships and the airplane led to increased survival of associated biota. Improvements in threshing and harvesting machinery led to decreased contamination of seed lots. The invention of specialized plastic coolers permitted the increased survival for longer distances of intentionally transported organisms—a modern Wardian case. The importation of used tires for retreading provided an additional vector for mosquitoes in the water held in such tires. Examples of sociopolitical factors noted by OTA (1993) include new patterns of immigration and tourism, globalization of trade and free trade agreements, and the expansion of the pet and horticultural industries.

Working synergistically with such dynamic factors are changes in both origin and destination areas (Carlton 1996, 1999). An important dynamic characteristic of origin areas is that the biotic diversity of such areas may be constantly changing because of new invasions and many other potential changes. Examples of changes in the destination area include alterations in water quality, biodiversity (including new invasions), or temperature regimes. Changes in destination areas may make such regions more or less hospitable to invasions. In turn, changes in the resistance or susceptibility to invasions may be highly species-specific responses.

The arrival of new invasions in an area can be described as a hub-and-spoke model: each new population (a new hub) interacts with an array of potential new routes (new spokes). Thus, a given species can be dispersed by a given number of vectors from its site of origin. But when that species is introduced to new site, forming what may be a distinct metapopulation, vector diversity may change, such that the species may be able to be dispersed by an increasing number or frequency of the same or new vectors to new regions to which its site of origin was not connected. When the species is introduced to these new sites, vector diversity may again change, and so on.

The rationale for limiting a species' spread therefore is not only to reduce its economic or environmental impact but also to limit its ability to interact with a greater number of hubs and a greater diversity of spokes.

Conclusion

The framework for vector science proposed here forms a critical and complex yet manageable matrix. In order to provide the scientific data needed to support management choices, broad vector research agendas should be implemented in most countries. Little or no information is available for most regions on the total diversity of vectors known to transport species to a given region, the diversity of species moved with and along such vectors, or the strength of any specific vector. In contrast, a great deal of speculation, assumption, estimation, and conjecture often is available, leading, not surprisingly, to diffuse strategies that leave numerous invasion windows open. Coupled with this is the need for adequate public information campaigns that describe in direct, simple terms the concerns about exotic species invasions and why new invasions must be prevented.

Numerous existing quarantine, control, and management efforts paint a clear picture that the situation is not hopeless. Indeed, public and political awareness of the impact of exotic species invasions at the beginning of the twenty-first century is unparalleled, suggesting that we can reduce species invasions dramatically if we implement powerful and dynamic vector management strategies.

Acknowledgments

We thank the many participants of the Vectors Workshop held in November 1999 at the Smithsonian Environmental Research Center in Edgewater, Maryland, whose presentations and discussions catalyzed many of the concepts presented here. Jeff Waage and Richard Everett provided valuable comments on the manuscript.

References

Andow, D. A., D. W. Ragsdale, and R. F. Nyvall, eds. 1997. *Ecological Interactions and Biological Control.* Westview Press, Boulder, CO.

Baskin, Y. 2002. *A Plague of Rats and Rubbervines.* Island Press/Shearwater Books, Washington, DC.

Carlton, J. T. 1979. "History, biogeography, and ecology of the introduced marine and estuarine invertebrates of the Pacific coast of North America," PhD dissertation, University of California, Davis.

———. 1985. "Transoceanic and interoceanic dispersal of coastal marine organisms: The biology of ballast water," *Oceanography and Marine Biology, an Annual Review* 23:313–371.

———. 1992. "Dispersal of living organisms into aquatic ecosystems as mediated by aquaculture and fisheries activities," in *Dispersal of Living Organisms into Aquatic Ecosystems,* edited by A. Rosenfield and R. Mann, 13–45. Maryland Sea Grant Publication, College Park, MD.

———. 1996. "Pattern, process, and prediction in marine invasion ecology," *Biological Conservation* 78:97–106.

———. 1999. "The scale and ecological consequences of biological invasions in the world's oceans," in *Invasive Species and Biodiversity Management,* edited by O. T. Sandlund, P. J. Schei, and A. Viken, 195–212. Kluwer Academic Publishers, Dordrecht, The Netherlands.

Carlton, J. T., and J. Geller. 1993. "Ecological roulette: The global transport and invasion of nonindigenous marine organisms," *Science* 261:78–82.

Carlton, J. T., and J. Hodder. 1995. "Biogeography and dispersal of coastal marine organisms: Experimental studies on a replica of a 16th-century sailing vessel," *Marine Biology* 121:721–730.

Carlton, J. T., D. M. Reid, and H. van Leeuwen. 1995. *Shipping Study. The Role of Shipping in the Introduction of Non-Indigenous Aquatic Organisms to the Coastal Waters of the United States (Other Than the Great Lakes) and an Analysis of Control Options.* Report Number CG-D-11-95. Government Accession Number AD-A294809. The National Sea Grant College Program/Connecticut Sea Grant Project R/ES-6. Department of Transportation, U.S. Coast Guard, Washington, DC, and Groton, CT.

Clark, A. H. 1949. *The Invasion of New Zealand by People, Plants and Animals. The South Island.* Greenwood Press, Westport, CT.

Claudi, R., and J. H. Leach. 2000. *Nonindigenous Freshwater Organisms. Vectors, Biology, and Impacts.* Lewis Publishers, Boca Raton, FL.

Cohen, A. N. 1997. "Have claw, will travel," *Aquatic Nuisance Species (ANS) Digest* (Freshwater Foundation, Excelsior, Minnesota) 2(3):1, 16–17, 23. Online at http://www.anstaskforce.gov/digest.htm.

Courtenay, W. R., Jr., and J. R. Stauffer, Jr., eds. 1984. *Distribution, Biology, and Management of Exotic Fishes.* Johns Hopkins University Press, Baltimore, MD.

Crosby, A. W., Jr. 1972. *The Columbian Exchange. Biological and Cultural Consequences of 1492.* Greenwood Press, Westport, CT.

Crosby, A. W., Jr. 1986. *Ecological Imperialism. The Biological Expansion of Europe, 900–1900.* Cambridge University Press, Cambridge, UK.

Crossman, E. J., and B. C. Cudmore. 2000. "Summary of fish introductions through canals and diversions," in *Nonindigenous Freshwater Organisms. Vectors, Biology, and Impacts,* edited by R. Claudi and J. H. Leach, 393–398. Lewis Publishers, Boca Raton, FL.

Darwin, C. 1859. *On the Origin of Species by Means of Natural Selection.* John Murray, London.

Davies, N., and G. K. Roderick. 2000. "Determining the pathways of marine bioinvasion: Genetical and statistical approaches," in *Marine Bioinvasions: Proceedings of the First National Conference,* edited by J. Pederson, 251–255. Massachusetts Institute of Technology, MIT Sea Grant College Program, MITSG 00-2, Cambridge, MA.

Desmond, R. 1995. *Kew. The History of the Royal Botanic Gardens.* Harvill Press, London, and the Royal Botanic Gardens, Kew.

Drake, J. A., H. A. Mooney, F. di Castri, R. H. Groves, F. J. Kruger, M. Rejmánek, and M. Williamson, eds. 1989. *Ecology of Biological Invasions: A Global Perspective.* Wiley, New York.

Ellis, J. 1770. *Directions for Bringing over Seeds and Plants from the East-Indies and Other Distant Countries in a State of Vegetation; Together with a Catalogue of Such Foreign Plants as Are Worthy of Being Encouraged in Our American Colonies, for the Purposes of Medicine, Agriculture, and Commerce, to Which Is Added the Figure and Botanical Description of a New Sensitive Plant, Called* Dionaea muscipula *or Venus's Fly-Trap.* L. Davis, London.

Elton, C. S. 1958. *The Ecology of Invasions by Animals and Plants.* Methuen, London.

Fuller, P. L., L. G. Nico, and J. D. Williams. 1999. "Nonindigenous fishes introduced into inland waterways of the United States," American Fisheries Society Special Publication 27.

Gollasch, S., J. Lenz, M. Dammer, and H.-G. Andres. 2000. "Survival of tropical ballast water organisms during a cruise from the Indian Ocean to the North Sea," *Journal of Plankton Research* 22:923–937.

Grosholz, E. D., G. M. Ruiz, C. A. Dean, K. A. Shirley, J. L. Maron, and P. G. Connors. 2000. "The impacts of a non-indigenous marine predator in a California bay," *Ecology* 81:1206–1224.

Guthrie-Smith, H. 1921. *Tutira. The Story of a New Zealand Sheep Station.* University of Washington Press, Seattle (1999 reprint).

Hehn, V. 1885. *The Wanderings of Plants and Animals from Their First Home.* Swan Sonnenschein & Co., London.

Hokkanen, H. M. T., and J. M. Lynch, eds. 1995. *Biological Control: Benefits and Risks.* Cambridge University Press, Cambridge, UK.

Howarth, F. G. 1991. "Environmental impacts of classical biological control," *Annual Review of Entomology* 36:485–509.

IUCN (International Union for the Conservation of Nature and Natural Resources). 1967. *Towards a New Relationship of Man and Nature in Temperate Lands.* Part III. *Changes Due to Introduced Species.* IUCN Publications, New Series no. 9.

Jervis, M., and N. Kidds, eds. 1996. *Insect Natural Enemies: Practical Approaches to Their Study and Evolution.* Chapman & Hall, New York.

Kew, H. W. 1893. *The Dispersal of Shells. An Inquiry into the Means of Dispersal Possessed by Fresh-water and Land Mollusca.* Kegan Paul, Trench, Trubner & Co., London.

Kruger, F. J., D. M. Richardson, and B. W. van Wilgen. 1986. "Processes of invasion by alien plants," in *The Ecology and Management of Biological Invasions in Southern Africa,* edited by I. A. W. Macdonald, F. J. Kruger, and A. A. Ferrar, 145–155. Oxford University Press, Cape Town, South Africa.

Lavoie, D. M., L. D. Smith, and G. M. Ruiz. 1999. "The potential for intracoastal transfer of non-indigenous species in the ballast water of ships," *Estuarine, Coastal and Shelf Science* 48:551–564.

Lever, C. 1992. *They Dined on Eland. The Story of the Acclimatisation Societies.* Quiller Press, London.

MacKenzie, D. R., C. S. Barfield, G. G. Kennedy, R. D. Berger, and D. J. Taranto. 1985. *The Movement and Dispersal of Agriculturally Important Biotic Agents.* Claitor's Publishing Division, Baton Rouge, FL.

Mann, R., ed. 1979. *Exotic Species in Mariculture.* MIT Press, Cambridge, MA.

McDowall, R. M. 1994. *Gamekeepers for the Nation. The Story of New Zealand's Acclimatisation Societies 1861–1990.* Canterbury University Press, Christchurch, NZ.

McKnight, W., ed. 1993. *Biological Pollution: The Control and Impact of Invasive Exotic Species.* Indiana Academy of Science, Indianapolis.

Mead, A. R. 1961. *The Giant African Snail. A Problem in Economic Malacology.* University of Chicago Press, Chicago, IL.

Mills, E. L., J. R. Chrisman, and K. T. Holeck. 2000. "The role of canals in the spread of nonindigenous species in North America," in *Nonindigenous Freshwater Organisms. Vectors, Biology, and Impacts,* edited by R. Claudi and J. H. Leach, 347–379. Lewis Publishers, Boca Raton, FL.

Mooney, H. A., and J. A. Drake, eds. 1986. *Ecology of Biological Invasions of North America and Hawaii.* Ecological Studies 58, Springer-Verlag, New York.

Munro, A. L. S., S. D. Utting, and I. Wallentinus, eds. 1999. *Status of Introductions of Non-indigenous Marine Species to North Atlantic Waters, 1981–1991,* Copenhagen, International Council for the Exploration of the Sea, Cooperative Research Report No. 231.

NRC Committee on Biological Diversity in Marine Systems. 1995. *Understanding Marine Biodiversity: A Research Agenda for the Nation.* National Academy Press, Washington, DC.

Olenin, S., S. Gollasch, S. Jonusas, and I. Rimkute. 2000. "En-route investigations of plankton in ballast water on a ship's voyage from the Baltic Sea to the open Atlantic coast of Europe," *Internationale Revue Hydrobiologie* 85:577–596.

OTA, United States Congress. 1993. *Harmful Non-indigenous Species in the United States.* OTA-F-565. U.S. Government Printing Office, Washington, DC.

Pereira, B. A. 1926. *Snail Pest at Deniyaya.* Year Book, Department of Agriculture, Ceylon, 62.

Por, F. D. 1978. *Lessepsian Migration. The Influx of Red Sea Biota into the Mediterranean by Way of the Suez Canal.* Springer-Verlag, Berlin.

Purseglove, J. W. 1965. "The spread of tropical crops," in *The Genetics of Colonizing Species,* edited by H. G. Baker and G. Ledyard Stebbins, 375–389. Academic Press, New York.

Rice, R. G. 1930. "Accidental dispersal by motor-car," *Journal of Conchology* 19:26–27.

Rosenfield, A., and R. Mann, eds. 1992. *Dispersal of Living Organisms into Aquatic Ecosystems.* Maryland Sea Grant Publication, College Park, MD.

Ruiz, G. M., and J. T. Carlton. 2003. *Bioinvasions: Vector Analysis and Management Strategies.* Island Press, Washington, DC.

Ruiz, G. M., and J. A. Crooks. 2001. "Biological invasions of marine ecosystems: Patterns, effects, and management," in *Waters in Peril,* edited by L. Bendell-Young and P. Gallaugher, 3–17. Kluwer Academic Publishers, Dordrecht, The Netherlands.

Ruiz, G. M., P. W. Fofonoff, J. T. Carlton, M. J. Wonham, and A. H. Hines. 2000. "Invasion of coastal marine communities in North America: Apparent patterns, processes, and biases," *Annual Review of Ecology and Systematics* 31:481–531.

Runnstrom, S., and P. Vivier, chairmen. 1953. *The Introduction of Foreign Species* [a collection of 10 papers]. Verhandlungen der Internationalen Vereinigung für Theoretische und Angewandte Limnologie, Mitteilugen 12.

Sandlund, O. T., P. J. Schei, and A. Viken, eds. 1999. *Invasive Species and Biodiversity Management.* Kluwer Academic Publishers, Dordrecht, The Netherlands.

Shotts, E. B., Jr., and J. B. Gratzek. 1984. "Bacteria, parasites, and viruses of aquarium fish and their shipping waters," in *Distribution, Biology, and Management of Exotic Fishes,* edited by W. R. Courtenay Jr. and J. R. Stauffer Jr., 215–232. Johns Hopkins University Press, Baltimore, MD.

Simberloff, D. 1986. "Introduced insects: A biogeographic and systematic perspective," in *Ecology of Biological Invasions of North America and Hawaii,* edited by H. A. Mooney and J. A. Drake, 3–26. Wiley, New York.

Skjerve, E., and Y. Wasteson. 1999. "Consequences of spreading of pathogens and genes through an increasing trade in foods," in *Proceedings of the Norway/UN Conference on Alien Species,* Trondheim, 1–5 July 1996, compiled by O. T. Sandlund, P. J. Schei, and A. Viken, 259–267. Directorate for Nature Management and Norwegian Institute for Nature Research, Trondheim.

Spongberg, S. A. 1990. *A Reunion of Trees. The Discovery of Exotic Plants and Their Introduction into North American and European Landscapes.* Harvard University Press, Cambridge, MA.

Vasuki, V. 1990. "Effect of insect growth-regulators on hatching of eggs of three vector mosquito species," *Proceedings of Indian Academy of Sciences* 99:477–482.

Welcomme, R. L. 1988. International introductions of inland aquatic species, *FAO Fisheries Technical Paper* No. 294, Food and Agriculture Organization of the United Nations (FAO), Rome, Italy, 318 pp.

———. 1992. "A history of international introductions of inland aquatic species." *ICES Marine Science Symposia* 194:3–14.

Wells, M. J., R. J. Poynton, A. A. Balsinhas, K. J. Musil, H. Joffe, E. van Hoepen, and S. K. Abbott. 1986. "The history of introduction of invasive alien plants to southern Africa," in *The Ecology and Management of Biological Invasions in Southern Africa,* edited by I. A. W. Macdonald, F. J. Kruger, and A. A. Ferrar, 21–35. Oxford University Press, Cape Town, South Africa.

Wilson, C. L., and C. L. Graham, eds. 1983. *Exotic Plant Pests and North American Agriculture.* Academic Press, New York.

Wittenberg, R., and M. J. W. Cock. 2001. *Invasive Alien Species: A Toolkit of Best Prevention and Management Practices.* CABI Publishing, Wallingford, UK.

Wonham, M. J., W. C. Walton, G. M. Ruiz, A. M. Frese, and B. S. Galil. 2001. "Going to the source: Role of the invasion pathway in determining potential invaders." *Marine Ecology Progress Series* 215:1–12.

Woods Hole Oceanographic Institution. 1952. *Marine Fouling and Its Prevention.* United States Naval Institute, Annapolis, MD.

WWF. 1998. *Resolving the DDT Dilemma: Protecting Biodiversity and Human Health.* World Wildlife Fund, Washington, DC.

4

The ISSG Global Invasive Species Database and Other Aspects of an Early Warning System

Maj de Poorter, Michael Browne, Sarah Lowe, and Mick Clout

Biological invasions can be enormously costly in both environmental and economic terms, and they are occurring with increasing frequency because of the rapid growth in global trade and travel. Invasive alien species (IAS)[1] are alien species that threaten ecosystems, habitats, or species. They occur in all major taxonomic groups, including viruses, fungi, algae, mosses, ferns, higher plants, invertebrates, fish, amphibians, reptiles, birds, and mammals. They have invaded and affected native biota in almost every ecosystem type and have caused hundreds of extinctions, especially on oceanic islands or in "ecological islands" such as freshwater ecosystems. The impacts of IAS are immense and insidious. The environmental costs include irretrievable losses of native species and ecosystems; the economic costs include damage to crops, livestock, property, and human health (McNeely et al. 2001). For these reasons, IAS are a major focus of international concern, including in the Convention on Biological Diversity (CBD), and are the subject of cooperative international efforts such as the Global Invasive Species Program (GISP).

Prevention of Biological Invasions

There are four basic strategies for dealing with IAS: prevention, early detection, eradication, and control. The aim of an early warning system (EWS) for IAS is to increase the chances of successful prevention of biological invasion. There are two aspects to prevention: a first line of defense, which consists of the prevention of introduction of a potential IAS, and a second line of defense, which consists of the prevention of a full-blown invasion if the potential IAS is introduced.

Prevention of introduction is the first and most cost-effective option (Wittenberg and Cock 2001; IUCN 2000). This lesson has been learned the hard way in several cases of highly destructive and costly invasive organisms such as the brown tree snake (*Boiga irregularis*) in Guam, water hyacinth (*Eichhornia Crassipes*) in Africa and elsewhere (Lowe et al. 2000), and brushtail possum (*Trichosurus vulpecula*) in New Zealand (Clout and Lowe 2000). By *introduction* we mean any movement (direct or indirect), by human agent, of a species outside its native range.[2]

The second line of defense against biological invasion is the early detection of an introduced potentially invasive alien species, allowing for rapid response (e.g., eradication before numbers have become too big or the area of spread too vast). Early detection and rapid response become important when border controls fail to intercept a species "hitchhiking" or being smuggled. It is also crucial in detecting the development of unforeseen invasive characteristics of a species that was mistakenly cleared for introduction. A quick response is also needed when a species has been deliberately introduced for use in a contained situation (e.g., a zoo or laboratory) and then escapes into the wild. Early detection coupled with rapid response is sometimes called preventive action. However, because it is aimed at preventing establishment or spread it is a second line of defense only, and it should never be relied on in lieu of preventive measures.

International Early Warning Systems

International EWSs have been developed in other fields. Examples of these systems include the following:

- The Early Warning Program of the International Decade for Natural Disaster Reduction (IDNDR) defines the objective of early warning as "to empower individuals and communities, threatened by natural or similar hazards, to act in sufficient time and in an appropriate manner so as to reduce the possibility of personal injury, loss of life and damage to property, or nearby and fragile environments." A set of guidelines was prepared, covering local, national, regional, and international early warning (UN 1997). For example, "Collaboration and coordination is essential between scientific institutions, early warning agencies, public authorities, the private sector, the media, and local community leaders to ensure that warnings are accurate, timely, meaningful and can result in appropriate action by an informed population" (UN 1997).
- In 1994 the Food and Agriculture Organization (FAO) of the United Nations established an Emergency Prevention System (EMPRES) for Transboundary Animal and Plant Pests and Diseases in order to minimize the risk of emergencies (including famines) and major losses caused by the spread of agricultural pests and diseases that migrate or spread across borders. EMPRES has a major focus on 15 diseases on an "A" list and several others of secondary importance on a "B" list. However, the system also acknowledges that unclassified diseases can have severe economic or trading implications (especially when there is a link to public health), as shown by recent events such

as the bovine spongiform encephalopathy epidemic in Europe and outbreaks of Nipah virus in Malaysia (FAO 2002).

- Infectious human disease has the potential to spread internationally. With the development of vaccines and the discovery of potent antimicrobial drugs, humanity has developed the ability to prevent many infectious diseases and cure many others. More recently, however, epidemics have spread around the globe unchecked through a combination of increased global travel by humans and disease vectors (e.g., mosquitoes), the emergence of newly recognized diseases (including those caused by microorganisms crossing the species barrier), the reemergence of known diseases (e.g., tuberculosis), and the emergence and spread of resistance to antimicrobial drugs. The World Health Organization (WHO) gathers global disease intelligence and coordinates the multifaceted response needed to contain outbreaks quickly and prevent their international spread (Heymann 2001).

Early Warning in the Context of IAS

An EWS in the context of IAS has to deal with a large number of species that could invade and a complex and diverse set of potential impacts on biodiversity, including predation, competition, hybridization, disease, and cumulative effects (McNeely et al. 2001). An EWS for IAS therefore must deal with a very complex set of circumstances. The improvement and development of risk prediction and assessment, especially relating to the prevention of IAS introduction, must be a priority for the foreseeable future.

IAS Early Warning in the First and Second Lines of Defense

Humans cannot take action to prevent natural disasters (earthquakes, tsunamis, cyclones), and an EWS for such natural events can aim only to reduce the damage or loss of life they cause. In the case of disease, the FAO and WHO systems reduce the impact of disease, but they can also prevent introduction into a new country. Thus, these systems contribute to the first and second line of defense.

In the case of IAS, the equivalent of a catastrophic event is the introduction of a potentially invasive alien species. By definition, this is the direct, indirect, or secondary result of human activity and decisions. Therefore, there is a window of opportunity to avoid the introduction altogether in addition to minimizing or mitigating the impact. This ability to strengthen both the first line of defense and the second line of defense is an important consideration in the development of an EWS for IAS.

Relationships with Other Management Components

Early warning is closely tied to other aspects of IAS management. Huge amounts of information are not useful if there is no ability to act, hence the need for awareness of the IAS problem, the political will to act, the legal ability to act, and frameworks in which to act (e.g., border control, risk analysis, precautions, and ecosystem approaches). Likewise, the ability to act is limited if there is insufficient information or if information is not accessible. Management tools are discussed further in Wittenberg and Cock (2001), McNeely et al. (2001), and PCE (2000).

Requirement for Global Coverage of IAS Early Warning

The Early Warning Program of the IDNDR states, "International bodies and regional organizations must work to maintain the vital importance of timely exchange and unrestricted access of observational data and other warning information between countries, particularly when hazardous conditions affect neighbouring countries" (UN 1997). Under the Convention of the World Meteorological Organization, rapid access to information from beyond the national borders of threatened countries allows the provision of forecasting and warning services for meteorological and hydrological phenomena (Zillman 1998). The FAO and WHO early warning systems likewise work across political boundaries.

IAS are a global problem. Moreover, trade pathways across political boundaries provide opportunities for secondary introduction and facilitate spread of species to other countries, and transboundary ecosystems allow natural mechanisms of spread to cause secondary introductions. Information sharing is crucial. Data from countries where a species has previously invaded can provide useful information on invasion rate and speed, land cover types prone to invasion, possible economic and ecological consequences, and management approaches (McNeely et al. 2001). Likewise, information on distribution, invasiveness, or pathways of species in other countries is crucial in preventing new or secondary biological invasions. It is clear that a global IAS EWS is needed, in addition to national and regional systems.

The Aim of an IAS EWS

In the IAS context, the aim of an EWS should be to prevent biological invasions. This can be achieved through timely prediction, identification, and warning of risks to strengthen the first and second lines of defense. Global, regional, and national systems will be needed in order to reach this goal.

Components of an EWS for IAS

The first component of an EWS for IAS is the ability to identify and predict certain risks. Management information that is relevant to prevention, early detection, and rapid response should also be accessible. The second component is the ability to ensure the completeness and timeliness of data and information collected. Finally, the third component is the ability to provide timely warning to those at risk.

Component 1: Risks to Be Identified and Predicted and the Nature of Related Information Needed

Risk: "This alien species is likely to be or to become invasive."

Any type of organism, from microorganisms to large mammals, may become invasive. The crucial question of invasion biology is to predict which introduced species will become problems and which will remain innocuous.

In first instance, ecological traits were investigated as discriminators, with limited success (Williamson 1996; Goodwin, McAllister, and Fahrig 1999; Lockwood 1999; Rejmánek 1999). However, some characteristics noted by Rejmánek and others later in this volume are considered key indicators of invasiveness in plants (e.g., those linked to reproduction and dispersal). Other indicators for plants include vegetative reproduction for terrestrial plants and especially for aquatic species; seed dispersal by vertebrates, characters favoring passive dispersal by humans; and efficient competitors for limiting resources, especially relevant for natural and seminatural ecosystems. Tolerance to environmental factors such as shade or salinity, particular life form or habitat (e.g., climbing vine, aquatic species), and adaptive mechanisms such as a plant's ability to fix nitrogen are also indicators (Wittenberg and Cock 2001). The size of primary (native) geographic ranges of plant species is a promising predictor of their invasiveness, but several important exceptions are known (Goodwin, McAllister, and Fahrig 1999). Members of taxonomic families that hold a disproportionately large number of IAS may pose a higher risk of becoming invasive (Lockwood et al. 2001).

Although these characteristics have some value in predicting the invasiveness for an alien species, there is overwhelming evidence that the most important factor identifying risk of invasiveness is a history of invasiveness. This is very strongly expressed in several expert publications, such as Wittenberg and Cock (2001): "Only one factor has consistently high correlation with invasiveness: whether or not the species is invasive elsewhere." In Chapter 6 of this volume, Rejmánek et al. note, "Extrapolations based on previously documented invasions are fundamental for predictions in invasion ecology," and Samways (1996) states, "The best approach is to be aware of which species have been local, regional or intercontinental invasives elsewhere and then be vigilant for their first appearance in a new area." "The best predictor of which species will become problematic is whether or not a species has proven to be invasive elsewhere, especially under similar (climatic and geographic) conditions and in related ecosystems" (Simberloff 1999).

Risk: "This is a location or area where an IAS may invade next."

Using information from an IAS's native range to predict where it may become invasive next has been favored by those predicting the vulnerability of areas to invasive alien invertebrates or vertebrate poikilotherms such as the cane toad. More complicated systems using a combination of climatological and ecological data are also in development (Peterson 2001; PGSF 2001). However, it has also been recognized that IAS have moved outside their native habitat range: for instance, the brush-tail possum in New Zealand has a wider habitat range than in its native Australia (Clout and Ericksen 2000). Similarly, it is also known that in the alien range, invasive plant species may show a wider climate tolerance than in their native range (B. Marambe, pers. comm., 2000). Therefore, predictions of where an alien species may become invasive next ideally should be based on both its native and invasive precedents.

Risk: "There are pathways in place here that are known or likely to be used by IAS."

Identifying the risk of an IAS getting somewhere requires information on the pathways it can use (e.g., shipping), where those pathways are (e.g., shipping routes), and where such a pathway connects a source of a particular IAS with a potential destination. Information about pathways used can be obtained from research, pathway risk assessment (including risk–goods assessment), and quarantine and border control reports on actual interceptions. Other sources of information on where those pathways are located can come from transport agencies and other sources that keep track of volumes, origins, and destinations of trade. Information on the distribution of IAS is needed to identify places where they could enter a pathway; interception reports from border control and quarantine agencies could further pinpoint the sources of a particular IAS introduction to particular national borders.

Islands and other vulnerable ecosystems

It has become clear to most people working on alien species threats to biodiversity that islands (and other geographically and evolutionary isolated places) are different from continental situations in a number of ways. They are more vulnerable to invasions (see Lever 1994; Elmqvist 2000; Courchamp, Chapuis, and Pascal 2003) and more likely to suffer catastrophic biodiversity loss as a result of invasions.[3] The CBD has repeatedly recognized the very urgent need to address IAS issues in isolated and vulnerable ecosystems. This vulnerability should be reflected in the EWS where possible.

Information management

In addition to data that allow prediction and identification of risks, key management information should also be part of the EWS. This may include methods that will assist with prevention, early detection, and rapid response (e.g., methods for treatment of risk goods, methods for eradication). Further information on tools that can be used to deal with IAS can be found in Wittenberg and Cock (2001).

Component 2: Information and Data Collection and Timely Availability

Sources of information and data

To ensure that as much relevant information as possible finds its way into the EWS, efforts must be made to seek out information from sectors as diverse as science and research, ecosystem management, transport and trade, and biosecurity agencies. The technical challenge will be to integrate it all and make it useful and to obtain sufficient resources to ensure the information is collected in sufficient detail. The political challenge will be to reach agreement on the need to have national IAS information of this kind available internationally and publicly.

Timeliness of information and data

The EWS needs to learn of new developments quickly in order to disseminate timely notifications or warnings, and there is an element of intelligence gathering involved. For instance, this intelligence would include reports of new incursions, the development of invasiveness after a lag phase, and newly discovered pathways and vectors.

International development of an EWS for IAS

In order to obtain all relevant information for an IAS EWS, international agreement is needed on the types of information that should be covered, whether there should be any requirements for mandatory notification, and who should provide the information and how.

Free exchange of information between nations is a general requirement for an international EWS (Zillman 1998). However, the provision of such information can have negative consequences if it leads to trade restrictions or a reduction in tourism. This creates a conflict of interest, which must be solved internationally. A similar conflict of interest has been dealt with by the WHO; in the past, countries have been reluctant to report outbreaks of human epidemics for fear of negative impacts on travel, trade, or tourism (Heymann 2001). However, Heymann reports that this traditional reluctance is beginning to change. The growth of the Internet and other electronic media has meant that official transparency about outbreaks and prompt reporting have become increasingly important because they are a defense against unverified rumors and mistakes and a precursor for quick availability of international aid.

Component 3: Timely Warnings to Those at Risk

Passive availability of information

Unlike most natural disasters, which have a discrete duration and then an absence, the threat of biological invasions is constant. Whenever someone wants to find out how likely a given species is to invade their territory, for instance, he or she must be able to access the EWS. Therefore, there is a continuous need for a passive access: easy and fast access for users who are looking for information and want to access the EWS capacity to identify risks or to find information on methods that can be used for rapid response.

Active provision of information and alerts

Whereas databases, networks, and Listservs can currently provide information to those looking for it, there is no system in place to alert those who may not be looking but who nevertheless should be forewarned of a new or spreading alien bioinvasion. An IAS EWS should provide alerts where a new risk has emerged. Such a new risk may include a new instance in which a particular species has been found to be an invasive alien, a new incursion, or a new pathway. The EWS should be able to assess what users are at risk of the new development and then actively notify them. Following the example of other international EWSs, a global IAS EWS would need to feed into regional or national systems, as appropriate, to further distribute the alert to those who need to take action.

Need for international agreement

International agreement must be reached on what level or type of risk warrants such active alert and on who in a country or region should be alerted (e.g., what agency has the responsibility to act on the international alert?). Such active dissemination of information is hampered by the piecemeal nature of agencies that are responsible for dealing with the management of risks to biodiversity (e.g., as opposed to agriculture or horticulture). Once such responsibilities are more generally established in most nations, a system of alerts can be more realistically envisaged, be they through the CBD clearinghouse, through expansion of existing agricultural (plant, animal) or human health EWSs, through a newly developed system specifically for IAS with a biodiversity impact, or a combination of all these options. Cooperation between the CBD, the International Plant Protection Convention, the Office Internationale Epizootique, the WHO, and other relevant entities with experience in global-scale early warning will be needed, and existing systems of international early warning for catastrophic events will be able to provide useful precedent.

There is likely to be an advantage in negotiating such an agreement for international warning and alerts in conjunction with negotiating international requirements for reporting and notification.

GISP Approach to a Global IAS EWS

As part of its contribution to the first phase of the GISP, the World Conservation Union (IUCN) Invasive Species Specialist Group (ISSG) set out to design and build the Global Invasive Species Database (GISD) to provide a global, searchable first resource for essential information on IAS management, including information needed for early warning purposes. Details of the development of this database are given in Appendix 4.1. In addition, in Phase II GISP is developing and coordinating a distributed network of invasive species information. The aim is to provide accessible information on scientific, technical, and other aspects of IAS, including IAS identification, prevention, and eradication. The output will be a distributed network of linked databases capable of providing scientific, technical, and institutional information relevant to IAS.

The Global Invasive Species Database and Early Warning System

The ISSG GISD (http://www.issg.org/database), developed as part of GISP Phase I, contains globally sourced information on IAS from all taxonomic groups and incorporates early warning aspects. It aims to answer the key question, "Is this species an IAS anywhere else in the world?" which is the best indicator for the likelihood of an alien species being or becoming invasive. Its predictive system is a first attempt to answer the question, "Where else could this IAS survive and invade?" The GISD contains information on ecological characteristics of IAS, pathways used, invasive characteristics, IAS distributions (native, alien, and alien invasive), ecological impacts, management information, references, and contacts. Like an encyclopedia, it is designed to be easy to explore and contains high-quality summarized information supplied by a network of invasive species experts, managed and updated by the ISSG database team. Database searches are simple and intuitive, and each species account is a ready source of useful information. It addresses the needs of a range of users from the well-informed to the naive, providing individuals, communities, agencies, policymakers, and scientists with access to information on IAS and links to other sources of information.

Priorities for further development of the GISD range from a focus on some of the world's worst invasive species to a focus on areas where information and resources are scarce, including small island developing states and other islands (Browne 2001). As a matter of principle, access to information is available at no cost. Another priority will be to provide access for those who have limited Internet access by providing printed reports or CD-ROMs.

The Role of Listservs

A helpful contribution to early warning can be achieved through the use of Listservs. For example, a message along the lines of "there is some deliberation about a plan to use alien species X for purpose Y in country or region Z" posted on the well-established Aliens-L Listserv[4] usually elicits several responses if the species in question has been an invasive alien elsewhere. Another Listserv with South Pacific regional range and more emphasis on agricultural weeds is Pestnet. Listservs or other information networks with similar aims can be found in the national contexts as well, but for IAS early warnings, international coverage is of utmost importance. Listservs cannot be as comprehensive as a distributed network or a global database and may lack the aspects of consistency, standardization, and quality control. However, Listservs have a role to play because of their ability to quickly deal with ad hoc issues, their flexibility, and the horizontal information transfer enabled by their design.

Conclusion

In the IAS context, the aim of an EWS should be to prevent biological invasions. This is accomplished by providing timely prediction, identification, and warning of risks in order to strengthen the first and second lines of defense. Global, regional, and national systems will need to be collaborative to maximize the effort.

The nature of the data and information collected and managed in an IAS EWS must allow users to identify and predict the following risks:

- Alien species that are likely to be or become invasive
- Area where an IAS could survive and invade next
- Occurrence of pathways that are known or likely to be used by IAS

In addition, practical management information that allows swift decisions and actions must be included, and the system must have predictive capacity.

Information and data on species that are useful in identifying and gauging prediction relevant risks include the following:

- History of invasiveness of the species
- Invasive tendencies of related taxa
- Geographic information on native range that includes habitat and climate preference
- Specific ecological characteristics
- Land cover type in all regions where the species is invasive and where it is native
- Climates and habitats where the IAS is invasive and native
- Distribution of the IAS (invasive, alien, native), including new incursions and early detections
- Interception data from quarantine and border control
- Pathways historically used
- Known impacts
- Practical management information (e.g., methods, reference, contacts)
- Occurrence of potentially invasive alien species on islands

All available and relevant information and data should be available to the EWS, including that from sources as diverse as

- Research
- Practical management (including reports on successful and failed eradication and control efforts, reports from surveys, monitoring, and new incursions)
- Implementation of rules and regulations (e.g., interception reports from quarantine or border control agencies, results from national import risk assessments, pathway risk assessments, species lists)

Global, regional, and national EWSs must dovetail with each other, both in the provision of information and data to the EWS and in the distribution of alerts, warnings, and information.

Information flow must be timely; new developments must be fed into the EWS so that warnings and alerts can be generated as soon as possible. In addition to voluntary cooperation, international agreements must be reached, including the following:

- Requirements for notification and reporting (e.g., what and how)
- Generation of alerts (e.g., whom to alert and how)
- Collaboration and interaction of national, regional, and global EWSs

EWSs are particularly important in the development of international agreements. Existing international EWSs are important resources in reforming existing programs and developing new ones. Cooperation within existing mechanisms is important to maximize success.

Acknowledgments

The GISD project has benefited from advice and support from Greg Sherley, Philip Thomas, Jean-Yves Meyer, Wendy Strahm, Laurie Neville, Fabio Corsi, Bob Meese, Jim Space, Jerry Cooper, Pam Fuller, Jim Carlton, Brian Steves, Jerry Cooper, Souad Boudjelas, and Shyama Pagad. Synergy International built the database according to ISSG's design specifications. Principal supporting organizations are the IUCN, the National Biological Information Infrastructure (NBII), the Manaaki Whenua–Landcare Research (New Zealand), and the University of Auckland. Additional support has come from the Scientific Committee on Problems of the Environment, Diversitas, Fondation d'Entreprise Total, New Zealand AID, and the Pacific Development and Conservation Trust. We are grateful to the dedicated invasive species specialists from around the world who contribute in a voluntary capacity by providing profiles of invasive species, images, reviews, and updates on current issues.

Appendix 4.1

Development of the GISD

The GISD was developed by the ISSG as a contribution to the GISP[5] Phase I. It provides global information on IAS to agencies, resource managers, decision makers, and interested individuals. The database focuses on invasive species that threaten biodiversity and covers all taxonomic groups from microorganisms to animals and plants. Species information is supplied or reviewed by expert contributors from around the world and includes biology, ecology, distribution, management information, references, contacts, links, and images.

Information in the GISD helps to empower practitioners, support training programs, and help build capacity. In order to get this information into the hands of more people, a CD-ROM version of the GISD will be produced, and the GISD will also become part of a distributed network of invasive species information databases: the Global Invasive Species Information Network.

Analysis of Requirements

Review of Existing Information Sources

The ISSG undertook a review of existing databases and other sources of information on invasive species and an analysis of potential users and their needs before embarking on the design of the GISD.[6] A wide variety of information on invasive species that threaten biological diversity is available on the Internet and elsewhere. Most resources have a local or regional focus rather than a global one, and many of them deal only with agricultural pests. They vary enormously in terms of their taxonomic and geographic coverage and the level of detail they contain. Most importantly, invasive species information is widely dispersed and difficult to access. Finding vital information often is a time-consuming and frustrating process.

Analysis of Users and Their Needs

The following generic issues underpinning the invasive species problems in the Pacific region were identified in the *Technical Review of Invasive Species in the Pacific* (SPREP 2000):

- A shortage and inaccessibility of scientific information
- A lack of awareness on the impacts of invasive species on biodiversity
- Insufficient dissemination of information to the relevant decision makers and government officials
- A shortage of technically trained personnel in Pacific island countries

The user-friendly design of the GISD is a response to the information needs of communities and decision makers all over the world who are facing pest and weed infestations and the threat of further invasions. The 250 profiles currently available in the GISD comprise core information elements that were developed in consultation with practitioners and program managers. Information they contain is presented in plain English.

Groups that were canvassed for their comments on users and their needs included delegates at several invasive species meetings, ISSG members and other groups within the IUCN, readers of the Aliens-L Listserv, groups and individuals working directly with invasive species, and other developers of invasive species databases.

Groups and individuals that were identified as potential users include

- Conservation and biodiversity program managers
- Land managers and land owners
- Border control and quarantine agencies and their staff
- Decision makers in all sectors of society
- Communities and nongovernment organizations
- Practitioners and volunteers
- Training agencies and trainees
- Educational institutions
- Researchers
- Invasive species specialists
- The media

There was general support for the creation of a comprehensive global database of invasive species that would gather and disseminate otherwise widely dispersed information. Our consultation process suggested that the database should capture and display synthesized information rather than raw experimental data. It became apparent that there was an urgent need for reliable and accessible information on IAS in developing countries, particularly those with islands whose unique, previously isolated biodiversity is especially threatened by these species (Clout 1999; McNeely 1999; SPREP 2000).

Choices to Be Made

Users

There were a variety of views on the level of complexity and the amount of detail that the database should be able to deal with. It was obvious that the database would not be able to provide all things to all people, so choices had to be made and priorities set. Our approach was to focus on the database as a management tool and an awareness-raising, educational tool rather than a research tool. The focus was on practitioners working on the front lines of IAS prevention and management. As a result, the language and format used cater to users who might have a secondary school level of education, with English often being a second or third language.

Providing Internet Access Is Not Enough

Many potential users were interviewed about their information needs and their access to the Internet. Many had no access to the Internet or very slow or unreliable access (e.g., the South Pacific and parts of Africa). Alternative methods of accessing information, such as hardcopy or CD-ROM, must be provided in order to maximize access to information. The ISSG is planning to produce a CD-ROM version of the GISD in 2004 and to distribute it using a variety of trusted networks. We will evaluate the response to the CD-ROM and update it annually.

Why a Centralized Database?

It was decided to create a central repository for global information because many countries and regions do not have access to information that is relevant to them, and in many cases no other repository for information about their particular invasive species problems is available. Access to *relevant* information is particularly important given that, for example, only one in five species that are invasive in the Pacific region are also invasive in the Caribbean. In addition, much of the information that is available has not been digitized, including hardcopy reports, expert knowledge in the form of unpublished notes or personal observation, and practitioners' management expertise, often unpublished or available only in obscure technical reports.

Finally, the target audience we identified will not find unstructured raw data very useful. The content of the GISD is information that has been synthesized, analyzed, or otherwise adapted in an attempt to fully meet users' needs. Data capture is based on core information elements, creating a standardized view, but the level of detail in fields is flexible because the source material is so variable. In addition, each species profile has links back to local, national, and regional resources where more detailed and locally specific information can be found. All information in the GISD is created or reviewed by invasive species experts and is presented in plain English. The result of this approach is a database that delivers authoritative scientific information in a user-friendly manner to a broad audience.

Why Global?

General

IAS are a global problem. Moreover, trade pathways that cross political boundaries provide opportunities for new or secondary introduction and facilitate spread of species to other countries, and transboundary ecosystems allow natural mechanisms of spread to cause secondary introductions into other nations. Information sharing is crucial. Data from countries where species have previously invaded can provide useful information on invasion rate and speed, possible economic and ecological consequences, and management approaches (McNeely et al. 2001), and information on distribution, vectors, and pathways is crucial for the prevention of new or secondary biological invasions.

Our goal is to create and disseminate core information on any species that has been identified as invasive anywhere. Realistically speaking, it will take some time for the GISD to reach a truly global scope; populating a database takes time. The initial focus for the GISD was on the list of 100 of the world's worst invasive species, IAS that are of relevance to islands, and those relevant to the South Pacific, but subsequent partnerships and funding have already resulted in a broader scope. For example, in 2002 a memorandum of cooperation, which covered database enhancements and the uploading of North American species profiles, was signed with the NBII.

Identification of Risk: Early Warning

A crucial question of invasion biology is to predict which introduced species will become problems and which will remain innocuous. A variety of ecological traits, tolerance to ecological factors, and other characteristics have been investigated in terrestrial plants, insects, and avifauna (e.g., Rejmánek 1996; Williamson 1996; Goodwin, McAllister, and Fahrig 1999; Lockwood 1999; Wittenberg and Cock 2001). Although several of the characteristics investigated have some value in the prediction of invasiveness for an alien species in well-defined circumstances, there is overwhelming evidence that the most important factor identifying risk of invasiveness is previous invasiveness (i.e., invasion history). This is very strongly expressed in several expert publications: "Only one factor has consistently high correlation with invasiveness: whether or not the species is invasive elsewhere" (Wittenberg and Cock 2001).

Species that appear in the GISD meet this criterion: they have all been described as invasive (i.e., they have had negative impacts on biodiversity) somewhere in the world. By assessing the risk of invasiveness and focusing on parts of the world where little information is currently available, the GISD plays an important role in early warning. This is especially crucial in analyzing risk before intentional introductions, preventing unintentional introductions, and planning and prioritizing management responses where prevention has failed. Additional information on species' current global distributions and the mechanisms by which their ranges expand provides further crucial information necessary for early warning and invasion prevention.

Overall Conclusion of the Need Analysis

The aims of the GISD are to facilitate effective prevention and management of IAS and to raise awareness by providing easy access to authoritative, globally sourced information and expertise. To achieve this it was decided that the database should

- Be global in scope and focused primarily on threats to biodiversity
- Have a particular focus on developing countries and islands
- Derive authority from the participation of IAS experts

- Provide direct access to essential information on invasive species in plain English (no superfluous information and unnecessary jargon)
- Have a particular focus on prevention and management information (i.e., strategies for prevention, containment, eradication, and control)
- Be easily accessible through the Internet (e.g., fast download)
- Be able to deliver its contents as CD-ROM or hardcopy pages and booklets
- Be noncommercial, nonprofit, and freely available
- Be independent
- Be suitable for education and awareness-raising purposes as well as management purposes

Description of the GISD

Information Elements in the GISD

The core information elements in the GISD are

- Taxonomy, common names in many languages, descriptions, and images (to identify and describe each invasive species)
- Distribution (to describe where it is)
- Pathways and vectors (to describe how it got there)
- Impacts (to describe what effect it has there)
- Prevention and management information (to describe how to deal with it)
- Contact details of specialists (for advice)
- Links to more information

Web Site

The attractive and user-friendly Web site design allows visitors to quickly find answers to specific questions, but novices also can easily acquaint themselves with invasive species issues. The Web site is uncluttered and fast to load, allowing satisfactory access even in regions with slow Internet access. The text is jargon-free so that users with English as a second or third language can access the information. Each species profile is made up of five pages of information: ecology (including impacts and expansion mechanisms), distribution records, management information, expert contacts, and references and links. (Note: examples of the different pages are reproduced in Appendix 4.2.)

Ecology Page

This page summarizes the basic characteristics of each species, its ecological relationships, and its range expansion mechanisms. It contains images and common and scientific names. Other information on the ecology page includes distinguishing features, habitat, food and nutrient needs, uses, methods and rate of reproduction, and notes on any special life cycle stages that may increase understanding of the invasion process. There are also brief statements about general impacts and the management of that species.

Distribution Page

This page displays the known global distribution of each species alphabetically by country name. For each country there are more detailed records. They describe its biostatus (e.g., occurrence, native vs. alien, invasive), any specific impacts of a species in that location, when and how it was introduced, and any relevant management information. All sources are documented.

Management Information Page

This page includes information about management of each invasive species from three different sources: general advice about management from the ecology page, location-specific management information, and any management-related references and links. All efforts are taken to access best practice information globally and make it available. Management information often resides in places where it is difficult to access (technical reports published internally in an agency or even inside a practitioner's head if she or he did not have time to write it up but instead went on to the next practical management task). Adding this information to the database is very time-consuming, but we believe that a key role for the GISD is to make such information widely available. This will help ensure more efficient use of limited resources and better biodiversity outcomes.

References and Links Page

References and useful links for each species are sorted into those that are general in nature and those that deal specifically with management information. As much as possible, references and readable PDF files are directly available on the Web site. Web sites with further relevant information are accessed by hyperlink. References identified as containing management information give users access to knowledge accumulated by invasive species management experts. They may describe different programs and techniques or the reasons for the success or failure of attempts at prevention, containment, eradication, and control. They may warn about unexpected side effects or give advice on best practice.

Contacts Page

The contacts page gives users access to expert advice or assistance. The database contains the name, telephone number, e-mail address, and affiliation or institution of experts who have volunteered to respond to such requests for specific advice and assistance.

Searches Available to the User

Users can search the database by

- Scientific name, synonym, or common name
- Taxonomic groupings (order = Homoptera or genus = *Euglandina*)

- Organism type (e.g., bird, fish, microorganism, aquatic plant)
- Country or region
- Habitat (e.g., natural forest, wetlands)
- Combination searches, such as organism type in a habitat type in a place (e.g., "land invertebrate in grasslands in Mississippi")

All searches return a list of species with its scientific and common names and a short summary of salient points. Clicking on a species name leads straight to the full species profile.

System Specifications

The database and Web site were built by Synergy International Ltd. in Auckland, New Zealand. The Microsoft Access database provides an interface directly to SQL Server using Open Database Connectivity. The data are stored and manipulated by Microsoft SQL Server 7 and Transact-SQL. The SQL Server database is the foundation for the Dynamic HTML Web interface, which uses Internet Information Server and Active Server Page to search and display database information. This arrangement optimizes user-defined queries and the maintainability and integrity of the data. It will accommodate increased capacity as its content grows, and it provides the potential for development of a remote data entry interface, with password access for expert contributors, if that option is pursued. The system is hosted on a data server provided by Landcare Research–Manaaki Whenua, a New Zealand Crown Research Institute. The GISD can be found at http://www.issg.org/database/welcome and is mirrored by the NBII at http://www.invasivespecies.net/database/welcome.

Future Developments for the GISD

The GISD will soon become a key contributor to the Global Invasive Species Information Network. Short-term goals for further development of the database include ongoing population with species information and development of a CD-ROM version of the database for use in places where Internet access is difficult or nonexistent. We are developing remote data entry, and we also intend to develop a global master list of invasive species. The master list will deliver a global picture of the invasive species problem by classifying and integrating the growing number of lists of problem species that are being published by diverse local, national, and regional programs, both online and in print.

User Feedback and Practical Support Received

In 2002, the ISSG signed a Memorandum of Cooperation with the NBII that allows a two-stage database enhancement program and will also result in a rapid increase in the number of species in the database.

The GISD has consistently attracted interest from conservation and biodiversity programs and invasive species information resources around the world. Material support for our work has come in the form of fact sheets, management information, predictive modeling results, and many other useful data sets, tools, and applications. This positive response to the GISD is very encouraging. The GISD has arrangements with many conservation programs to share invasive species information and to link back to each other. The GISD project began with, and remains indebted to, invasive species specialists around the world who voluntarily create or review information for the GISD. They contribute expert advice and valuable support.

Conclusion

The GISD contains core information in plain English on all types of IAS with a focus on those that threaten biodiversity. The information is either created or reviewed by invasive species specialists and is disseminated freely in response to the growing need for prompt and easy access to accurate information and expert advice. Core content of the database includes descriptions, distribution and pathways information, management information, references, links to other information sources, and the contact details of experts who can offer specific advice.

The GISD was created within tight funding constraints but was designed to be flexible and allow expansion. The GISD project has been remarkably successful, thanks to the goodwill and generosity shown by experts and supporters around the world. After more than 3 years of operation, we can point to a growing public profile, endorsements from experts, and strong support from partners such as the NBII and the New Zealand Agency for International Development as evidence of the value of this work.

The GISD (http://www.issg.org/database) aims to raise awareness about IAS by providing a broad audience with easy access to authoritative information and to facilitate effective prevention and management of invasive species problems by disseminating specialists' knowledge and experience globally.

Appendix 4.2

GISD Sample Pages

Figure 4.1. GISD home page at http://www.issg.org/database/welcome/

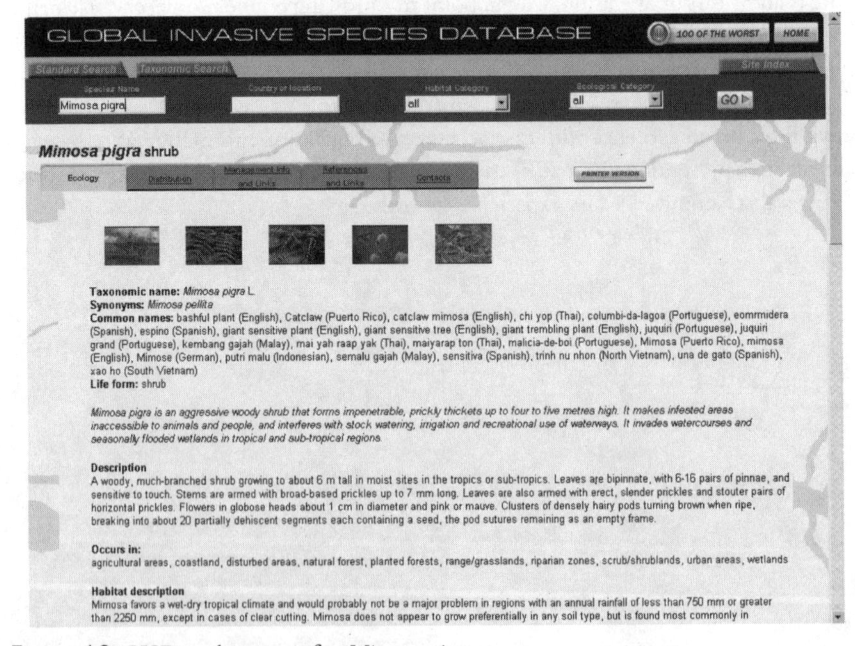

Figure 4.2. GISD ecology page for *Mimosa pigra*

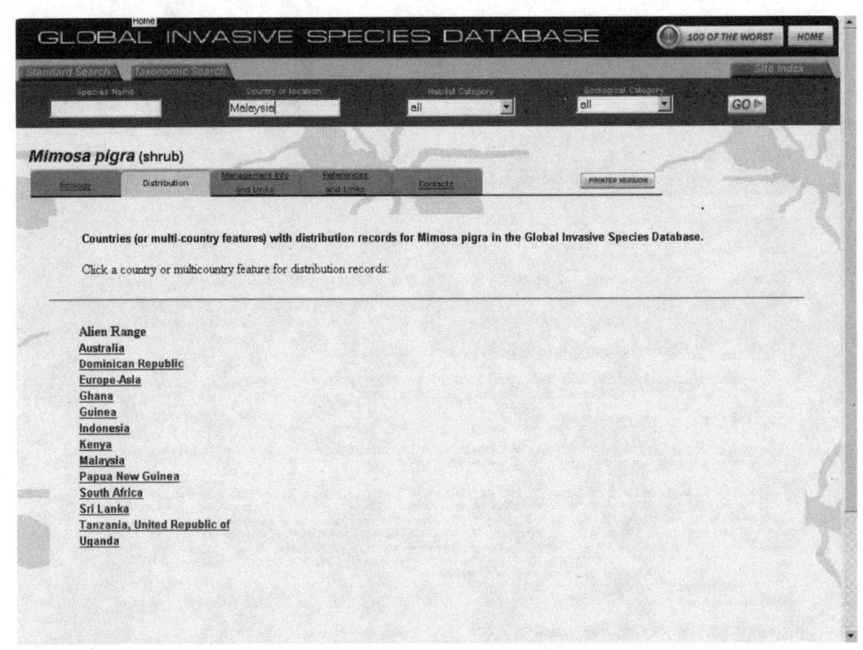

Figure 4.3. GISD distribution page for *Mimosa pigra*

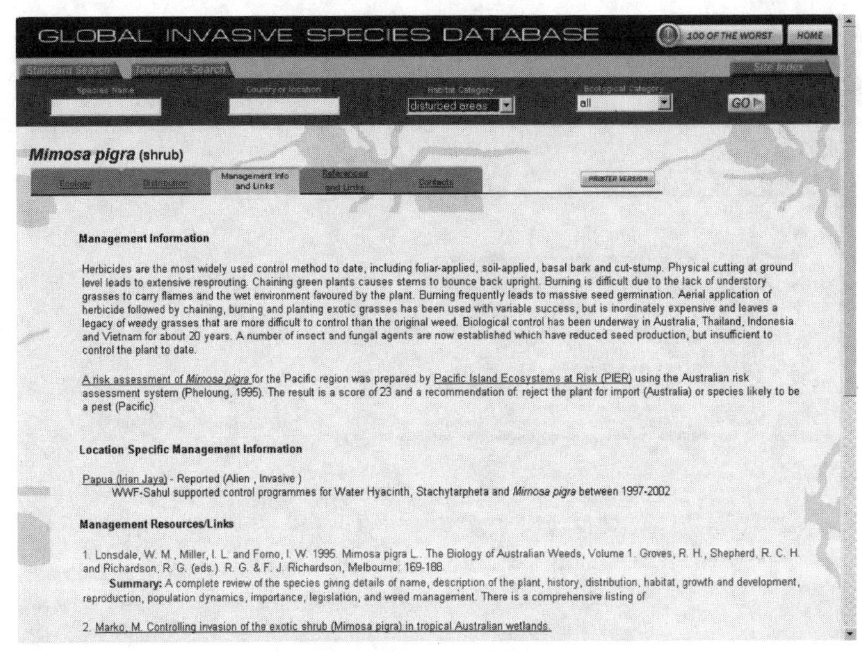

Figure 4.4. GISD management information page for *Mimosa pigra*

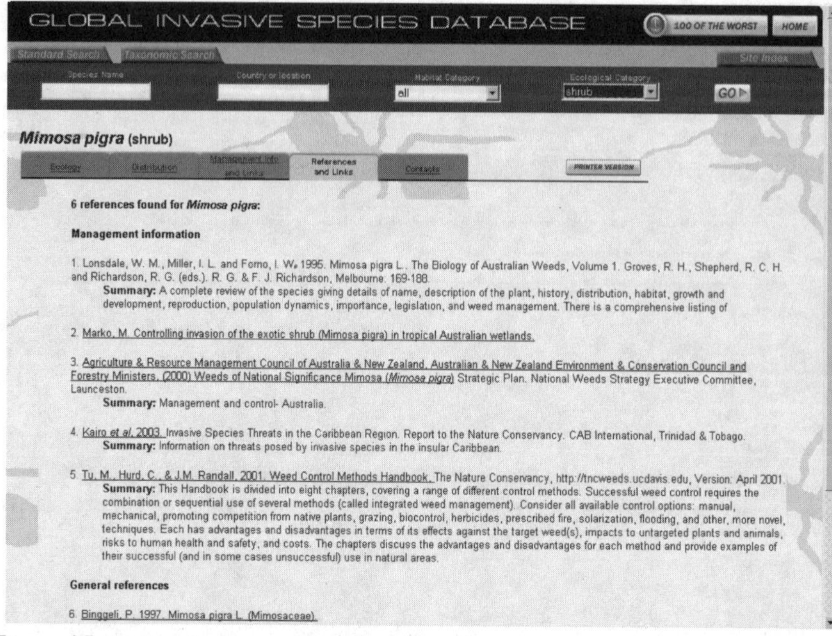

Figure 4.5. GISD references and links page for *Mimosa pigra*

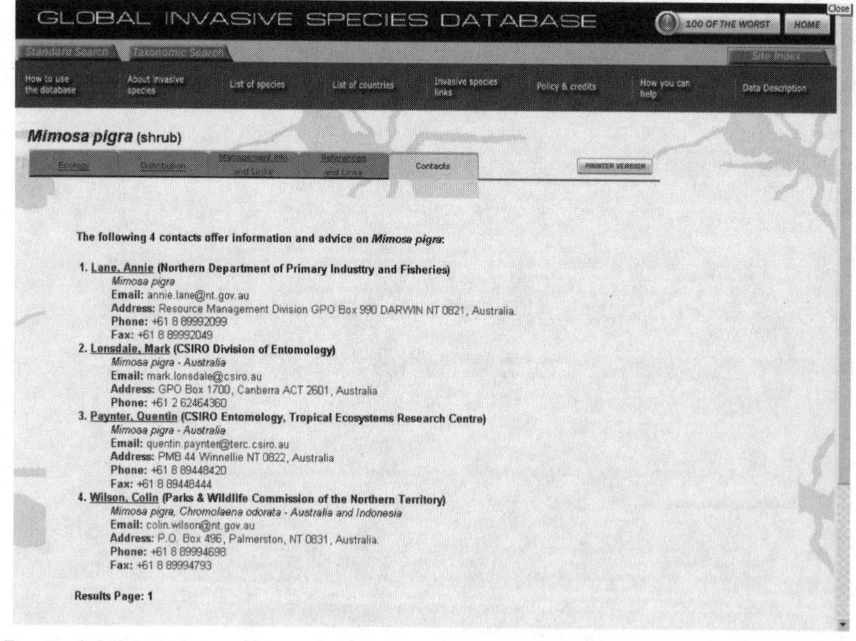

Figure 4.6. GISD contacts page for *Mimosa pigra*

Notes

1. An alien species (nonnative, nonindigenous, foreign, or exotic species) is a species, subspecies, or lower taxon occurring outside its natural range (past or present) and dispersal potential (i.e., outside the range it occupies naturally or could not occupy without direct or indirect introduction or care by humans) and includes any part, gamete, or propagule of such species that might survive and subsequently reproduce (IUCN 2000). Invasive alien species are alien species whose establishment and spread threaten ecosystems, habitats, or species with economic or environmental harm (McNeely et al. 2001). In the IUCN context they are described as follows: "Invasive alien species are alien species that become established in natural or semi-natural ecosystems or habitat, are agents of change, and threaten native biological diversity" (IUCN 2000).
2. *Introduction* means the movement, by human agency, of a species, subspecies, or lower taxon (including any part, gamete, or propagule that might survive and subsequently reproduce) outside its natural range (past or present). This movement can be either within a country or between countries.
3. They are also more likely to respond to successful eradication and border control methods to reduce or remove threats. The CBD has repeatedly recognized the very urgent need to deal with IAS issues in isolated and vulnerable ecosystems.
4. Aliens-L is a Listserv dedicated to invasive species and maintained by the IUCN. It allows users to freely seek and share information on IAS and the threats they pose to the earth's biodiversity. To subscribe, send an e-mail to listadmin@indaba.iucn.org with no subject and the message "subscribe aliens-L" in the body of the text.
5. See http://jasper.Stanford.EDU/GISP/ for more information.
6. Results of a review of Internet-based resources are available at http://www.issg.org/database/reference/index.asp, and several non-Internet resources (e.g., *The Proceedings of the International Conference on Eradication of Island Invasives,* publication features, and newsletters) are listed at http://www.issg.org.

References

Browne, M. 2001. "Update from the Global Invasive Species Database." *Aliens* 14:7–8. Online: http://www.issg.org/database.

Clout, M. N. 1999. "Biodiversity conservation and the management of invasive animals in New Zealand," in *Invasive Species and Biodiversity Management* (based on a selection of papers presented at the Norway/United Nations Conference on Alien Species, Trondheim, Norway, July 1–5, 1996), edited by O. Sandlund, P. Schei, and A. Viken. Kluwer Academic Publishers, Dordrecht, The Netherlands. Pp. 349–359.

Clout, M., and K. Ericksen. 2000. "Anatomy of a disastrous success: The brushtail possum as an invasive species," in *The Brushtail Possum: Biology, Impact and Management of an*

Introduced Marsupial, edited by T. Montague, 1–9. Landcare Research, Lincoln, New Zealand. Pp. 1–9.

Clout, M. N., and S. J. Lowe. 2000. "Invasive species and environmental change in New Zealand," in *Invasive Species in a Changing World,* edited by H. A. Mooney and R. J. Hobbs. Island Press, Washington, DC. Pp 369–384.

Courchamp, F., J. Chapuis, and M. Pascal. 2003. "Mammal invaders on islands: Impact, control, and control impact." *Biological Reviews* 78:347–383, Cambridge University Press.

Elmqvist, T. 2000. *Invasive Species on Islands: Consequences and Management Options.* Paper presented at the EC Portuguese Presidency Meeting on Biodiversity Research, May 14–16, 2000, Azores. Online: http://cimar.org/biodiversity/papers/emqvist.htm.

FAO. 2002. "Emergency Prevention System for Transboundary Animal and Plant Pests and Diseases." Online: http://www.fao.org/EMPRES/default.htm.

Goodwin, B., A. McAllister, and L. Fahrig. 1999. "Predicting invasiveness of plant species based on biological information," *Conservation Biology* 13(2):422–426.

Heymann, D. 2001. *Strengthening Global Preparedness for Defense against Infectious Disease Threats.* Statement by Dr. David L. Heymann, Executive Director for Communicable Diseases, WHO, to the Committee on Foreign Relations, U.S. Senate Hearing on the Threat of Bioterrorism and the Spread of Infectious Diseases, September 5, 2001. Online: http://www.who.int/emc/surveill/index.html.

IUCN. 2000. "IUCN guidelines for the prevention of biodiversity loss caused by alien invasive species." Published by the ISSG as special lift-out in *Aliens* 11. Online: http://iucn.org/themes/ssc/pubs/policy/invasivesEng.htm.

Lever, C. 1994. *Naturalized Animals.* Poyser Natural History, London.

Lockwood, J. 1999. "Using taxonomy to predict success among introduced avifauna: Relative importance of transport and establishment," *Conservation Biology* 13(3):560–567.

Lockwood, J., D. Simberloff, M. McKinney, and B. Von Holle. 2001. "How many, and which, plants will invade natural areas?" *Biological Invasions* 3:1–8.

Lowe, S., M. Browne, S. Boudjelas, and M. De Poorter. 2000. "100 of the world's worst invasive alien species: A selection from the Global Invasive Species Database." Published by the ISSG as special lift-out in *Aliens* 12(December 2000).

McNeely, J. A. 1999. "The great reshuffling: How alien species help feed the global economy," in *Proceedings of the Norway/UN Conference on Alien Species, the Trondheim Conferences on Biodiversity, July 1–5, 1996,* O. T. Sandlund et al., eds., Directorate for Nature Management/Norwegian Institute for Nature Research, Trondheim, Norway, 1996, p. 53.

McNeely, J. A., H. A. Mooney, L. E. Neville, P. J. Schei, and J. K. Waage, eds. 2001. *Global Strategy on Invasive Alien Species.* IUCN, Gland, Switzerland, on behalf of the Global Invasive Species Programme.

PCE. 2000. *New Zealand Under Siege: A Review of the Management of Biosecurity Risks to the Environment.* Office of the Parliamentary Commissioner for the Environment. Wellington, New Zealand.

Peterson, T. 2001. "Modeling species invasions: New methods and new data from biodiversity informatics," in *Assessment and Management of Alien Species That Threaten Ecosystems, Habitats and Species.* CBD Technical Series, No. 1, Montreal.

PGSF. 2001. "Parliamentary reports." Public Good Science Fund, Wellington, New Zealand.

Rejmánek, M. 1996. "A theory of seed plant invasivness: The first sketch." *Biological Conservation*, 78:171–181.

Rejmánek, M. 1999. "Invasive plant species and invasible ecosystems," in *Invasive Species and Biodiversity Management* (based on a selection of papers presented at the Norway/United Nations Conference on Alien Species, Trondheim, Norway, July 1–5, 1996), edited by O. Sandlund, P. Schei, and A. Viken, 79–102. Kluwer Academic Publishers, Dordrecht, The Netherlands.

Samways, M. 1996. *Managing Insect Invasions by Watching Other Countries.* Paper presented at the Norway/United Nations Conference on Alien Species, Trondheim, Norway, July 1–5, 1996.

Simberloff, D. 1999. *The Ecology and Evolution of Invasive Nonindigenous Species.* Paper presented at the Global Invasive Species Program Workshop on Management and Early Warning Systems, Kuala Lumpur, Malaysia, March 22–27, 1999.

Soberón, J., J. Golubov, and J. Sarukhan. 2001. "Predicting the effects of *Cactoblastis cactorum* (Berg) on the Platyopuntia of Mexico: A model on the route of invasion," in *Assessment and Management of Alien Species That Threaten Ecosystems, Habitats and Species.* Abstracts of keynote addresses and posters presented at the sixth meeting of the Secretariat of the Convention on Biological Diversity, Subsidiary Body on Scientific, Technical, and Technological Advice, held in Montreal, Canada, March 12–16, 2001. Montreal, SCBD, 123 p. CBD Technical Paper No. 1.

SPREP. 2000. *Invasive Species in the Pacific: A Technical Review and Draft Regional Strategy.* South Pacific Regional Environmental Program, Apia, Samoa.

Sutherst, W., R. Floyd, and G. Maywald. 1995. "The potential geographical distribution of the cane toad, *Bufo marinus* L. in Australia," *Conservation Biology* 9(6):294–299.

UN. 1997. *Guiding Principles for Effective Early Warning by the Convenors of the International Expert Groups on Early Warning of the Secretariat of the International Decade for Natural Disaster Reduction.* United Nations IDNDR Secretariat, Geneva.

Williamson, M. 1996. *Biological Invasions* (Population and Community Biology Series 15). Chapman & Hall, London.

Wittenberg, R., and M. J. W. Cock, eds. 2001. *Invasive Alien Species: A Toolkit of Best Prevention and Management Practices.* CABI Publishing, Wallingford, UK.

Zillman, J. W. 1998. *Meteorological and Hydrological Early Warning Systems.* Paper presented at the International Conference on Early Warning Systems for Natural Disaster Reduction, Potsdam, Germany.

5

Characterizing Ecological Risks of Introductions and Invasions

David A. Andow

Risk characterization is a synthesis and summary of information about a potentially hazardous situation that addresses the needs and interests of decision makers and of interested and affected parties. Risk characterization is a prelude to decision making and depends on an iterative, analytic–deliberative process.

—NRC (1996: 27)

The characterization of the environmental risks associated with species invasions and introductions is in its early stages. Characterization of risk is decision driven (NRC 1996) rather than analytic and summarizing (NRC 1983). It involves both analysis and deliberation, which widens the scope of concerns to include all interested and affected parties. Although it remains convenient to refer to a risk assessment and a risk management process that are linked by risk characterization (NRC 1983), for the complex and poorly defined issues related to species invasions and introductions, it will be useful to apply the model of risk characterization. Figures 5.1 and 5.2 contrast the 1983 and 1996 National Research Council (NRC) models. The main advantage of the risk characterization model is that it can be adapted to complex risk situations involving potentially controversial issues and conflicting interests. Characterizing environmental risks of biological invasions and introductions is still in a rudimentary state, primarily because the decisions are not clearly formulated, decision options are poorly characterized, and environmental science is not focused on informing the decision-making process. Here I apply some of the concepts of risk characterization to the problem of biological invasions and introductions, focusing on the role of ecological science in the process.

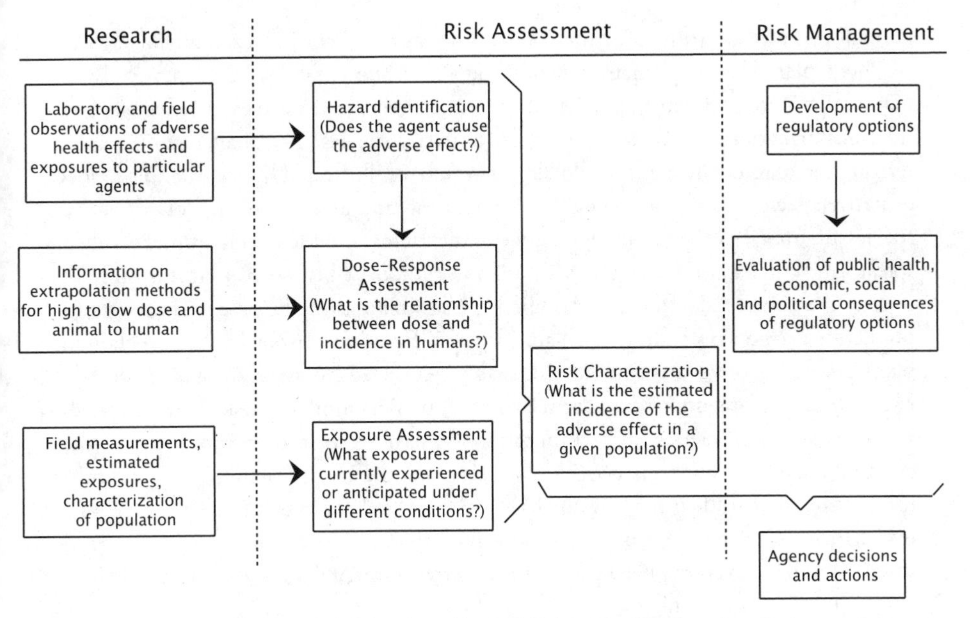

Figure 5.1. Risk analysis model developed by NRC (1983), showing the relative independence of risk assessment and risk management. Risk characterization is a summarizing activity concerned solely with scientific results. Figure from NRC (1996: 15).

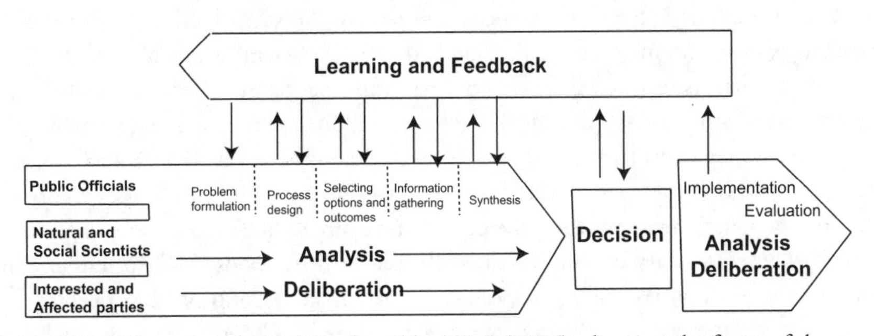

Figure 5.2. Risk analysis model developed by NRC (1996), showing the focus of the process on decision making. Risk characterization is a synthetic process integrating analysis and deliberation.

Formulating the Risk Problem

Identifying the actual risks of biological invasions and introductions remains a complex problem. Early in the 1900s the most widely recognized risk was plant pest risk: that the invading or introduced organism would cause some plant to grow less vigorously. Many national quarantine acts were established because some alien plant pest harmed an important plant. In the United States, white pine blister rust was instrumental in the establishment of the Federal Quarantine Act of 1912 (Horsfall 1983). Although plant pest risk remains one of the key environmental risks of alien organisms, many other potential detrimental environmental effects have been established, such as the transformation of ecosystems and loss of native species. The challenge in characterizing these other environmental risks is twofold, as illustrated by the example of loss of native species. First, the biological mechanisms that give rise to the loss of native species are many, so the association between the increase in the population of an alien species and the loss of a native species often is tenuous. This means that it is difficult to map the actual risks to particular alien species in anything but a descriptive, non-predictive way. Second, these risks are in part a function of social values. Although plant pest risk is a well-recognized risk that society needs to avoid or manage, loss of native species is less universally appreciated as an environmental risk. Not all species are treated by all societies as equal in terms of the need for preservation, despite the extraordinary efforts of conservation enthusiasts. In the United States, a lycaenid butterfly or any nematode is not as valued as the American bald eagle or the monarch butterfly. In Japan, some rare insects are so valued that a pet trade has built up around them. Moreover, even the perception of these environmental risks of alien species is in part a function of cultural values.

The articulation between the scientifically identified risks and culturally identified risks of alien species remains rudimentary. This articulation between science and cultural values must be done so that the risks can be adequately characterized. There are far too many permutations between alien species, potentially invaded habitats and probable abundance, distribution and ecological effect to treat the scientific issues independently from the cultural issues. It will be critical to involve people who are affected by alien species, those with interests in controlling them, and those who may be affected by control efforts. Three considerations that are often overlooked at this risk characterization stage are fairness, prevention, and rights (NRC 1996). Which groups of people, organisms, or ecosystems are exposed to the greatest risks? In addressing the issue of fairness, it may become necessary to consider the total risks of these exposed groups rather than only the incremental risk of alien species. For example, a group of poor people may depend on the qualities of certain waterways for their livelihood. Although the incremental risk associated with alien species may be small for these people, its absolute effect may be devastating compared with that experienced by wealthy people. At present no scientific research has characterized the distributional effects of alien species on different human groups. In contrast, much research has been conducted to show that some habitats are at greater risk than others. For example, small oceanic islands are at particular risk from alien species, although the reasons for this pattern are not certain (Simberloff 1986). Regardless, the decision-making process must treat these unequal risks fairly.

What can be done to avoid risks from alien species? The simplistic answer to this question is to stop the movement and introduction of alien species. When ecologists critiqued the release of alien natural enemies in biological control, their recommendation was to stop all releases until it could be proved that the release would cause no environmental harm (Simberloff and Stiling 1996). This is certainly a possible option for avoiding risks from alien species, but much more scientific thought is needed to develop additional options for avoiding these risks. It is not possible to prove that a human action will not cause environmental harm, so stopping movement and introduction cannot be sustained indefinitely as a scientific argument. It may be possible to disaggregate the problem and develop additional risk procedures targeted at a narrower scope of problems. For example, some alien insect egg parasitoids released in northern temperate climates may entail a much narrower range of risks that are quantifiable (Andow, Lane, and Olson 1995) than any kind of release on oceanic islands. Moreover, it is not possible to stop international trade.

Types of Invasions and Introductions and Some Decision Options

Biological invasions and introductions can be grouped in several ways. From the perspective of risk characterization, however, these scientific definitions and distinctions are useful only insofar as they contribute to framing or informing decisions. Consequently, the classification outlined here will explicitly link to decision making. This outline is rudimentary and provides only nominal insights on decision making; much more work in this area is needed.

The distinction is between species and genotypes that invade on their own dispersive powers (invading species) and species and genotypes that are introduced by humans (introduced species). Less can be done to respond to invading species than to introduced species. For invading species, it might be possible to stop or slow dispersal, as is being done for the gypsy moth (Sharov and Liebhold 1998), but most decisions focus on managing the environmental consequences of the invading species. For introduced species it may be possible to manage the pathways of introduction to reduce the risks of the alien species in addition to managing the environmental consequences of the species. The rest of this chapter concentrates on introductions.

Intentional introductions should be distinguished from accidental (unintentional) introductions of the alien organism or genotype. Intentional introductions provide many more decision opportunities for reducing risk than do accidental ones. If the species or genotypes are to be released intentionally, it should be possible to deliberately not release them, release them differently, or manage them to mitigate risk. Horticultural introductions, biological control introductions, and genetically engineered crops are examples. Halting or modifying accidental introductions entails more indirect interventions, such as controlling dispersal vectors. Such introductions include those associated with dry or wet ballast, packing materials, and plant pests associated with their host plants. The risk characterization processes differ between intentional introductions and accidental introductions.

At present, risk management focuses on restricting the importation or movement of potentially invasive introductions. This is implemented through various plant pest and quarantine laws and agreements, such as the International Plant Pest Convention (IPPC). As discussed earlier, however, risk characterization may need to be broadened to allow additional dimensions of the risk problem to be evaluated.

Presumption of Risk or Presumption of Safety?

In any risk characterization process, one of the initial issues to be considered is burden of proof. Do we presume there is a risk until proven otherwise, or do we presume there is no risk until proven otherwise? In actual risk analysis, an intermediate position often is taken, but a presumption of risk or safety underlies all risk analysis.

The presumption of risk or safety is not a scientific issue. Rather, it is a social issue determined by how risk averse a society is. Scientific information can reduce uncertainty and thereby influence the perception of risk and how risk averse a society might become, and science can provide a rationale for a presumption of risk or safety, but science does not determine this presumption.

Precautionary Principle and Principle of Substantial Equivalence

The precautionary principle holds that when a human activity raises the threat of harm to human health or the environment, precautionary measures should be taken even if some cause and effect relationships are not fully established scientifically. It derives its power from the same source at the aphorism "Better safe than sorry" and is in many ways difficult to fault. This principle embraces a presumption of risk but is sufficiently tentative to allow many perspectives. How great a threat? How strong a measure? How much scientific uncertainty? Answers to these three questions determine the strength of the precautionary principle and situate the principle along a spectrum of strong presumption of risk to weak presumption of risk.

The principle of substantial equivalence suggests that when a human activity is similar to a widely accepted activity with well-characterized effects (or lack thereof) on human health or the environment, then the two activities are substantially equivalent, and no additional risks need be considered. If it looks like a duck, acts like a duck, and quacks like a duck, then it is a duck. This principle embraces a presumption of safety but is also sufficiently tentative to allow many perspectives. How similar? How widely accepted? How well characterized? Answers to these three questions determine the strength of the principle of substantial equivalence and situate it along a spectrum of strong presumption of safety to weak presumption of safety.

At the outset, it should be clarified that different classes of introductions may be held to different presumptions of risk. Not all classes of introductions are equally risky, and social or cultural demands for safety may vary widely. As a cursory review shows (Figure 5.3), the application of these principles to risks of alien introductions is highly variable. For accidental introductions of plant pests, almost every country balances a presumption of risk and a presumption of safety and has some form of quarantine to reduce the risk. Introductions of weed biological control agents are held to a strong standard of presumption of risk. It can be very difficult to obtain permission to introduce these biological control agents in some countries, such as Japan, and others, such as the United States and Australia, require significant testing to demonstrate a degree of safety before release. In most countries horticultural releases are considered safe, and little is done to manage potential environmental risks of these introductions, despite abundant scientific evidence that these introductions cause environmental harm (Mack 1991; Mack and Lonsdale 2001). Most countries do not intercept potential horticultural introductions very effectively. Clearly, science is only one of several criteria that are used to set a presumption of risk or safety for a class of alien introductions.

Three categories of introductions deserve special mention. Introductions of genetically modified (GM) crops have engendered significant controversy worldwide. In the United States, the Food and Drug Administration (FDA) presumes that these crops are

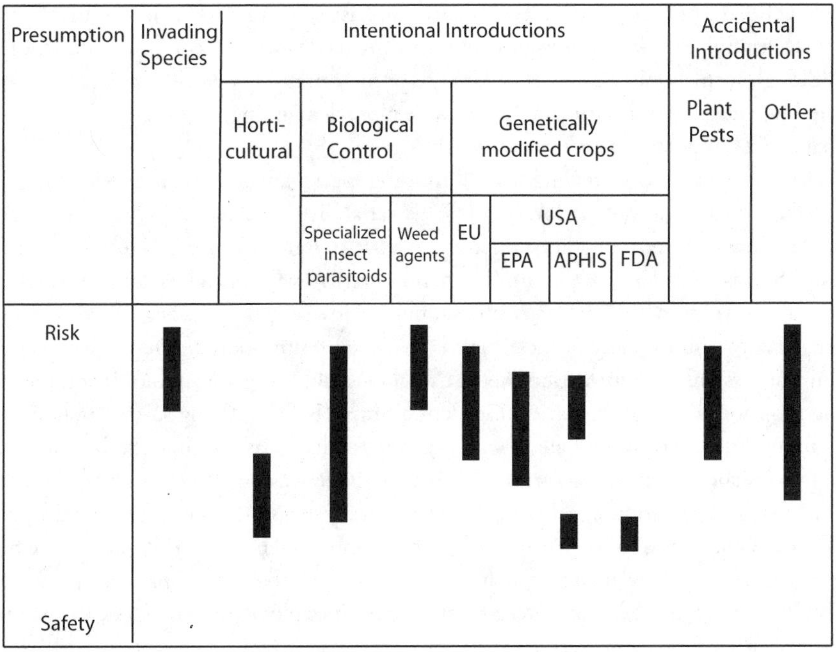

Figure 5.3. Presumption of risk or safety associated with various classes of biological invasions and introductions, with particular emphasis on the presumptions inherent in the U.S. regulatory system. This is not a measure of the actual risk of the species as determined when risk analysis is completed but a measure of the presumed risk of any species as risk analysis commences. APHIS, Animal and Plant Health Inspection Service; EPA, Environmental Protection Agency; FDA, Food and Drug Administration.

substantially equivalent to standard, non–genetically engineered foods (FDA 1992) and takes a presumption of safety. The U.S. Department of Agriculture (USDA) Animal and Plant Health Inspection Service (APHIS) regulates GM crops under two systems. The first is a permitting system in which they presume a plant pest risk until shown otherwise, and the other is a notification system in which they presume safety. The Environmental Protection Agency (EPA) presumes that resistance evolution will occur to *Bacillus thuringiensis* crops, they test nontarget effects without a strong presumption of risk or safety, and they consider the human health effects to be substantially equivalent to those of other *Bacillus thuringiensis* insecticide products, which are considered safe. In contrast, the EU has presumed that GM crops are risky until proven otherwise. It is in part these differences in the presumption of risk or safety that fuel the controversy over the commercial use of these organisms.

Biological control of arthropods by introducing alien specialist insect parasitoids has been regulated in the United States by the USDA APHIS. Until the early 1980s, APHIS used a presumption of safety. Specialized parasitoids were believed to have a stable host range, which was usually restricted to a particular insect pest and its relatives, which often were also pestiferous. In the early 1980s, the potential detrimental nontarget effects from these introductions were emphasized (Pimentel et al. 1984; Howarth 1991), leading the USDA to reconsider its oversight of biological control introductions. Although formal policy changes have yet to occur, obtaining permission to release certain biological control organisms has become more demanding, requiring scientific data that addresses potential environmental risks.

Other kinds of accidental releases of invading species and genotypes also have highly variable presumptions of safety or risk. This variability is related in part to the variety of ways in which accidental introductions can occur. Release of vertebrate pets and fish bait often is considered safe (without scientific basis), whereas release of dry ballast is considered risky (with much scientific support). However, the most significant characteristic common to many of these cases is that the presumption of safety or risk is still fluid, and scientific information can be useful in establishing a firm foundation for the presumption of safety or risk. An excellent example is the evolving science policy surrounding the use of water ballast. Initially, water ballast was presumed to be safe with respect to species invasions, largely because the issue was not thoroughly considered. With the accumulation of scientific information on invasions caused by the transport of water ballast around the world (e.g., Carlton and Geller 1993), this perspective has gradually shifted. Now it appears that movement of water ballast is considered risky, and several changes in the use of water ballast have been implemented to reduce the potential risk of alien species invasions.

These three categories are similar in that there is not yet agreement on the presumption of risk or safety. Without this agreement, it is very difficult to implement a scientifically consistent risk analysis protocol. On some of these issues it may take several years before a consensus can be reached, but on others, such as water ballast and biological control, resolution may come soon.

Standards of Evidence and Uncertainty

Closely coupled to the presumption of risk or safety is the scientific issue of standards of evidence. Assuming that the risks (endpoints) are identified, how they are demonstrated to exist and how their severity is measured for any particular case are scientifically problematic. The normal standard of science is to presume that an event does not occur until it can be conclusively demonstrated to occur, usually through multiple lines of evidence. When translated into the language of risk analysis, normal science would presume safety until proven otherwise. This standard of evidence is too exacting to be useful in a policy environment. Although some scientists might assert that because safety cannot be proved, a presumption of risk is scientifically invalid, this is an irresponsible position.

The role of science in risk characterization is less about proof and more about dealing with uncertainty. There are two major sources of uncertainty in risk analysis. The first is uncertainty in outcome; ecological events are not easily predicted even with abundant knowledge and information. Science uses probabilistic methods to treat this kind of uncertainty, and although it can become complicated, it entails estimating the probability of occurrence of particular outcomes and weighting these outcomes. Although none of this is trivial, the conceptual framework is reasonably well developed.

However, the more critical source of uncertainty occurs in the absence of scientific knowledge. This enters into the analysis as uncertainty because it is necessary to make scientific assumptions to bridge these uncertainties. These assumptions are all conditional on a presumption of risk or safety. For example, suppose debarked logs from northeast Asia were being used as dunnage in trans-Pacific shipping, and we have no information about the insect pests that feed on the tree in Asia. Debarking removes the serious risk of introducing alien species of bark beetles (scolytids), which live in the bark and vascular tissue just underneath the bark. We could assume that once these logs are debarked, there is no additional significant risk (presumption of safety) associated with the heartwood, or we could assume that other important alien pests might inhabit the remaining wood (presumption of risk). If we rely on previous experience and consider the time constraints and competing demands to reduce the risk of alien species invasions, we might conditionally presume safety because invasive wood-boring beetles in heartwood (e.g., some cerambycids and buprestids) are uncommon. This decision would consequently expose us to the accidental introduction of the alien invasive Asian long-horned beetle. Clearly, we cannot always assume that any risk of an introduced alien species is significant, but we also have little scientific basis to readily inform these kinds of decisions. Ecological science must focus more on the scientific dimensions of uncertainty in the decision-making process related to invasions and introductions.

This second kind of uncertainty is central to risk analysis. In its simplest form, the precautionary principle states that the lack of scientific knowledge should not stop society from taking action to avoid potential risks. All effective risk analysis protocols will indicate how uncertainty can or should be treated because risk analysis is always conducted with incomplete scientific knowledge. To evaluate the attitude of a risk analysis model or protocol toward uncertainty, several dimensions are relevant. Can the full range of decision options be chosen under extreme uncertainty and under little uncertainty? Are the decisions restricted in some way under uncertainty, either temporally, spatially, or otherwise conditioned? How is new information integrated after a decision is made? These issues prove to be central in a comparison of the international agreements that can affect risk analysis of invasive introductions.

Ex Ante and Ex Post Analysis

Ex ante risk analysis occurs before a decision is made and helps to inform the decision-making process. Ex post analysis occurs after the decision and is useful in evaluating the decision or decision-making process. Most of the ecological research on species invasions has not been aimed to inform either type of analysis.

Williamson (1999) makes this point in part when he says that prediction is harder than explanation. Most ecological research has focused on describing alien invasions and introductions and making statistical summaries of previous invasions and introductions in the search for patterns. These efforts have led to hypotheses about what makes a species more invasive and what makes a habitat more prone to invasion (e.g., Mooney and Drake 1986). These and related efforts have been useful in identifying risks, delineating the extent of the issue, and motivating the political and scientific desire to reduce the risks associated with species invasions. They also demonstrate that the decision to do nothing to reduce the risks of alien invasions and introductions is costly indeed. However, they have not been very useful for decision making to reduce these risks.

Several insights from this literature are useful as ex post risk analysis to inform the decision-making process. Williamson's (1996) tens rule is quite helpful in the decision-making process. The tens rule suggests that most potential invaders will either not establish or not cause any recognizable ecological harm. Consequently, any decision-making process must be one that efficiently winnows the many inconsequential aliens from the few harmful ones.

Alien species establishment is positively correlated with the amount and mode of international trade (Sailer 1978; OTA 1993). The increasing volume of trade has increased the opportunity for accidental introductions. The changing modes of trade have provided new groups of organisms the opportunity to invade. For example, the development of rapid trans-Atlantic steam shipping in the late nineteenth century enabled economic shipping of live plant material, which led to the introduction of many scale insects (Sailer 1978). This implies that trade should be examined to reduce the risks of alien introductions. This must be a rapid, efficient process to focus examination on the few potentially risky trade vectors, related in some way to the mode of transport, the habitat of origin, and the receiving habitat. Based on previous experience, any significant changes in the mode of transport should be examined for its potential for accidental introduction of new alien species.

Another important pattern is that species that have been invasive in one locale are more likely to be invasive in other locales than other species (Williamson 1996). This historical information provides a powerful sieve for isolating several harmful species for more concentrated risk characterization efforts. Research to identify invasive species worldwide will be helpful for developing decision processes to characterize and manage these risks.

Though useful, historical information is insufficient for identifying risky alien species. Many ecologists suggest that a focus of research should be to develop the ability to predict invasions (an ex ante analysis). In a certain sense this is correct, but in the usual scientific sense this may not be necessary. Usually ecologists frame the prediction problem as follows: given any particular species and an identified risk (e.g., probability of establishment, probability of adverse ecological effect), use ecological data to predict the ecological risk associated with that species. Although numerous schemes have been proposed, none yet are widely accepted. Rejmánek and Richardson (1996) propose a set of ecological characteristics that accurately ranks historical invasiveness patterns in *Pinus* spp. This is one of the more thorough analyses of a taxonomic group, but their scheme must be tested independently of the historical record before it could be used for decision making.

This detailed level of scientific prediction may not be necessary for effective risk characterization and management. As mentioned earlier, scientific contributions to reducing uncertainty are key for developing effective and efficient risk analysis. Consequently, the problem might not be predicting the invasiveness of any particular species. Instead, the problem might be winnowing down the possible introductions to a much smaller subset to be evaluated without inadvertently eliminating some that are risky. Risk analysis of invasiveness could then be restricted to this smaller group of higher-risk introductions (those retained by the winnowing process). The first sieve could be history. Any species that has been invasive somewhere in the world could cause ecological harm somewhere else in the world. It would be necessary to determine the likely pathways and vectors these identified invasive species might take to colonize other parts of the world. Additional sieves must be developed, and a couple of approaches can be suggested, both of which would be expected to retain a very low proportion of the species for detailed analysis.

Of the established alien species that have caused no discernible ecological harm in their new habitats, what factors can predict the species that could cause ecological harm in another invaded habitat? Of the alien species that have not established anywhere beyond their native range, what factors can predict the species that could cause ecological harm in another invaded habitat? Neither of these conditional problems has been addressed ex post in the ecological literature, but scientific work on either would contribute to an effective decision process. If additional sieves, such as the two suggested here, are developed, the ex ante prediction problem becomes narrower: of the species that could cause ecological harm were they to invade or be introduced, which are likely or not likely to cause actual harm? Using alien pests as an example, the problem could be framed as two issues. Given that the species is an established alien nonpest somewhere else, under what conditions is it likely to become a pest here? Given that the species is a native pest somewhere else, is it likely to become a pest here? Again, these problems have not been addressed beyond the species level in the ecological literature.

Some Risk Analysis Approaches

As mentioned earlier, risk analysis of potential introduced species focuses on restricting the importation or movement of potentially invasive introductions. This is implemented through various plant pest and quarantine laws and agreements. Internationally, the Agreement on the Applications of Sanitary and Phytosanitary Measures (SPS Agreement) of the World Trade Organization (WTO) forms one umbrella under which risks of alien invasive species are evaluated. The SPS Agreement specifies the IPPC and its secretariat as the vehicle to implement risk analysis. Another, independent international agreement that addresses invasive species risk is the Convention on Biodiversity (CBD). These approaches will be compared in the following sections, and specific risk analysis models that can be used under these agreements will be examined, taking into account some of the theoretical issues discussed earlier.

Approved and Prohibited Lists

The use of approved and prohibited lists of species for introduction illustrates the contrast between the presumption of risk and safety. Approved lists allow introductions of listed species; all others are presumed risky and not allowed. The U.S. government uses an approved list for restricting import of specified fruits and vegetables from specified countries (7 CFR 319.56), and Hawaii allows import only of animals and microorganisms on a conditionally approved list. Prohibited lists presume that listed species cannot be safely introduced. The Lacey Act prohibits import of specified fish and wildlife taxa, and the Federal Noxious Weed Act prohibits import of 93 listed plant species.

Although some ecologists might support an approved list approach to the introduction of alien species, scientific debates over the present issue are a red herring. Because this discussion is about deliberate introductions, there are many options for management in addition to prohibiting all such introductions. Moreover, many of these introductions involve powerful vested interests, such as the pet and horticultural trade, pet owners, and gardeners. Consequently, risk characterization is exceedingly complex, and science must consider itself merely a partner to the many diverse interests that will determine the outcome of the risk analysis. As discussed earlier, ecological science might broaden its concerns and address scientific issues related to fairness and novel approaches to avoiding risk. Lists will not be enforceable for the large class of accidental introductions. For these, an entirely different approach must be developed.

Qualitative Risk Assessment

A particularly vexing problem is to analyze risks associated with accidentally introduced species. Indeed, they are somewhat refractory to quantitative analysis because the information on the possible alien species is insufficient.

The U.S. National Park Service developed a qualitative risk assessment model for managing nonindigenous plants and animals (Whiteaker and Doren 1989). These models use a point scale to rank the current impact of the alien species, its potential for harm, control feasibility, and the consequences of delay. These are combined to determine which species should be targets for management. Similar approaches were adopted by the State of Minnesota Exotic Species Task Force to rank exotic plants and animals as benign, neutral, and threatening (Minnesota Interagency Exotic Species Task Force 1991).

These efforts predated a major improvement in qualitative risk assessment of potential plant pest risk at the USDA (Orr, Cohen, and Griffin 1993). This model focuses on the commodity that would serve as an importation vector for potential plant pests. The protocol involves listing all alien potential plant pests that would be associated with the commodity in the exporting habitat and determining qualitatively for each pest species the probability of pest establishment and the economic and environmental consequences of establishment. These two qualitative measures are then combined into a single qualitative unmitigated plant pest risk for each species. An example is the assessment of plant pest risk of importing unprocessed pine and spruce logs from Mexico (Thacz et al. 1998). The protocol uses an expert panel to assign the qualitative measures to potential plant pests and consultation with a large group of additional experts to create consensus around these measures. This and other qualitative assessment models feed into a conventional risk management decision-making process. Several of these assessments have been successfully completed, so the adequacy of the assessment process probably can be evaluated scientifically. For example, are the measures of economic damage and environmental damage independent? Have the assessments led to effective management changes?

The SPS Agreement

International agreements currently in force will have great influence on the development of appropriate risk analysis procedures for invasive species. The most significant of these in relation to invasive species is the SPS Agreement of the WTO and the standards, guidelines, and recommendations that it identifies as necessary to ensure adequate protection.

On April 15, 1994, the Final Act of the Uruguay Round of Multilateral Trade Negotiations was signed, putting into force the Agreement on the Applications of Sanitary and Phytosanitary Measures on January 1, 1995 (the SPS Agreement of the WTO). This agreement allows members to restrict international trade to protect human, animal, or plant life or health from pests and diseases as long as the restriction is necessary and scientifically justified. These restrictions are called sanitary or phytosanitary measures, and apply to all plants and animals within the territory of member states. Clearly, these measures could be used to protect against the adverse effects of invasive species because all such effects directly or indirectly harm animal or plant life or health. Because restrictions can be applied only when necessary and scientifically justified, the SPS Agreement makes a presumption of safety and places the burden of proof on demonstrating risk. In addition, because the agreement is a trade agreement, members must use measures that minimize trade restrictions while achieving an acceptable level of protection. This means that when sufficient scientific information exists, the sanitary or phytosanitary measures can restrict trade, but methods that have the smallest effect on trade must be used.

Under conditions of uncertainty, members may restrict trade (Article 5, paragraph 7), but this is a provisional decision, and they are obligated to obtain the scientific information to justify their actions in a reasonable period of time. The time frame is not specified in the agreement, and the possibility that a country may not have the capacity to obtain the information is not discussed. These implicit ambiguities could increase the risk of species invasions, but the interpretation of this clause probably will occur through case decisions and has not yet been completely settled.

One of the key provisions of the agreement is to designate certain international standards, guidelines, and recommendations as ones necessary to protect human, animal, or plant life or health. The agreement identifies three of them and provides a means to add additional ones. For food safety the agreement identifies the Codex Alimentarius Commission, for animal health it identifies the International Office of Epizootics, and for plant health it identifies the IPPC. Several significant but unresolved issues regarding this framework are

- Whether all potentially invasive species will be ensnared by one of these international agreements
- Whether any invasive species will be ensnared by more than one of them, triggering double oversight or necessitating additional agreements addressing the scope of each
- Whether the various agreements are harmonized to treat similar risks from different taxa similarly
- Whether international capacity to develop and implement procedures under each agreement is sufficient to ensure equivalent protection against invasive species under all the other agreements

Each of these international agreements, with the possible exclusion of the Codex, should be evaluated for how they would protect against harm from invasive species.

Here I provide a general overview of the IPPC. This evaluation is by necessity provisional, but it may provide a basis for future discussion.

The IPPC is administered within the Food and Agriculture Organization (FAO) and was established to provide uniform, scientific approaches to controlling the international spread of pests of plants and plant products. It has been revised several times, most recently in November 1997, as a consequence of the SPS Agreement. The convention covers all plant pests, which are defined as organisms that cause an unacceptable (potential) economic impact to plants in some area. Pests may be excluded or their movement managed to reduce the economic risks to plants. It is up to the importing country to demonstrate that a potential pest is of economic concern, so there is a presumption of safety. Pest status is demonstrated by a pest risk analysis (PRA). The PRA guidelines are under revision, but one of the most recent is the 2003 book on *Pest Risk Analysis for Quarantine Pests.* PRAs can be initiated in three ways: identification of a pathway that presents a potential pest hazard, identification of a pest that may necessitate action, or a review or revision of policies or priorities. Next, the potential pests are evaluated to determine whether they need to be regulated. If they are determined not to need regulation, the process stops. Otherwise, the potentials for entry, establishment, and spread are evaluated and the potential economic impact is assessed. Both direct and indirect pest effects are assessed. Noncommercial impacts, which are inadequately measured in terms of prices in established markets, can be approximated with an appropriate nonmarket valuation. The economic evaluation should be in terms of a monetary value wherever possible. Risk management is next proposed to reduce the risk to an acceptable level. The acceptable level of risk may be expressed in a number of ways, including reference to existing phytosanitary requirements, indexed to estimated economic losses, expressed on a scale of risk tolerance, and compared with the level of risk accepted by other countries.

Potential Strengths

- The FAO has a significant program for building the capacity of countries to conduct PRAs and implement effective management to reduce the risks from exotic plant pests. It would be wasteful of limited financial and human resources to attempt to establish a parallel risk analysis structure for invasive species.
- The connection of the IPPC to the SPS Agreement means that using the IPPC to evaluate and manage invasive species would have widespread and rapid adoption.
- The flexibility to initiate a PRA based on pathway, organism, or policy is excellent.
- The initial focus of the PRA on problem identification is very useful.

Potential Concerns

- The presumption of safety embedded in the IPPC is potentially problematic and may not be consistent with the use of the precautionary principle. However, if all countries were to have the capacity to conduct technically justified PRAs, then the initiation of a PRA can itself be considered a precautionary act. Greater attention should be paid to developing procedures that trigger the initiation of a PRA. Moreover, because PRAs are expensive, shifting the cost of the PRA to the importing party may also act as a precautionary measure. Additional precautionary measures could be added by modifying the PRA guidelines.

- Invasive species risks probably will not fall uniformly on all people. People with greater dependency on locally available natural resources are more likely to be adversely affected than those with access to external resources. Attention to the distribution of risk among more vulnerable peoples must be added to the PRA process.
- The sole focus on economic costs will undervalue many of the ecological effects of invasive species. The definition of risk must be expanded to include other, specific noneconomic costs, especially those that are not well characterized by nonmarket valuations.
- The use of uncertainty in the PRAs must be integrated into the analysis process more completely instead of being added on as an additional consideration. Precaution can be implemented into the PRA by creating a bias in the interpretation of uncertainty.
- The sole reliance on the present PRA model for the risk analysis of invasives may be limiting. As mentioned earlier, some invasive species may be so harmful that they can be listed on a prohibited list. A possibly suitable precautionary measure would be to prohibit importation to certain ports of consignments carrying certain cargoes from specified locations because they are likely to be carrying one of these prohibited species.

These suggested changes to the IPPC are all consistent with the 1997 convention, although some of them appear to challenge its underlying philosophy. These changes might take many years to implement into the IPPC, but the science base that underlies the need to make these changes is still incompletely organized, so it will be necessary to take the time to mobilize the scientific expertise. Strategically, effective implementation of oversight and procedures to reduce the risks of invasive species may entail close coordination with present quarantine and plant protection structures and personnel. Consequently, rapid tactical engagement may be needed to ensure that risks of invasive species are reduced worldwide.

Convention on Biodiversity

The CBD was drafted in 1992 and has been ratified by several countries, not including the United States. The CBD is an international agreement to conserve and sustainably use biological diversity for the benefit of present and future generations. It identifies numerous measures to enable the signatories to accomplish this goal, including protection of indigenous biodiversity and conditions on restricting its use or protecting it against external threat. It directly relates to invasive species risks by specifying in Article 8(h) that the contracting parties shall, as far as possible and as appropriate, prevent the introduction of, control, or eradicate alien species that threaten ecosystems, habitats, or species. In addition, the preamble of the CBD includes a commitment to the precautionary principle. It specifically states that where there is a threat of significant reduction or loss of biological diversity, lack of full scientific certainty should not be used as a reason for postponing measures to avoid or minimize such a threat.

On January 29, 2000, the Cartagena Protocol on Biosafety to the CBD was adopted by consensus and went into force on September 11, 2003. The protocol applies to genetically engineered organisms that will be intentionally introduced into the environment, such as seeds. From the perspective of international trade, the protocol applies to a narrow range of trade items. Food and feed commodities that might be accidentally introduced are excluded from the protocol.

The protocol provides two clauses that have immediate relevance to invasive species risk: the savings clause and the precautionary principle (Stewart and Johanson 2000). The savings clause indicates how the protocol relates to other international agreements, most critically the WTO and the SPS Agreement. The clause is in the preamble and states,

> *Recognizing* that trade and environment agreements should be mutually supportive with a view to achieving sustainable development,
>
> *Emphasizing* that this Protocol shall not be interpreted as implying a change in the rights and obligations of a Party under any existing international agreements,
>
> *Understanding* that the above recital is not intended to subordinate this Protocol to other international agreements.

This is clearly contradictory, stating on one hand that parties must abide by the rights and obligations under previous international agreements, specifically WTO and the SPS Agreement, while asserting that the protocol is not subordinate to these very same agreements. It will probably take great effort to resolve this issue, but as Stewart and Johanson (2000) argue, later agreements (e.g., the Biosafety Protocol) usually take precedence over prior agreements (e.g., WTO and the SPS Agreement) when disputing parties have agreed to both. Because the United States has not signed and ratified the CBD, much uncertainty will remain about the relationship between the CBD and the WTO.

The protocol also provides a clear statement of the precautionary principle. Lack of scientific certainty regarding the extent of potential adverse effects of a genetically engineered organism on the conservation and sustainable use of biological diversity, taking into account risks to human health, shall not prevent a party from making a decision to avoid or minimize such potential adverse effects.

Although the CBD and Biosafety Protocol provide strong principles under which invasive species can be restricted and regulated, they lack the detailed specificity of the IPPC and the PRA protocols developed to implement it. Moreover, the capacity to implement these principles is weak or absent in most countries, including the United States. Consequently, much effort is needed before widespread use of these principles for reducing and managing risks from potential invasive species will be realized.

Comparison of the SPS Agreement and the CBD for Alien Invasives

Burden of Proof

This is perhaps the most significant difference between the SPS Agreement and the CBD. The SPS Agreement allows restriction of movement or regulation of invasive aliens only when necessary and scientifically justified. If the Biosafety Protocol were extended to invasive aliens, the CBD would allow restriction of movement or regulation as appropriate to avoid or minimize potential adverse effects even in the absence of scientific certainty. Although this difference appears stark and unresolvable, there is room for interpretation. As indicated earlier, all risk analyses of potential invasive species, whether conducted under the SPS Agreement or not, will lack scientific certainty. Thus, this difference between the agreements may hinge on an interpretation of what constitutes scientific justification and how it is to be realized within the PRAs of the IPPC under the SPS Agreement.

Uncertainty

There appears to be a big difference between the agreements on how uncertainty affects decision making. Under the SPS Agreement decisions can be made when inadequate scientific knowledge is available, but such decisions are provisional. It is necessary to acquire the needed scientific information within a reasonable period of time. Under the Biosafety Protocol, decisions made under scientific uncertainty can be final. Although some disputes have provided interpretation about what is a reasonable period of time to acquire the missing information under the SPS Agreement (a developed country with the resources to gather the information must start to gather it within a couple of years of making the provisional decision), it is clear that countries lacking in capacity will not be penalized because they cannot afford to gather the information. On the other side, although the Biosafety Protocol indicates that precautionary decisions can be final, in practice, if new contradictory information were to become available, the decision probably would be reversed. This would make all precautionary decisions provisional on future contradictory information. Thus, the main difference between the two agreements may be in the requirement to gather the supporting scientific information and the timeline for gathering it under the SPS Agreement, both of which are absent from the CBD.

Implementation

The SPS Agreement is implemented through the IPPC, which evaluates plant pest risk associated with alien invasives. The CBD has not yet been implemented, but it will evaluate risks of alien invasives to biological diversity and sustainable use of natural resources. This will include ecosystem properties and habitat quality, neither of which is explicitly evaluated under the IPPC PRAs. Clearly the mandate under the IPPC does not exclude evaluating ecosystem and habitat risks, but the detailed methods have not yet been developed. It is also clear that the training and capacity-building efforts to conduct risk analysis under the IPPC cannot be matched with a parallel effort under the CBD. The unavoidable conclusion is that some accommodation between the SPS Agreement and the CBD must be found within the IPPC and its PRA protocols. This will probably entail substantial compromise under both agreements and the IPPC. It will be necessary to develop risk analysis models for evaluating ecosystem and habitat risks before such an accommodation can be expected.

Some Future Needs

- The diversity of local cultures requires that alien species risks be characterized in those cultural contexts, involving people who may be affected by the alien species, involved in controlling the species, or affected by those controls. Specifically, scientific research should be initiated that characterizes the effects of alien species on different human groups. Additional research is needed to delineate approaches or methods for avoiding the risks of alien species. Research should be continued to identify habitats or ecosystems of concern.
- Consensus must be reached regarding the presumption of risk or safety for many kinds of introductions. For cases in which this is presumption is still fluid and not crystallized around particular interests, scientific information is likely to have a significant effect.
- There is a pressing need to develop a risk analysis model for biological introductions, concentrating on accidental introductions and their effects on ecosystems and native habitats. It is premature to suggest the structure of such a model, but it should have several properties.
  - It should blend approaches based on biological species with those based on the pathway of introduction.
  - It should rapidly and efficiently winnow attention to the few risky activities without creating excessive costs for less risky ones.
  - It should effectively characterize all environmental risks associated with the few identified potentially risky activities and enable decisions that are supported by all interested parties.

- –It should adapt to changes in transportation mode or changes in the structure of the originating or receiving landscapes.
- –Some detailed suggestions are provided in the text, but the overall structure of the biological introduction risk analysis model remains to be outlined and developed.

- There is a need to evaluate the adequacy of the SPS Agreement for reducing the ecological risks of introductions. Specifically, some changes to the operating practices of the IPPC probably are needed.

References

Andow, D. A., C. P. Lane, and D. M. Olson. 1995. "Use of *Trichogramma* in maize: Estimating environmental risks," in *Biological Control: Benefits and Risks,* edited by H. M. T. Hokkanen and J. M. Lynch. Cambridge University Press, Cambridge, UK, pp. 101–118.

Carlton, J. T., and J. B. Geller. 1993. "Ecological roulette: The global transport of nonindigenous marine organisms," *Science* 261:78–82.

FDA. 1992. "Statement of policy: Foods derived from new plant varieties; Notice." *Federal Register* 57:22984–23005.

Horsfall, J. G. 1983. "Impact of introduced pests on man," in *Exotic Plant Pests and North American Agriculture,* edited by C. L. Wilson and C. L. Graham, 2–14. Academic Press, New York.

Howarth, F. G. 1991. "Environmental impacts of classical biological control," *Annual Review of Entomology* 36:485–509.

Mack, R. N. 1991. "The commercial seed trade: An early disperser of weeds in the United States," *Economic Botany* 45:257–273.

Mack, R. N., and W. M. Lonsdale. 2001. "Humans as global plant dispersers: Getting more than we bargained for," *BioScience* 51:95–102.

Minnesota Interagency Exotic Species Task Force. 1991. "Report and recommendations" submitted to the Natural Resources Committees of the Minnesota House and Senate, July 1991.

Mooney, H. A., and J. A. Drake, eds. 1986. *Ecology of Biological Invasions of North America and Hawaii.* Springer-Verlag, New York.

NRC. 1983. *Risk Assessment in the Federal Government: Managing the Process.* National Academy Press, Washington, DC.

———. 1996. *Understanding Risk: Informing Decisions in a Democratic Society.* National Academy Press, Washington, DC.

Orr, R. L., S. D. Cohen, and R. L. Griffin. 1993. *Generic Non-Indigenous Pest Risk Assessment Process (for Estimating Pest Risk Associated with the Introduction of Non-Indigenous Organisms).* U.S. Department of Agriculture, Animal and Plant Health Inspection Service, Washington, DC.

OTA. 1993. *Harmful Non-indigenous Species in the United States.* U.S. Congress, OTA-F-565. U.S. Government Printing Office, Washington, DC.

Pimentel, D., C. Glenister, S. Fast, and D. Gallahan. 1984. "Environmental risks of biological pest controls," *Oikos* 42:283–290.

Rejmánek, M., and D. M. Richardson. 1996. "What attributes make some plant species more invasive?" *Ecology* 77:1655–1660.

Sailer, R. I. 1978. "Our immigrant insect fauna," *Bulletin of the Entomology Society of America* 24:3–11.

Sharov, A. A., and A. M. Liebhold. 1998. "Bioeconomics of managing the spread of exotic pest species with barrier zones," *Ecological Applications* 8:833–845.

Simberloff, D. 1986. "Introduced insects: A biogeographic and systematic perspective," in *Ecology of Biological Invasions of North America and Hawaii,* edited by H. A. Mooney and J. A. Drake, 3–26. Springer-Verlag, New York.

Simberloff, D., and P. Stiling. 1996. "How risky is biological control?" *Ecology* 77:1965–1974.

Stewart, T. P., and D. S. Johanson. 2000. "A nexus of trade and the environment: The relationship between the Cartagena Protocol on Biosafety and the SPS Agreement of the World Trade Organization." *Agriculture Sanitary & Phytosanitary and Standards Report,* 1–45.

Thacz, B. M., H. H. Burdsall Jr., G. A. DeNitto, A. Eglitis, J. B. Hanson, J. T. Kliejunas, W. E. Wallner, J. G. O'Brian, and E. L. Smith. 1998. *Pest Risk Assessment of the Importation into the United States of Unprocessed* Pinus *and* Abies *Logs from Mexico.* General Technical Report FPL-GTR-104. USDA Forest Service, Forest Products Laboratory, Madison, WI.

Whiteaker, L. D., and R. F. Doren. 1989. *Exotic Plant Species Management Strategies and List of Exotic Species in Prioritized Categories for Everglades National Park.* Research/Resources Management Report SER-89/04, Southeast Regional Office, National Park Service, U.S. Department of the Interior. Holmstead, Florida.

Williamson, M. 1996. *Biological Invasions.* Chapman & Hall, New York.

Williamson, M. 1999. "Invasions," *Ecography* 22:5–12.

6

Ecology of Invasive Plants: State of the Art

Marcel Rejmánek, David M. Richardson, Steven I. Higgins, Michael J. Pitcairn, and Eva Grotkopp

More than a century ago De Candolle (1855), Darwin (1859), Hooker (1864), Franchet (1872), and Goeze (1882) made important contributions to the understanding of plant invasions. There were also useful studies in the first half of the twentieth century (e.g., Allan 1936, 1937). However, only after the publication of Elton's (1958) classic *The Ecology of Invasions by Animals and Plants* did invasion ecology emerge as a new discipline, first slowly (Salisbury 1961; Baker and Stebbins 1965) and later explosively (Drake et al. 1989; Protopopova 1991; Lohmeyer and Sukopp 1992; Williamson 1996; Cox 1999; Sandlund, Schei, and Viken 1999; Mooney and Hobbs 2000; Groves, Panetta, and Virtue 2001). The scope of contemporary invasion biology is inevitably broad, ranging from essentially theoretical studies (Shigesada and Kawasaki 1997; Metz, Mollison, and van den Bosch 2000) to practical recommendations on how to deal with particular species and ecosystems (Groves, Shepherd, and Richardson 1995; Bossard, Randall, and Hoshovsky 2000). In this chapter we summarize what is known about plant invasions and, particularly, what is useful to managers.

Five questions constitute a backbone of invasion ecology:

- Which taxa invade?
- How fast?
- What makes ecosystems invasible?
- What is the impact?
- How can we control or eradicate harmful invaders?

The overriding frustration of invasion ecology has been that very general answers (i.e., those that explain patterns across a wide range of systems) to these questions have been elusive. Some authors argue that there are no generalizations to be made other than trivial correlations that have very limited value for management. The most important factor preventing the formulation of generalizations at a scale useful to managers is the effect of the environment. Therefore, although we will discuss the progress made toward answering each of the five questions, it should be kept in mind that the domain of any generalizations is limited. For each question we need to add, "In which habitats?" Our discussion deals almost exclusively with vascular plants, although some conclusions also apply to other invasive taxa.

Invasive taxa (spreading where they are not native) are a subset of naturalized taxa (non-native, forming sustainable populations without direct human help but not necessarily spreading). This distinction is critical because not all naturalized taxa reported in floras and checklists are invasive. Not all naturalized plant taxa, and not even all invaders, are harmful invaders; the last mentioned can be called exotic weeds or exotic pest plants (Rejmánek 2000a; Richardson et al. 2000c). The majority of weedy taxa in Europe, Malaya, Mexico, and Taiwan are native (Espinosa and Sarukhán 1997; Hsu 1975; Turner 1995; Williamson 1996; Villasenor and Espinosa 1998), whereas the majority of weedy taxa in Australia, the United States, Chile, South Africa, New Zealand, and many other islands are nonnative, invasive taxa (Bromilow 1995; Haselwood and Motter 1983; Lorenzi and Jeffery 1987; Matthei 1995). Because introduced plant taxa differ in their reproduction, rates of spread, and impacts, we must be able to prioritize management actions directed at excluding, monitoring, containing, or eradicating some alien plants.

Which Taxa Invade?

By definition, invasive taxa spread where they are not native. It follows that addressing the question "Which taxa will invade?" implicitly includes the notion of spread rate. This obviously means that there is some overlap between the questions of "Which taxa?" and "How fast?" However, for convenience we reserve the issue of "How fast?" for the next section.

To answer the question of "Which taxa?" invasion ecology offers five groups of largely complementary approaches:

- Stochastic (roles of initial population sizes, residence times, and numbers of introduction attempts)
- Empirical, taxon-specific (does a particular taxon invade elsewhere?)
- Evaluation of biological characters responsible for or associated with invasiveness
- Evaluation of habitat compatibility
- Experiments

Stochastic Approach

The most robust but admittedly rather trivial generalization in invasion ecology is that the probability of invasion success increases with initial population size and with the number of introduction attempts. This was documented by analyses of success rates in biological control of insects by insects in Canada (Williamson 1989) and successful introductions of birds to New Zealand (Green 1997) and Australia (Newsome and Noble 1986; Duncan et al. 2001). The same "gambling opportunity" (Egler 1983) seems to apply for alien plants as well. In southeastern Australia, Mulvaney (2001) found a strong correlation between the extent of planting and the probability that a woody taxon had become naturalized. The usefulness of this approach seems to vary between the genera for which we have relevant data. The contrast between *Eucalyptus* and *Pinus* (the two genera of trees with the greatest extent of planting as aliens in the Southern Hemisphere) is particularly useful for illustrating some important differences. For eucalypts, there is a weak but still highly significant correlation between the number of records of spontaneous occurrences and the number of plantations of 57 *Eucalyptus* taxa introduced to southern Africa (Figure 6.1). The extent of planting of eucalypts (propagule pressure) is much better correlated with invasive success than anything we have been able to find that distinguishes invasive from noninvasive taxa on the basis of life history or any other features of biology. As we show later in this chapter, the extent of invasiveness for different taxa of pines (measured as the area invaded) is better explained by life history considerations than by the residence times or extent of plantations (Richardson 1989). Sufficiently detailed data are lacking for other genera that have enjoyed similar levels of

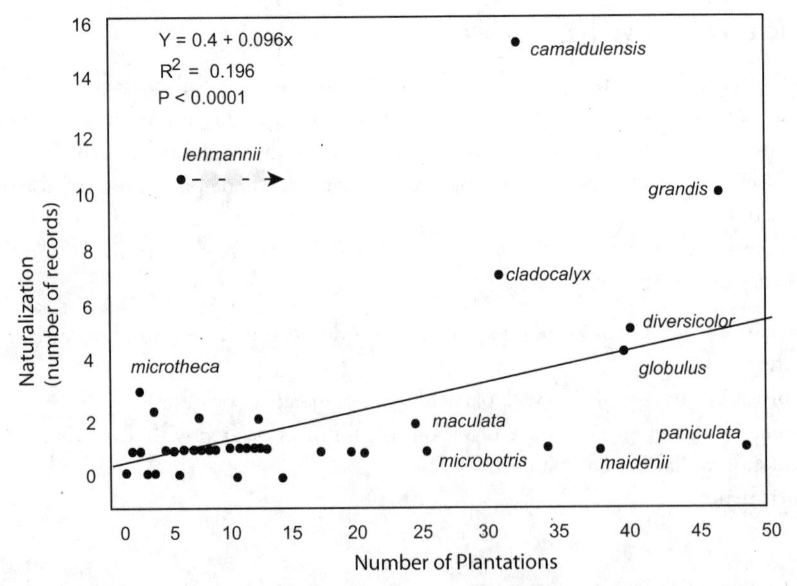

Figure 6.1. Relationship between the number of records of spontaneous occurrence and the number of plantations (Poynton 1979) of 57 *Eucalyptus* species in southern Africa. The point with an arrow represents *E. lehmannii*, a species that is commonly planted as an ornamental but, because of the negligible value of its timber, has been introduced in very few plantations (M. Rejmánek and D. M. Richardson, unpublished).

introduction outside their natural ranges, but we suspect that similar contrasts are common and that intermediate situations may be even more widespread. Both propagule pressure and inherent taxon-specific invasiveness are clearly important.

Residence time is also important. In Taiwan, for example, the number of discrete localities of naturalized legumes is significantly dependent on minimum residence time (years since the first record in the country; Figure 6.2). Similarly, in the northwestern United States, the number of invaded counties depends significantly on the minimum residence time of alien species recorded for the first time after 1950 (see data in Toney, Rice, and Forcella 1998). In Venezuela there is a highly significant correlation ($r = .67, p < .001$) between the log of the total number of known localities and minimum residence time for 116 alien grass taxa (M. Rejmánek, unpublished). The most widespread invaders in New Zealand are those that were introduced early (Allan 1937; Rejmánek 2000a). Rozefelds and Mackenzie (1999) make the same point for plant invasions in Tasmania. It is unlikely that there is some constant proportion of invasive taxa within pools of introduced taxa (a suggestion made recently by some ecologists). The reasonably constant proportion of alien taxa that invade across a wide range of systems (e.g., as expounded in the tens rule; Williamson 1996) is at least partly a result of the similar mean residence times of species in alien floras.

In summary, we may expect that stochastic invasion effects (which depend on initial inoculum size, residence time, and the number of introduction events and their spa-

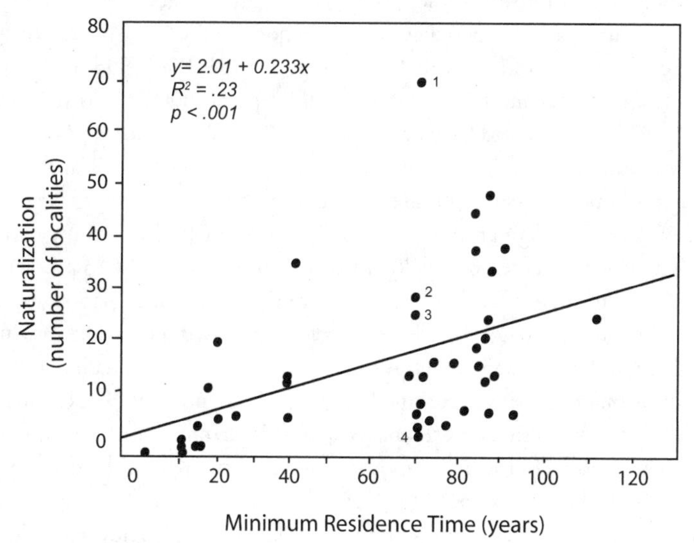

Figure 6.2. Number of discrete localities as a function of the minimum residence time (time since the first record or collection) for naturalized legumes (Fabaceae) in Taiwan. Four *Crotalaria* species with the same residence time but different degrees of invasiveness are numbered: 1, *C. zanzibarica;* 2, *C. micans;* 3, *C. incana;* 4, *C. bialata.* Figure from Wu, Chaw, and Rejmánek (2003).

tial distribution) will increase in importance as life history similarity between taxa of interest increases (as in *Eucalyptus*, as discussed earlier). However, it is important to realize that the stochastic approach allows us to make only certain probabilistic predictions about establishment and spread; it provides us with no information on particular taxa. Nevertheless, outcomes of probabilistic predictions can serve as multipliers for predictions made by the other approaches.

Empirical, Taxon-Specific Approach

Knowing whether a particular species is invasive elsewhere can help managers make practical decisions. This approach could be called tautological, but global transplant "experiments" unquestionably are an extremely rich source of valuable information in invasion ecology. These experiments comprise thousands of taxa introduced in a very wide range of habitats subjected to different types and intensities of natural and human-induced disturbance. Extrapolations based on previously documented invasions are fundamental for predictions in invasion ecology. With the development of relevant databases, this approach should lead to immediate rejection of imports of many invasive taxa (prevention) and prioritized control of those that are already established. This knowledge is very powerful, and the U.S. Department of Agriculture has recognized its importance for more than two decades (Reed 1977). Using data from previously invaded regions, we can learn much, not only about the degree of invasiveness of given taxa but also their possible ecological and economic impacts. There may even be some valuable management experience to draw upon. Many local or biome-specific databases (e.g., Swarbrick and Skarratt 1994; Binggeli 1998; Rice 1998; Ridgway et al. 1999; Kartesz and Meacham 1999; Randall 2000; Space 2000), floras (Sharma and Pandey 1984; Webb, Sykes, and Garnock-Jones 1988), biological floras (Poschlod et al. 1996), and weed manuals of donor countries (e.g., Spain for California and Chile; see also Maillet and Lopez-Garcia 2000) are helpful, even if incomplete.

Obviously, data of this kind can be used safely only for making one-way predictions. It is safe to say which taxa are potentially invasive, but it is very risky to predict which taxa are "safe" (noninvasive). Reichard and Hamilton (1997) found a poorer correlation between invasion and "does not invade elsewhere" than between invasion and "invades elsewhere" (see also Williamson 2001). A major problem is the lag phase that has been documented for plant invasions worldwide (Crooks and Soulé 1999 and Richardson 2001 review examples and suggest explanations for this phenomenon). The question is, when (how many years after arriving in a new region) can a taxon be declared to be "safe" (noninvasive), bearing in mind that lag phases of many decades are not unusual? Clearly, premature rating of a taxon as noninvasive based on its performance in one or a few regions can be misleading. An example is *Metrosideros excelsa*, which horticulturists hailed in the late 1960s as a safe replacement hedge plant for the highly invasive *Leptospermum laevigatum* in South African fynbos. A few decades later, the former species is a serious invader in some areas (Richardson and Rejmánek 1998).

Extrapolations based on the performance of taxa elsewhere have been used either on their own or in combination with an evaluation of biological characters associated with invasiveness in several regional screening procedures (Panetta 1993; Scott and Panetta 1993; Williams 1996; Owen 1997; Reichard and Hamilton 1997; Walton and Ellis 1997; Pheloung, Williams, and Halloy 1999). Daehler and Carino (2000) recently assessed the predictive abilities of some of these procedures using a set of known invaders and noninvaders in Hawaii. Again, the knowledge of whether a species is invasive elsewhere turned out to be the key information for correct predictions.

Regrettably, the major obstacle to efficient implementation of this approach is the lack of long-term commitment of any international agency to create and update a truly global database of invasive plant species. After many discussions with interested colleagues, we are inclined to believe that this task should be a financial responsibility of nothing short of the United Nations. Ideally, a continually updated database should be associated with some major botanical institution with a large herbarium and extensive library (e.g., Royal Botanic Gardens in Kew, Missouri Botanical Garden in St. Louis, or Muséum National d'Histoire Naturelle in Paris). The professional staff of such an "International Invasive Plant Data Center" should include at least one plant taxonomist familiar with taxonomy of cultivated plants and difficult weedy genera (e.g., *Polygonum* s.l., *Prosopis, Senna, Solanum, Tamarix*), a plant ecologist specialized in demography of invasive taxa, a geographic information system–oriented plant biogeographer, an open-minded weed scientist, and a database manager. A basic information unit processed in this center will be a taxon locality record. Essential information from the individual records will be continuously summarized for individual taxa.

The proposed center is realistic and, considering potential benefits, inexpensive. However, the whole task is substantially more demanding than what many newcomers in this field are ready to admit. If it is not properly designed by professionals, it could be not only useless but even an obstacle to real progress ("Somebody is already working on it!"). Unfortunately, there will be no shortcuts in this area. Suggestions that the whole problem can be solved just by linking existing (in fact, mostly nonexistent) databases (Ricciardi et al. 2000) are unrealistic. Doing only that would just postpone the real solution. However, the center would certainly profit from a system of regional units associated with major herbaria on all continents. Simultaneously developing technologies for plant pathology and biodiversity inventories (Bridge et al. 1998; Oliver et al. 2000) will serve as an important source of inspiration. The need for such an information center is becoming incontestable. However, two important questions still remain: Who will fund it? Who will be responsible for professional guidance of the whole project?

Basic taxonomic units used in plant invasion ecology usually are species or subspecific taxa. Are higher units useful in invasion ecology? Genera are certainly worth considering. Species belonging to genera notorious for their invasiveness or "weediness" (e.g., *Amaranthus, Echinochloa, Ehrharta, Erodium*) should always be treated as highly suspicious. However, a continuum from invasive to noninvasive species is also common in many genera. What is actually more typical has to be rigorously tested. Recently, some attention has been paid to taxonomic patterns of invasive plants (Daehler 1998; Pysek 1998; Rejmánek and Richardson 2003). In terms of relative numbers of invasive species, some families seem to be consistently overrepresented: Amaranthaceae, Brassicaceae, Chenopodiaceae, Fabaceae, Gramineae, Hydrocharitaceae, Papaveraceae, Pinaceae, and Polygonaceae. Among large families, the only one that is conclusively underrepresented is Orchidaceae. There are also other currently underrepresented families, such as Acanthaceae, Podocarpaceae, Rubiaceae, and Zamiaceae. However, the recent widespread invasions of the Rubiaceae species *Cinchona pubescens* in the Galápagos and *Timonius timon* in Palau remind us of the danger of making conclusions based only on taxonomic affiliation.

Evaluation of Biological Characters

Transregional, taxon-specific extrapolations are very useful in many situations, but our lack of mechanistic understanding makes them intellectually unsatisfying. Understanding how and why certain biological characters promote invasiveness in taxa is a vital component in our toolbox of methods because even an ideal whole-Earth database will not cover all (or even most) potentially invasive taxa. In New Zealand, for example, Williams, Nicol, and Newfield (2001) report that 20 percent of alien weedy species collected for the first time in the second half of the twentieth century had never been reported as invasive outside New Zealand. Rapoport (1991, 1992) estimated that at least 10 percent of the earth's 260,000 vascular plant species are potential invaders; if this is true, then about 85 percent of them have yet to be recognized as such. For these reasons, several attempts have been made to find differences in biological characteristics of noninvasive and invasive taxa or, at least, between native taxa and nonnative invasive taxa in particular floras (Baker 1974; Forcella, Wood, and Dillon 1986; van Wilgen and Siegfried 1986; Richardson, Cowling, and Le Maitre 1990; Reichard 1994; Trepl 1994; Pysek, Prach, and Smilauer 1995; Thompson, Hodgson, and Rich 1995; Tucker and Richardson 1995; Baruch 1996; Binggeli 1996; Crawley, Harvey, and Purvis 1996; Rejmánek 1996b, 1999; Rejmánek and Richardson 1996; Thébaud et al. 1996; Williamson and Fitter 1996; Pysek 1997; Kolar and Lodge 2001; Rejmánek and Reichard 2001).

Baker (1965, 1974) made his predictions about the biology of "an ideal weed" a priori, using characters that he, based on his experience, believed were important for invasive taxa. Predictions of others are constructed a posteriori, based on comparisons of characters of invasive and noninvasive taxa. In a sense, these are again extrapolations but character-specific and, to a large extent, taxon-independent extrapolations. Among these, Rejmánek and Richardson (1996, see Table 6.1) derived the first region-independent screening procedure for woody species based exclusively on the biology of the species and some interactions with the environment. A problem with this approach is finding truly noninvasive taxa (incomplete information and, again, time lags pose problems). For example, in their analyses of woody invaders in North America, Reichard (1994) and Reichard and Hamilton (1997) used as "noninvasive" not only species that are invasive elsewhere (e.g., *Acacia decurrens, Cotoneaster microphyllus,*

Cryptomeria japonica, Duranta erecta, Pinus banksiana, Viburnum tinus) but also species that are invasive in North America (*Acer pseudoplatanus, Berberis darwinii, Gleditsia triacanthos*). Even in a much more carefully screened data set (Rejmánek and Richardson 1996), one "noninvasive" species (*Pinus caribaea*) is indeed in Australia. An alternative approach is to divide alien species into highly invasive and less invasive on the basis of analyses such as those in Figures 6.1 and 6.2 (e.g., *Eucalyptus camaldulensis* vs. *E. paniculata* or *Crotalaria zanzibarica* vs. *C. bialata*) (Richardson and Rejmánek 2004).

Discriminant analysis, multiple logistic regression, path analysis, and classification

Table 6.1. General rules for detecting invasive woody seed plants based on values of the discriminant function Z^*, seed mass values, and presence or absence of opportunities for vertebrate dispersal

		Opportunities for vertebrate dispersal	
		Absent	*Present*
$Z > 0$	Dry fruits and seed mass > 3 mg	Likely to be invasive[1]	Very likely to be invasive[2]
	Dry fruits and seed mass < 3 mg	Likely to be invasive in wet habitats[3]	
	Fleshy fruits	Unlikely invasive[4]	Very likely invasive[5]
$Z < 0$	All seed/fruit mass values	Noninvasive unless dispersed by water[6]	Possibly invasive[7]

Source: Modified from Rejmánek and Richardson (1996).

$^*Z = 23.39 - 0.63\sqrt{M} - 3.88\sqrt{J} - 1.09S$, where M = mean seed mass (in milligrams), J = minimum juvenile period (in years), and S = mean interval between large seed crops (in years). This discriminant function (Z) was based on a priori defined groups of invasive and noninvasive *Pinus* species. The function was later successfully applied on other gymnosperms and, as a component of this table, even on woody angiosperms (Rejmánek 1996b; Richardson and Rejmánek 2004). Note that parameters in this discriminant function are somewhat different from those in Rejmánek and Richardson (1996). This is because *Pinus caribaea* was excluded from the data set used to estimate the parameters.

[1]For example, *Acer platanoides, Cedrela odorata, Clematis vitalba, Cryptomeria japonica, Cytisus scoparius, Pinus radiata, Pittosporum undulatum, Pseudotsuga menziesii, Robinia pseudoacacia, Senna* spp., and *Tecoma stans.*

[2]Species with large arils (*Acacia cyclops*) are dispersed by birds.

[3]For example, *Alnus glutinosa* in New Zealand, *Eucalyptus globulus* in California, *Melaleuca quinquenervia* in southern Florida, *Tamarix* spp. in the southwestern United States, *Cinchona pubescens* in Galápagos, and *Baccharis halimifolia* in Australia.

[4]*Feijoa sellowiana* and *Nandina domestica* are frequently cultivated but noninvasive species in California. However, the second species is dispersed by birds and water in the southeastern United States.

[5]For example, *Berberis* spp., *Clidemia hirta, Crataegus monogyna, Lantana camara, Lonicera* spp., *Myrica faya, Passiflora* spp., *Psidium guajava, Rubus* spp., *Schinus terebinthifolius,* and *Solanum mauritianum.*

[6]*Nypa fruticans* is spreading along tidal streams in Nigeria and Panama. *Thevetia peruviana* can be dispersed over short distances by surface runoff in Africa.

[7]Examples of invasive species in this group are *Pinus pinea, Melia azedarach,* and *Maesopsis eminii* in Africa, *Quercus rubra* in Europe, *Mangifera indica* in the Neotropics, and *Persea americana* in Galápagos.

Table 6.2. Comparisons of invasive and noninvasive pine (*Pinus*) species based on phylogenetically independent contrasts (see Figure 6.3) between values of variables used in the discriminant function (see Table 6.1)

Variable and hypothesis	*No. of disagreements/total no. of contrasts*	p *Value binomial test*	p *Value* t *test*
√Minimum juvenile period	0/6	.016	.0063
Invasive < noninvasive			
√Mean seed mass	1/6	*ns*	.04
Invasive < noninvasive			
Mean interval between large seed crops (SCV)	2/6	*ns*	*ns*
Invasive < noninvasive			
√Seed mass + SCV	0/6	.016	.024
Invasive < noninvasive			

The computer program CAIC 2.0 was used to construct the independent contrasts and analyze the data. Data from Grotkopp, Rejmánek, and Rost (2002).

and regression trees (Huberty 1994; Christensen 1997; Breiman et al. 1984; De'ath and Fabricius 2000) are the most promising statistical tools in the assessment of biological characters responsible for invasiveness. Ideally, any comparison of species should be based on a reliable phylogeny and phylogenetically independent contrasts. This is desirable because individual species may not represent independent data points if the species traits analyzed are correlated strictly to phylogenetic history. Returning to the discriminant function for *Pinus* (Table 6.1), we can use an available phylogeny for the genus (Figure 6.3) to calculate six phylogenetically independent contrasts between invasive and noninvasive species (Table 6.2). Here, the results are consistent with phylogenetically uncorrected cross-specific comparisons. Even if this is very often the case, investigating the relationships of present-day species traits to phylogeny certainly can be beneficial and is to be encouraged (Crawley, Harvey, and Purvis 1996; Kotanen, Bergelson, and Hazlett 1998; Westoby et al. 1998; Freckleton 2000).

The major predictions made by an emerging theory of plant invasiveness based on biological characters include the following.

Fitness Homoeostasis Is Important

The ability of an individual or population to maintain constant fitness over a range of environments can be called fitness homoeostasis (Hoffman and Parsons 1991). Individual fitness homoeostasis (supported by phenotypic plasticity) seems to be equivalent to Baker's (1974, 1995) general purpose genotype. *Ageratum conyzoides, Eupatorium*

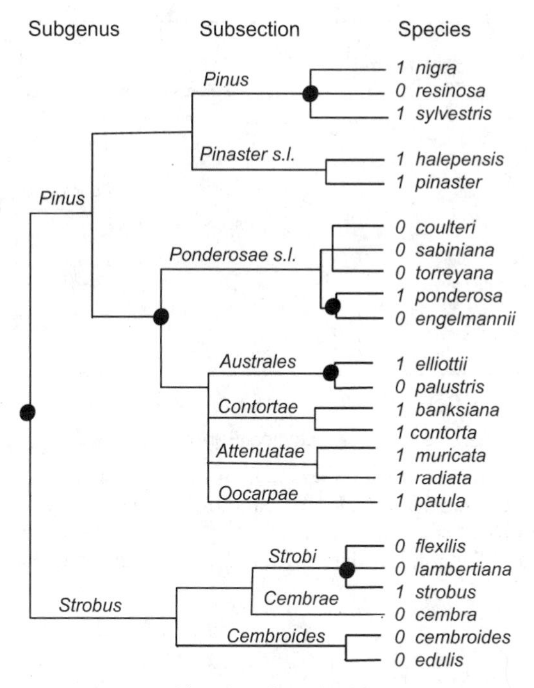

Figure 6.3. Cladogram showing the phylogenetic relationships between 12 invasive and 11 noninvasive *Pinus* species (Govindaraju, Lewis, and Cullis 1992; Liston et al. 1999) used for derivation of the discriminant function presented in Table 6.1. Phylogenetically independent contrasts are nonoverlapping groups of at least one invasive (1) and one noninvasive (0) species. Results are presented in Table 6.2.

adenophorum, and *Cortaderia jubata* are Baker's classic examples. As a result of both individual fitness homoeostasis and population genetic polymorphism, population fitness homoeostasis can contribute to the invasiveness of a taxon (Rejmánek 1999). Unfortunately, population fitness homoeostasis is not a readily quantifiable variable. Quantities such as mean relative physiological performance and mean relative ecological performance across environmental gradients (Austin 1982; Austin et al. 1985) can be used as surrogates for population fitness homoeostasis (Figure 6.4). A more direct measure would be the geometric mean of population growth rates (λ) across realistic ranges of environments (Venette 1997). However, reliably estimating λ (even for one population in one environment is a demanding task) (Bierzychudek 1999; Rejmánek 2000b). Clonal plants represent a special challenge for development of proper fitness measures (Winkler and Fischer 1999).

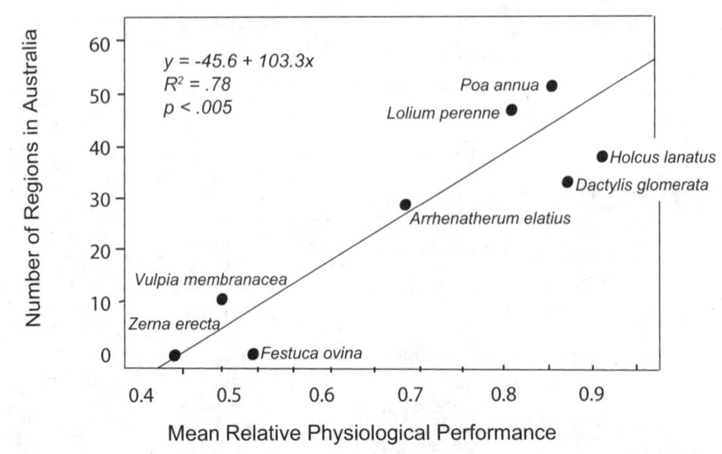

Figure 6.4. Relationship between the number of regions in Australia (out of 97) invaded by 8 European grass species (after Hnatiuk 1990) and their mean relative physiological performance over 15 nutrient concentrations. Calculated from results of Austin's (1982) experiments conducted in Bangor, UK. Figure from Rejmánek (2000a), reproduced with permission.

Genetic Change Can Facilitate Invasions, but Many Species Have Sufficient Phenotypic Plasticity to Exploit New Environments

Genetic change may facilitate the expansion of species (Purps and Kadereit 1999), but it is certainly not a prerequisite for species invasion. It seems that the success of invaders is based most often on the use of phenotypic plasticity rather than genetic change to exploit novel environments (Levin 2000; Willis, Memmott, and Forrester 2000; Thébaud and Simberloff 2001). Niche breadth of invaders along environmental gradients does not seem to be correlated with the level of genetic variation within populations and is rather a result of phenotypic plasticity (Sultan 1987; Sultan and Bazzaz 1993; Wilen et al. 1995; Williams, Mack, and Black 1995; Hermanutz and Weaver 1996). However, genetic change in ecologically marginal populations would allow the exploitation of habitats that otherwise would be off limits and thus promote invasion. Gene flow from ecologically central populations tends to constrain species expansion when selection pressures are weak (Levin 2000). On the other hand, current anthropogenic disturbance regimes may contribute to high genetic variability by increasing gene flow and thereby promoting the adaptability of invasive taxa to unpredictable conditions of disturbed habitats (Dietz, Fischer, and Schmid 1999). Introgressive hybridization (either between alien species or between alien and native species) may accelerate the ongoing invasions (Panetsos and Baker 1968; Abbott 1992; Ayres et al. 1999; Ellstrand and Schierenbeck 2000; Milne and Abbott 2000; Vilá, Weber, and

D'Antonio 2000; Daehler and Carino 2001). Ecological consequences of polyploidy are still far from clear. However, in general polyploids show more genetic diversity and occur across a large range of habitats with respect to stress, competition, and disturbance (Rothera and Davy 1986; Brochmann and Elven 1992). In the context of invasion biology, agamic polyploid taxa may be particularly important because their populations usually form complex mixtures of sexuals, facultative apomicts, and obligate apomicts (Bayer 1999; Campbell 1999). The relationship between invasiveness and the increase of within-population genetic variation due to multiple introductions (Novak and Mack 1993, 2001) certainly deserves more attention. A remarkably high genetic diversity of rapidly spreading populations of *Echium plantagineum* in Australia (Forcella, Wood, and Dillon 1986) is in part the result of hybridization among multiple introductions from Europe (Burdon and Brown 1986). For several reasons, multiple introductions (often undetected) can underlie many mysterious time lags between introduction and rapid spread of alien taxa.

Small Genome Size Is a Useful Indicator of Relative Invasiveness for Closely Related Taxa

The amount of DNA in the unreplicated haploid nuclear genome (genome size) of vascular plants is known to vary more than 800-fold from about 0.1 pg to more than 89 pg. Small genome size seems to be a result of selection for short minimum generation time, and because it is also associated with small seed size, high leaf area ratio, and high relative growth rate (RGR) of seedlings in congeners, it may be an ultimate determinant or at least an indicator of "weediness" (Grime, Hodgson, and Hunt 1988:33; Bennett, Leitch, and Hanson 1998) or invasiveness of plants in disturbed landscapes (Rejmánek 1996b, 1999; Grotkopp et al. 1998). However, because genome size apparently increased and decreased many times independently in many phylogenetic lineages (Bachman, Chambers, and Price 1985; Leitch, Chase, and Bennett 1998), the meaningful comparisons probably are limited to species within genera or within families. Invasive *Pinus* species, for example, have significantly smaller genome size. An extensive databank on plant genome sizes is available in the Royal Botanic Gardens at Kew (http://www.rbgkew.org.uk/cval/database1.html).

Several Characters Linked to Reproduction and Dispersal Are Key Indicators of Invasiveness

Reproduction and dispersal are key issues. Consistent seed production in new environments usually is associated with simple or flexible breeding systems (Hiscock 2000). For example, rare and endangered taxa in the genus *Amsinckia* (e.g., *A. furcata, A. grandiflora*) are heterostylic, whereas derived invasive taxa (*A. menziesii, A. lycopsoides*) are homostylic and self-compatible (Ray and Chisaki 1957; Pantone, Pavlik, and Kelley 1995; Schoen et al. 1997; but see Hamilton 1990). Self-pollination has been consistently identified as a mating strategy in colonizing species (Brown and Burdon 1987). Nevertheless, not all sexually reproducing successful invaders are selfers. Pannel and Barrett (1998; see also Barrett 2000; Larson and Barrett 2000) examined the benefits of reproductive assurance in selfers and outcrossers in model metapopulations. Not surprisingly, their results suggest that an optimal mating system for a sexually reproducing invader in a heterogeneous landscape should include the ability to modify selfing rates according to local conditions. In early stages of invasions, when populations are small, plants should self to maximize fertility. However, later, when populations are large and pollinators or mates are not limiting, outcrossing is more beneficial, mainly because of increasing genetic polymorphism.

Invasiveness of woody taxa in disturbed landscapes is associated with small seed mass (less than 50 mg), short juvenile period (less than 10 years), and short intervals between large seed crops (1–4 years) (Richardson and Rejmánek 2004). These three attributes contribute, directly or indirectly, to higher values of three parameters that are critical for population expansion: net reproduction rate (R_0), reciprocal of mean age of reproduction ($1/\mu$), and variance of the marginal dispersal density (σ^2) (see van den Bosch, Hengeveld, and Metz 1992) (Figure 6.5). For wind-dispersed seeds, the last parameter is negatively related to terminal velocity of seeds (Greene and Johnson 1989), which is

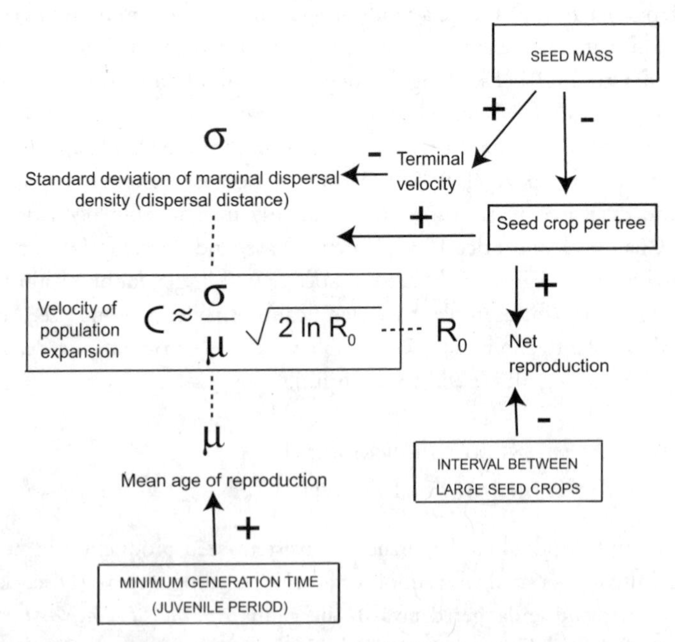

Figure 6.5. Positive (+) and negative (–) causal relationships between minimum generation time, seed mass, interval between large seed crops, and velocity of population expansion approximated by the van den Bosch, Hengeveld, and Metz (1992) equation.

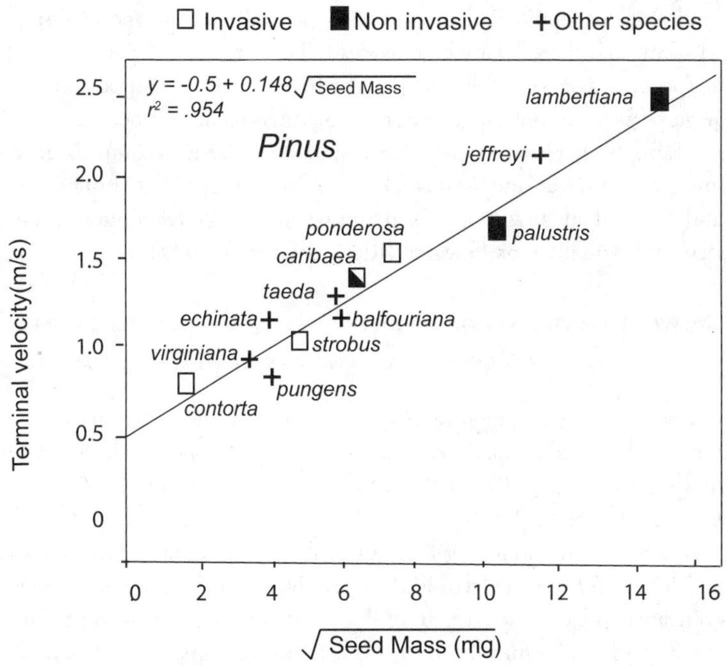

Figure 6.6. Dependence of terminal seed velocity on seed mass in 12 *Pinus* species (based on data in Siggins 1933; Krugman and Jenkinson 1974).

positively related to $\sqrt{\text{seed mass}}$ (Figure 6.6). Because of the trade-off between seed number and mean seed mass, small-seeded taxa usually produce more seeds per the same biomass (Eriksson and Jacobsson 1999; Guo et al. 2000; Leishman 2001). Long flowering and fruiting periods also seem to be associated with invasiveness of both woody and herbaceous taxa (Reichard 1994; Gerlach 2000; Cadotte and Lovett-Doust 2001). However, invasions of woody species with very small seeds (less than 3 mg) are limited to wet and preferably mineral substrates (Rejmánek and Richardson 1996). Based on invasibility experiments with herbaceous species, it seems that somewhat larger seeds (3–10 mg) extend species habitat compatibility (Burke and Grime 1996). Because seed mass seems to be positively correlated with habitat shade (Hodkinson et al. 1998), large-seeded aliens may be more successful in undisturbed, successionally more mature plant communities.

Seed Dispersal by Vertebrates Is Implicated in Many Plant Invasions

Seed dispersal by vertebrates is responsible for the success of many invaders in disturbed as well as "undisturbed" habitats (Strasberg 1995; Binggeli 1996; Rejmánek 1996a; Rejmánek and Richardson 1996; Richardson et al. 2000a). Even some very large-seeded alien species such as *Mangifera indica* can be dispersed by large mammals

(Fragoso and Huffman 2000). The proportion of naturalized plant species dispersed by vertebrates seems to be particularly high in Australia: more than 50 percent (Tables 2.1 and 7.3 in Specht and Specht 1999). Assessing whether there is an opportunity for vertebrate dispersal is an important component of the screening procedure for woody plants (see Table 6.1). However, vertebrate seed dispersal is a complicated process (Wilkinson 1997; Brewer and Rejmánek 1999; Andersen 2000; Kollmann 2000; Nathan and Muller-Landau 2000). When and where vertebrates promote plant invasions deserve substantially more research (Richardson et al. 2000a).

Low Relative Growth Rate of Seedlings and Low Specific Leaf Area Are Good Indicators of Low Plant Invasiveness in Many Environments

Many ecologists assume that a high RGR should be an important characteristic of invasive plant taxa in disturbed or open areas, especially in resource-rich environments (Baker 1974; Grime and Hunt 1975; Bazzaz 1986; Maillet and Lopez-Garcia 2000). Only a few studies have demonstrated this experimentally (Pattison, Goldstein, and Ares 1998; Baruch, Hernández, and Montila 1989). An analysis of seedling growth rates for 29 *Pinus* species (Figure 6.7) revealed that RGR of invasive species is significantly higher than that of noninvasive species, differences in RGR are determined primarily by leaf area ratio (LAR, leaf area per plant biomass), and LAR is determined primarily by specific leaf area (SLA, leaf area per leaf biomass). Consequently, invasive species have significantly higher SLA. Moreover, there is a highly significant ($R^2 = .685$, $p < .001$) positive relationship between RGR and invasiveness of 29 examined species expressed as the discriminant function Z (see Table 6.1). Identical results were obtained using phylogenetically independent contrasts (Grotkopp, Rejmánek, and Rost, 2002). Baruch (1996; Baruch and Goldstein 1999) implied that high SLA was an important factor associated with invasiveness of grasses and other plants. In general, SLA less than 90 cm^2/g (dry leaf mass) seems to be a good indicator of noninvasive or at least less invasive plant taxa.

Large Native Range Is an Indicator of Potential Invasiveness

The size of primary (native) geographic ranges of plant species is a promising predictor of their invasiveness (Forcella and Wood 1984; Rejmánek 1995, 1996b, 1999; Goodin, McAllister, and Fahrig 1998). Both population fitness homoeostasis and dispersal potential (see Rejmánek 1996b; Guo et al. 2000) seem to underpin this generalization. However, several important exceptions are known (Williamson 1996; Rejmánek 1999). Williamson (2001) believes that range is useful for predicting the probability of establishment but not the eventual range in the new region.

Vegetative Reproduction Is Responsible for Many Plant Invasions

Vegetative reproduction is responsible for an increase of habitat compatibility and therefore for successful establishment and spread of many species in terrestrial envia

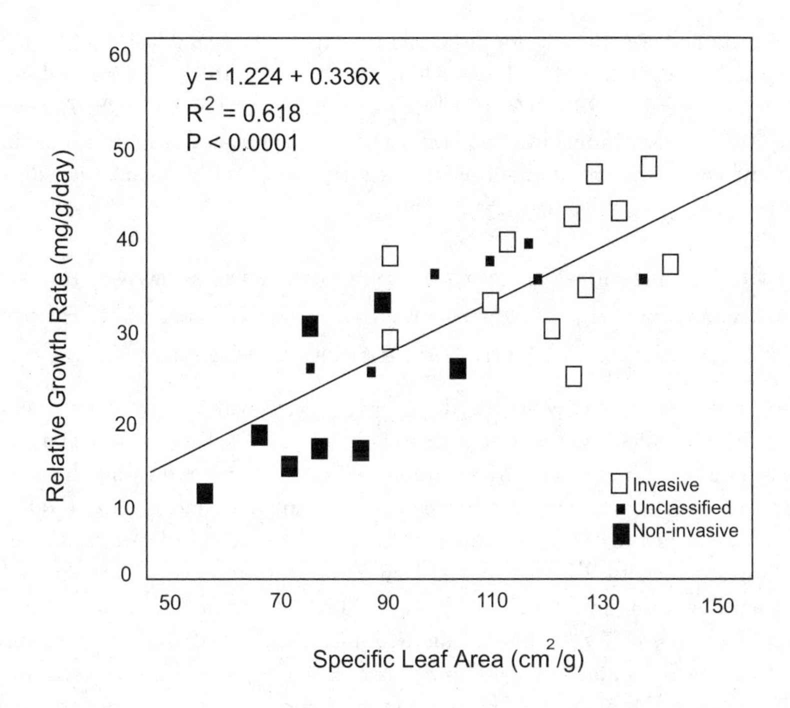

b

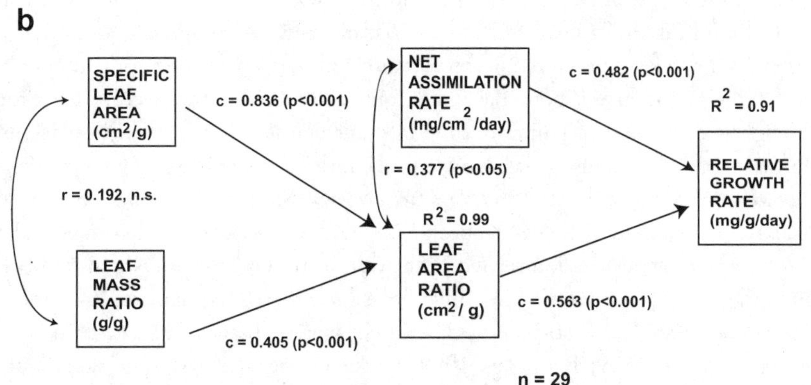

Figure 6.7. (**a**) Relationship between relative growth rate and specific leaf area in seedlings of 29 *Pinus* species. (**b**) Path diagram and path coefficients (*c*) showing strengths of causal links between 5 growth analysis variables for the same seedlings. Data from Grotkopp, Rejmánek, and Rost (1998).

ronments and even more so for dispersal in aquatic and wetland habitats (Auld, Hosking, and McFadyen 1983; Horak, Holt, and Ellstrand 1987; Pieterse and Murphy 1990; Henderson 1991; Barrett 1992; Ceccherelli and Cinelli 1999; Aptekar and Rejmánek 2000). Clonal invasive plants are more represented than nonclonal in wetter and colder than in drier and warmer climatic areas and in natural, less disturbed rather than artificial habitats (Pysek 1997).

Alien Taxa Are More Likely to Invade a Given Area if Native Members of the Same Genus (and Family) Are Absent, Partly Because Many Herbivores and Pathogens Cannot Switch to Phylogenetically Distant Taxa

Most plant taxa have coevolved in their native habitats with herbivores and pathogens (e.g., insects, mites, and fungi) that are somewhat specialized and limit their attack to a small number of host species. Many introduced species occur in their new habitat without these specific herbivores and pathogens. For example, Fenner and Lee (2001) compared the incidence of seed-attacking insects in flowerheads of 13 species of Asteraceae in Britain (where they are native) and New Zealand (introduced). They reported 6.24 percent flowerhead infection in Britain and 0.013 percent in New Zealand (one seed-attacking larva in 7,800 flowerheads examined). Strong, Lawton, and Southwood (1984) reviewed studies comparing the insect fauna of plants alien to North America (field crops, herbs, ornamental bushes, and trees) with the fauna in their native habitat. They found that most insects on introduced plants were less-specialized, polyphagous species. The lack of specialized herbivores and pathogens may allow some taxa to better realize their reproductive potential and become naturalized or invasive. Evidence for this is suggested by the substantial decline in plant density after introduction of insects, mites, or pathogens during classic biological control programs (e.g., *Hypericum perforatum,* Huffaker and Kennett 1959; *Senecio jacobaea,* McEvoy, Cox, and Coombs 1991; *Opuntia inermis* and *O. stricta,* Debach 1974; and *Salvinia molesta,* Room et al. 1981). An important generalization related to the discussed phenomenon is that taxa belonging to genera not represented in native floras (and therefore possessing traits different from those of resident taxa) are more likely to be invasive than alien taxa with native congeners (Darwin 1859; Rejmánek 1999; Lockwood et al. 2001). The success of taxa belonging to nonnative genera or higher taxa may result partly from the limited number of resident herbivores and pathogens able to switch to taxa phylogenetically distant from their native hosts (Futuyma, Keese, and Funk 1995; Becerra 1997).

The Ability to Use Generalist Mutualists Greatly Improves an Alien Taxon's Chances of Becoming Invasive

Plant species depending on nonspecific mutualisms (root symbionts, pollinators, and seed dispersers, including vertebrates) are more likely to overcome many abiotic and biotic barriers in new environments (Baker 1974; Richardson et al. 2000a). Because of the increased distribution and abundance of many generalist mutualists due to human activities, many ecosystems are becoming less resistant to invasion by alien plants (Richardson et al. 2000a; "invasional meltdown" *sensu* Simberloff and Von Holle 1999).

Efficient Competitors for Limiting Resources Are Likely to Be the Best Invaders in Natural and Seminatural Ecosystems

Undisturbed (natural and seminatural) plant communities in mesic environments are more likely to be invaded by tall plant species (Egler 1983; Gaudet and Keddy 1988; Pysek, Prach, and Smilauer 1995; Crawley, Harvey, and Purvis 1996; Williamson and Fitter 1996). The most prominent examples are new, taller, life forms (*Pinus* spp. and *Acacia* spp. in South African fynbos, *Cinchona pubescens* in shrub, fern, and grassland communities of Galápagos highlands). Undisturbed plant communities in semiarid habitats seem to be invasible especially by environmentally compatible species that rapidly develop deep root systems (e.g., *Bromus tectorum* or *Centaurea solstitialis;* Hulbert 1955; Roché, Roché, and Chapman 1994). In short, in undisturbed plant communities, efficient competitors for limiting resources are very likely to be successful invaders and the worst environmental weeds. Theoretically, given a set of R_i^* values (R_i^* is a level of resource below which an *i*th species cannot survive), for a pool of potential invaders it should be possible to predict the average likely success of each invading species in undisturbed communities (Tilman 1999). However, if seasonality, senescence, or even very low levels of natural disturbance allow establishment of shade-intolerant taxa that are taller than resident vegetation at maturity, then such taxa still can be successful and influential invaders despite their high R^* for light.

Characters Favoring Passive Dispersal by Humans Greatly Improve an Alien Plant Taxon's Chance of Becoming Invasive

The spread of many alien species depends heavily on human activities (Panetta and Scanlan 1995). Increasing volumes of soil are moved around, for example in topsoil, in mud on cars, or with horticultural stock. Species with numerous, small, soil-stored seeds are preadapted for this kind of dispersal (Hodkinson and Thompson 1997; UCPE 1996). Vigorous vegetative reproduction of some plants can trigger a positive feedback, which is why some plants used as fast cover later end up beyond garden or cemetery fences (e.g., many *Aizoaceae, Apocynaceae, Convolvulaceae, Crassulaceae, Oxalidaceae,* and *Trapeolaceae*).

Causal and correlative relationships between many of the factors discussed earlier are summarized and further discussed in Rejmánek (1999). Not surprisingly, some causal chains in this area cross several levels of biological organization. For example, the sequence "genome size + ⇒ nucleus volume + ⇒ parenchyma cell volume – ⇒ SLA + ⇒ LAR + ⇒ RGR + ⇒ invasiveness in disturbed environments" seems to be one of the reasons why we usually find a negative relationship between genome size and invasiveness among congeners (Grotkopp, Rejmánek, and Rost, 2002).

Evaluation of Habitat Compatibility

Recipient habitat compatibility usually is treated as a necessary condition for all invasions (Lindsay 1953; Sanders 1976; Chicoine, Fay, and Nielsen 1985; Panetta and Dodd 1987; Beerling, Huntley, and Bailey 1995). Knowledge of habitat compatibility is important both for prevention and exclusion of taxa that are not yet present in the country and for assessment of control and eradication priorities of established alien taxa. The match of primary and secondary environments is not always perfect (Michael 1981; Wilson et al. 1992) but usually reasonably close (Hultén and Fries 1986; Hickman 1993; Hügin 1995). In North America, for example, latitudinal ranges of naturalized European plant species in families Poaceae and Asteraceae are 15° to 20° narrower than their native ranges in Eurasia and North Africa (Rejmánek 2000a). These differences essentially reflect the differences in the position of corresponding isotherms and major biomes in Eurasia and North America (Rumney 1968; Walter 1968).

Several computer-based systems have been developed for determining the potential distribution of alien pests by assessing the similarity between climates of the species' natural range and regions outside this range (e.g., Jones and Gladkov 1999; Sutherst et al. 1999; Kriticos and Randall 2001). Such systems are very useful, but they should be used with caution because features other than climate often shape both natural and adventive distributions.

Major discrepancies between primary and secondary ranges have been found for aquatic plants, where secondary distributions often are much less restricted than their primary distributions (Cook 1985). Vegetative reproduction of many aquatic species seems to be the most important factor. For example, the native latitudinal range of the aquatic fern *Salvinia molesta* is just 8° (24°S to 32°S; southeastern Brazil; Forno 1983), whereas its secondary distribution ranges between 35° to the south and 30° to the north of the equator. In most places it occupies a greater diversity of environments than in its native range (Room and Julien 1995). This species is sterile and completely dependent on vegetative reproduction. Obviously, secondary ranges, if already known from other invaded continents (e.g., Cox et al. 1988), should be used in any prediction of habitat compatibility.

As for plants introduced (or considered for introduction) from Europe, several useful summaries of their ecological behavior are available. Ellenberg's (1974; Ellenberg et al. 1992) indicator values for about 3,000 vascular plants in central Europe are one example. Each species is characterized by its light (L), temperature (T), continentality (K), moisture (F), soil reaction (R), nitrogen (N) "figures" on a scale of 1–9, and salt and heavy metal tolerance "figures" on a scale of 1–3. Each species is also classified into a Raunkiaer life form and classified according to its leaf persistence and anatomical structure. Dealing with environments comparable to those of central Europe, reasonable predictions on habitat compatibility can be made. For example, figures for *Bromus tectorum* (L8, T6, K7, F3, R8, N4) can be interpreted as an occurrence in half–full light conditions, intermediate–warm and subcontinental–continental climate, dry and neutral–basic soils, and nitrogen poor–intermediate soils. The distribution of this invader in North America certainly agrees with this assessment. Ellenberg's indicator values are useful but not without problems. Several modifications have been proposed recently (Lindacher 1995; Diekmann and Lawesson 1999; Hill, Roy, and Mountford 2000; Schaffers and Sykora 2000).

Positions of 281 mostly herbaceous European species in Grime et al.'s (Grime, Hodgson, and Hunt 1988) Competitors-Stress tolerators-Ruderals (C-S-R) triangular matrix ordination of plants according to their established phase strategies provide another tool for habitat compatibility predictions. For example, species that are classified as stress tolerators (**S** corner of the triangle; e.g., *Nardus stricta* or *Rumex acetosella*) would not be successful invaders in highly disturbed environments (e.g., arable fields) or fertile, highly competitive environments (e.g., lowland alluvial meadows). On the other hand, ruderals such as *Senecio vulgaris* and *Tripleurospermum inodorum,* which are situated in the **R** corner of the triangle, are excellent invaders in periodically disturbed mesic environments. *Aira praecox* has an intermediate position between ruderals and stress tolerators. Not surprisingly, we find this species in open plant communities on sandy soils in North America. Grime et al.'s triangular ordination was also developed for 36 European tree species by Brzeziecki and Kienast (1994). Unfortunately, the stress corner of Grime et al.'s triangle is a mixed bag of several kinds of resource shortages. Recently, Grubb (1998) suggested how to decompose this complex mixture into nutrient, water, and light stresses and corresponding tolerance mechanisms. In the meantime, Hodgson et al. (1999) developed a simple method for extrapolating the C-S-R system to species that have not been the subject of previous ecological investigations. A combination of Ellenberg's indicator values and Grime et al.'s functional types (strategies) can be a powerful tool for predicting habitat compatibility of European species.

The strength of affiliation with vegetation syntaxa (associations, alliances, orders, classes) is very well known for almost all European plant taxa (Bolòs and Vigo 1995; Oberdorfer 1994). Because at least qualitative descriptions of environmental conditions (microclimate, soils, disturbance, management) of all syntaxa are available, potential habitat compatibility of individual plant taxa can be extracted from European literature. Also, this knowledge of the phytosociological behavior of individual taxa allows predictions about compatibility with analogous (vicarious) vegetation types. For example, communities of the alliance Trollio–Ranunculion in Japan probably would provide suitable habitats for species typical for the European alliance Adenostylion (e.g., *Doronicum austriacum* or *Petasites kablikianus*). Unfortunately, as the last example illustrates, these predictions are not a straightforward task for plant ecologists who are not familiar with Braun–Blanquet phytosociological system. Nevertheless, with the development of European vegetation databases (Rodwell et al. 1995) this wealth of knowledge about behavior of European plants should become more accessible.

By definition, the notion of habitat compatibility includes all the factors embraced in the concept of habitat. Most effort in assessments of habitat compatibility has been devoted to climatic and substrate compatibility, although it is well known that many other factors influence range limits. Managers should note that invasions result from interactions between habitat compatibility and propagule pressure. This is well illustrated using results of a study of the invasion dynamics of the New Zealand tree *Metrosideros excelsa* (Myrtaceae) in South African fynbos (details in Richardson and Rejmánek 1998). In the course of this study, we developed a new method for assessing the effects of habitat compatibility and spatially explicit dispersal potential on plant invasions with point sources (mature, seed-producing trees). Regressions of the number of *Metrosideros* saplings on a potential seed rain index revealed that wet habitats (seepage fynbos) are almost exactly 10 times more invasible than dry habitats (respective slopes are 16.36 and 1.59; Figure 6.8). This example clearly shows that classifica-

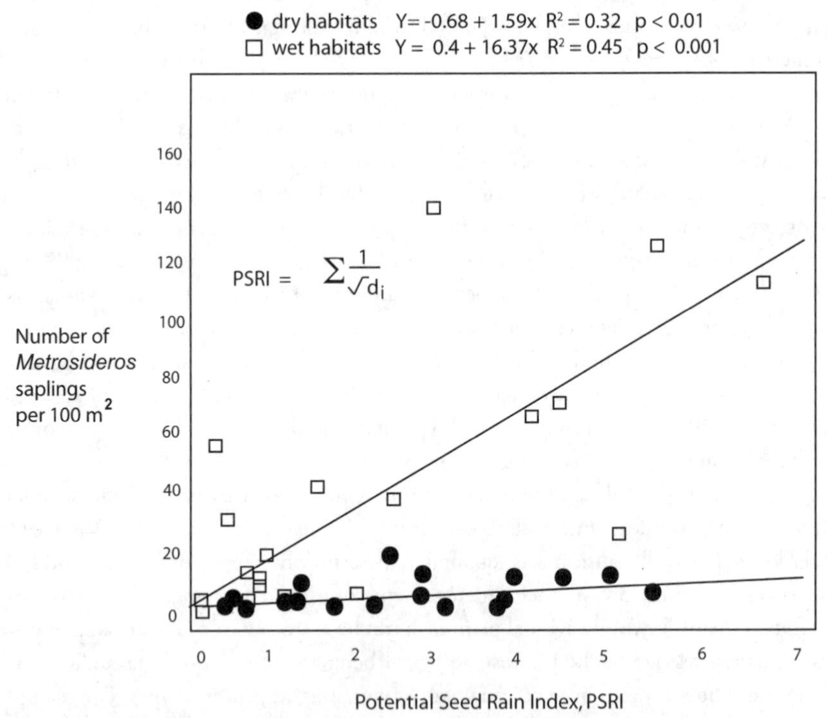

Figure 6.8. Dependence of *Metrosideros excelsa* sapling density on potential seed rain index (PSRI, d_i = distance to the *i*th seed-producing tree in meters within the radius 300 m) in dry and wet fynbos, Betty's Bay, Western Cape, South Africa. Slopes of the 2 regression lines are significantly different ($p < .01$). Fynbos plots were classified as dry (dominated by *Erica ericoides, Elegia filacea,* and *Metalasia muricata*) and wet (dominated by *Cliffortia hirsuta, Osmitopsis asteriscoides, Gleichenia polypodioides,* and *Erica perspicua*) based on results of average linkage clustering using Jaccard coefficient as a measure of similarity. Data from Rejmánek et al. (2005).

tion of habitats or communities into invasible and noninvasible cannot be absolute in many situations. Propagule pressure (mass effect) must always be considered. D'Antonio, Levine, and Thomsen (2001) presented a conceptual model to illustrate this (Figure 6.9). Needless to say, propagule pressure has profound implications for management. For example, habitats that are currently unaffected (or only slightly affected) by plant invasions may be deemed resistant to invasion. However, as populations of alien plants build up and propagule pressure increases outside or within such areas, invasions could well start or increase. Soil nitrogen reduction (through carbon amendment treatment) has been used as a remediation technique in invaded areas (Morgan and Seastedt 1999). The value of this method is likely to decrease as propagule pressure increases.

Many environmental factors influence invasions, and understanding them is important for management. For example, computer simulations by Higgins et al. (1999) showed that topography, vegetation flammability, and soil fertility all influenced the potential distribution of woody alien plants at the landscape scale. In a more mechanistic sense, disturbance regimes also mediate range limits but in complex, context-specific ways (Higgins and Richardson 1998). These results suggest that macroscale climatic factors, microclimatic factors, soils, and emergent properties of ecosystems such as disturbance regimes together define inherent habitat compatibility but that, in some cases and to some extent, propagule pressure can override these factors. Different types of information are needed to accurately model invasions at different spatial scales (e.g., local, landscape, regional, national) (Rouget and Richardson 2003).

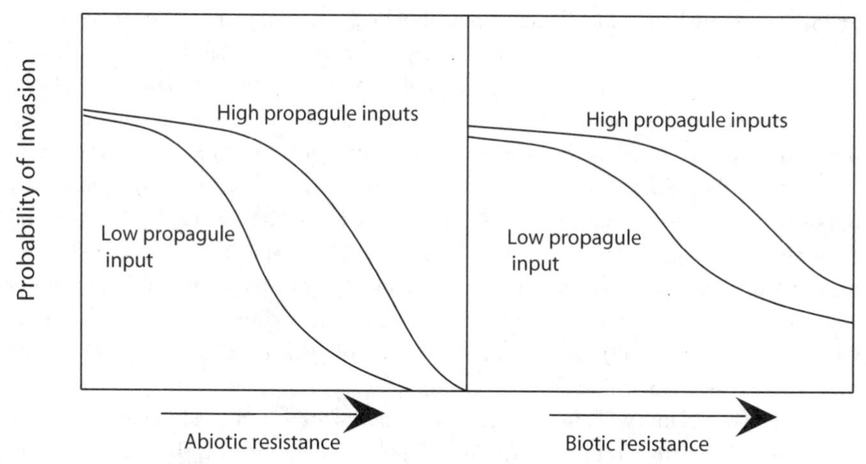

Figure 6.9. Probability of a taxon invasion depends on propagule supply (expressed, for example, as a number of propagules imported per square meter per year) and both abiotic and biotic resistance of particular ecosystems. Increasing propagule pressure increases probability of invasion at intermediate levels of abiotic resistance, and it substantially increases this probability at both intermediate and high levels of biotic resistance. Figure modified from D'Antonio, Levine, and Thomsen (2001).

Experiments

To some extent, predictions made on the basis of the first four approaches can be tested in quarantine field trials. This can be time-consuming and expensive but fruitful when dealing with limited numbers of herbaceous taxa (Hulbert 1955; Lonsdale 1994; Austin et al. 1985; Turnbull, Crawley, and Rees 2000). Deliberate introductions beyond a species' current range coupled with simultaneous manipulation of the environment may be the most powerful approach (Mack 1996). However, again, mostly unexplained time lags in invasions of some taxa (Kowarik 1995; Pysek and Prach 1995) make experimental testing less appealing as a tool for providing guidelines for management. Extensive experimental testing probably will be conducted more often when introduction of a taxon under question is highly desirable (i.e., when there is a good reason to consider introducing a species even if it is thought to have good potential of invading and where a detailed understanding of its invasion dynamics is needed to implement effective management). Kareiva, Parker, and Pascual (1996) provide an insightful analysis of the limitations of short-term invasion experiments.

Performance of introduced taxa in the presence of other species can be elucidated in classic two- or few-species competition experiments (Jolliffe 2000; Rejmánek, Robinson, and Rejmánková 1989) or, perhaps more realistically, in diffusion competition experiments (Austin 1982; Austin et al. 1985; Wilson and Keddy 1986) and individual-based, spatially explicit experiments (Silvertown and Wilson 2000). Parker and Reichard (1998) review some experiments on competition between native and nonnative plant species. It is essential to continue competition experiments long enough to reveal the eventual outcome. For example, only by the fourth year did *Lythrum salicaria* turn out to be a stronger competitor than *Typha angustifolia* (Mal, Lovett-Doust, and Lovett-Doust 1997). A realistic range of environmental conditions should be always considered because results of competition experiments can be substantially different under different conditions. For example, Schmidt (1981) showed that competition between the alien *Solidago canadensis* and the native *Urtica dioica* can have completely opposite outcomes depending on types of soils used for experiments. Such considerations and the issue of when (after how many years) to evaluate such experiments for meaningful results (because germination and short-term survival of seedlings do not necessarily translate into establishment and long-term occupation of sites; Turnbull, Crawley, and Rees 2000) greatly limit the value of such experiments for providing practical management guidelines.

Many new insights will also emerge from experimental studies addressing potential effects of global change on plant invasions. Unfortunately, it seems that many invasive taxa will profit from many of the expected global trends, including atmospheric nitrogen fertilization, elevated CO_2, and more extreme climatic events (Dukes and Mooney 1999; Mooney and Hobbs 2000; Smith et al. 2000; Chapter 12, this volume). For long-lived plants such as trees, manipulative experiments for gaining information on response to different elements of global change are impractical. For such plants, much useful information can be gleaned from natural (*sensu* Diamond 1986) experiments. For example, Richardson et al. (2000b) used information gathered from many sources and from many parts of the world on known responses of the various life history stages of *Prosopis* spp. to many interactions, conditions, resources, and disturbances associated with global change. This information was used to construct likely scenarios for invasive *Prosopis* spp. in South Africa. Similarly, many useful insights emerge on likely responses of invasive *Pinus* spp. in the Southern Hemisphere from comparisons of the performance of the same taxa in different areas (where different combinations of factors interact). Thus, geographically isolated samples can be used (with care) as surrogates for samples from a time series at one site (Richardson et al. 2000b). Given the spatial and temporal scales of invasions and global change phenomena, natural experiments such as these probably will be our only source of information for most cases (Richardson, Rouget, and Rejmánek 2004).

From this discussion it is clear that the most powerful predictions demand the simultaneous use of several approaches. The combination of resulting predictions can be affirmative, additive, or multiplicative, depending on their nature. However, the arithmetic here is complicated. All-encompassing indices and scores from screening exercises may be helpful, but really relevant information may be suppressed. Although the lure of global generalizations will surely keep ecologists busy for decades to come, careful attention to developing detailed habitat- or ecosystem-specific protocols is essential. Where prediction protocols exist for landscapes comprising mosaics of ecosystems, predictions for the most vulnerable system in the landscape should dictate management decisions.

How Fast?

Knowing how fast a taxon could spread is important because it defines how rapidly the impacts of the invader could escalate and how difficult it will be to control such a taxon in the future. A huge range of spread rates (five orders of magnitude: meters to tens of kilometers per year) have been reported in the literature. Can we explain this variance? Theoretical studies of alien plant spread provide several important insights here. Fundamentally, they show that spread is a function of both reproduction and dispersal. The most influential theoretical models of spread are the reaction–diffusion models; these are based on familiar population growth and interaction terms linked to diffusion terms that determine the population dispersal in space (for an example of an approximate solution, see Figure 6.5). Reaction–diffusion models have proved very robust for predicting spread rates of animals but tend to underestimate spread rates for plants; Clark et al. (1998) call this Reid's paradox of rapid plant migration. Fortunately, recent extensions to reaction–diffusion theory have resolved the problem to some extent by considering rare long-distance dispersal events. Whereas earlier models considered only local dispersal, the new models incorporate both local and long-distance dispersal and are therefore called stratified diffusion models (Shigesada, Kawasaki, and Takeda 1995). It is now clear that predicting spread rates for plants entails a quantitative knowledge of rare, long-distance dispersal events. The diffusion models provide several taxon-independent insights into spread rates.

The simple diffusion models of spread identify two stages of spread: a lag phase in which the invasive species is below the detection threshold and a phase of spread at a constant velocity. The stratified diffusion models by contrast predict a lag phase followed by a phase of accelerating spread that clearly more closely mirrors observed invasion patterns. Because the only difference between the stratified diffusion and the simple diffusion model is the way in which dispersal is modeled, we can conclude that the type of dispersal can have a qualitative impact on the spread predictions, whereas the population growth terms have only a quantitative influence. Although the diffusion models provide taxon-independent insights into spread rates for specific studies, it becomes important to understand how life history traits of the invader, disturbance, and distribution of habitat influence spread rates. It is in this domain that spatially explicit simulation models become powerful. For instance, in a study of pine invasions in South African fynbos, Higgins and coworkers (Higgins, Richardson, and Cowling 1996; Higgins and Richardson 1999) quantified how observed variance in factors such as fecundity, age at reproductive maturity, disturbance frequency, habitat loss, and rare long-distance dispersal interact to influence the spread rate. These results suggest that although dispersal had the largest effect on predicted spread rate, the other factors all interacted to influence the spread rate in meaningful ways.

The most valuable insight from studies of spread rates is that rare (less than 5 percent of seeds) long-distance (an order of magnitude greater than the modal dispersal distance) dispersal events have an almost overwhelming influence on the spread rate. Unfortunately, our knowledge of the mechanisms of long-distance dispersal is mostly anecdotal; quantitative studies are limited by the lack of data and have historically been hampered by inadequate statistical techniques. The few quantitative studies on long-distance dispersal suggest some unsettling facts for those interested in predicting spread rates. First, it seems that most existing data sets are inadequate for estimating rare, long-distance dispersal (Clark et al. 1999; Higgins and Richardson 1999). Moreover, it appears that the local dispersal method (i.e., dispersal by wind, ants, or mammals) tells us little about the ability of a taxon to accomplish long-distance dispersal and hence about its potential rate of spread (Wilkinson 1997; Clark et al. 1998). It would be very interesting and valuable to know whether future research on long-distance dispersal will support the implication that biology matters only little in the real world, where freak events mediate long-distance dispersal events.

The fastest consistent, natural long-distance dispersal is by running water or sea currents. This is largely a predictable process (Aptekar and Rejmánek 2000; Thébaud and Debussche 1991). However, as mentioned earlier, human-mediated long-distance dispersal is becoming more and more important. Inevitably, human-assisted dispersal can be the fastest of all dispersal modes: Singapore daisy (*Wendelia trilobata*) was able to cover 2,500 km of the Queensland coastline in 15 years, averaging some 167 km per year (Batianoff and Franks 1997). As in so many human-related affairs, human-mediated dispersal of alien plants is unpredictable to a large extent. Nonetheless, even a partial understanding of this process will enhance early detection of new invaders. Beside classic targets (e.g., railway stations, ports, nurseries, introduction stations, botanic gardens, arboreta, cotton factories), nontraditional spatially isolated foci should be examined regularly as well (e.g., campgrounds, paragliding launching sites, gardens of private hotels, butterfly gardens, surroundings of gardens of plant collectors who are known for their notorious weakness for alien taxa).

What Makes Ecosystems Invasible?

Analyses of ecosystem invasibility based just on one-time observations (a posteriori) usually are unsatisfactory (Rejmánek 1989). In most of the cases we do not know anything about the quality, quantity, and regime of introduction of imported propagules. Usually, it is impossible to separate the resistance of biotic communities from resistance determined by abiotic environments. Nevertheless, available evidence indicates that only very few alien species invade successionally advanced plant communities (Rejmánek 1989, 1996a). Here, however, the quality of usual species pools of introduced alien species (mostly r-strategists) probably is an important part of the story. Plant communities in mesic environments seem to be more invasible than communities in extreme environments (Rejmánek 1989). Apparently xeric environments are not favorable for germination and seedling survival of many introduced species (abiotic resistance), and wet terrestrial habitats do not provide resources for invaders because of fast growth and high competitiveness of resident species (biotic resistance). However, open water is notoriously open to all kinds of exotic aquatic plants (Ashton and Mitchell 1989). In general, disturbance, nutrient enrichment, slow recovery rate of resident vegetation, and fragmentation of successionally advanced communities promote plant invasions (Rejmánek 1989, 1999; Hobbs and Huenneke 1992; Huston 1994; Cadenasso and Pickett 2001).

Experiments on invasibility of different types of ecosystems have been gaining momentum in recent years (Hobbs 1989; Burke and Grime 1996; Knops et al. 1997; Tilman 1997; White, Campbell, and Kemp 1997; Crawley et al. 1999; Lavorel, Prieur-Richard, and Grigulis 1999; Rachich and Reader 1999; Smith and Knapp 1999; Levine 2000; Symstad 2000). These studies may provide robust, nontrivial generalizations in the future, but currently available evidence from such experiments offers little practical assistance to managers. A general theory of invasibility was put forward recently by Davis, Grime, and Thompson (2000). They suggest that intermittent resource enrichment (eutrophication) or release (due to disturbance) increases community susceptibility to invasions. Invasions occur when this situation coincides with availability of suitable propagules. The larger the difference between gross resource supply and resource uptake, the more susceptible the community is to invasion. Alpert, Bone, and Holzapfel (2000) make a similar point. This was anticipated by Vitousek and Walker (1987; Figure 6.10). In another experiment, Davis and Pelsor (2001) manipulated resources and competition and showed that fluctuations in resource availability of as little as 1 week in duration could greatly enhance plant invasion success (survival and cover of alien plants) up to 1 year after such events.

Is invasibility of a plant community related to the number of species present in that community? Davis, Grime, and Thompson (2000) suggest that there is no necessary relationship (see also Lonsdale 1999: 1533). Other studies show that such a relationship exists: positive at the landscape scale (Levine and D'Antonio 1999; Smith and Knapp 1999; Stohlgren et al. 1999; Stadler et al. 2000) and negative at the square centimeter and square meter scales ("neighborhood scales" *sensu* Levine 2000) (Tilman 1997; Knops et al. 1999; Lonsdale 1999: 1533; Levine 2000; Naeem et al. 2000; Lyons and Schwartz 2001). These studies relate the number of resident plant species to the number of alien plant species that establish or become invasive. But there is good evidence that the diversity of a wide range of other organisms in the receiving environment is as important as, if not more important than, the number of plant species currently occupying the site. Many invasions are contingent on the services of mutualists (pollinators, seed dispersers, and microbiota that form symbioses with plant roots). We

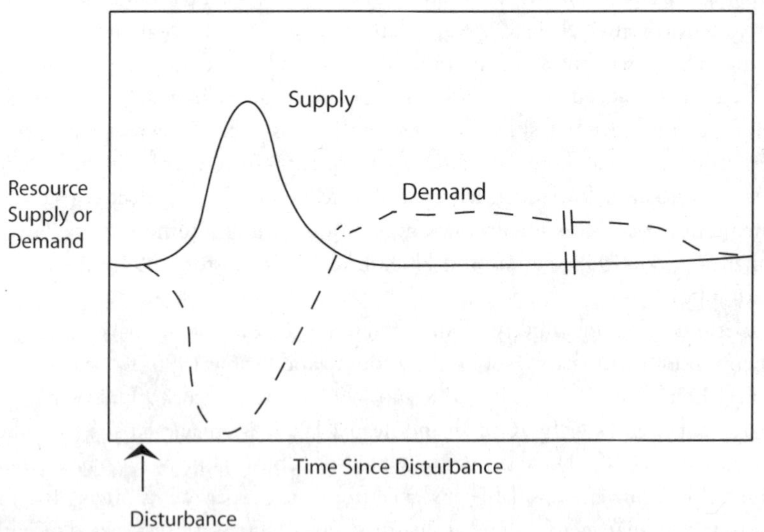

Figure 6.10. Changes in supply and demand of resources after disturbance in terrestrial ecosystems. Resource availability generally is at its maximum shortly after disturbance, although conditions of bare ground can inhibit seedling establishment in some sites. Figure modified from Vitousek and Walker (1987).

could therefore expect that diverse assemblages of mutualists would promote invasibility (Richardson et al. 2000a; Parker 2001).

We have summarized insights from many studies to produce a conceptual cause–effect diagram that captures all the fundamental aspects of the ongoing debate on the issue of invasibility (Figure 6.11). The fact that both invasibility and species diversity of residents are regulated in a similar way by the same set of factors (microclimate, spatial heterogeneity, and long-term regime of available resources) explains why there are so many reports of positive correlation between numbers of native and nonnative species when several different communities or areas are compared. Fast postdisturbance recovery of residents may be a key factor making the wet tropics more resistant to plant invasions (measured as a number of invading species; Rejmánek 1996a). However, there is probably one extra factor that is currently poorly understood: the historical and prehistoric degree of resident taxa exposure to other biota (Figure 6.11). Is this the reason why islands are more vulnerable and Eurasia least vulnerable to invasions (Huston 1994; Simberloff 1995)? Is instability of so many human-made monocultures a result

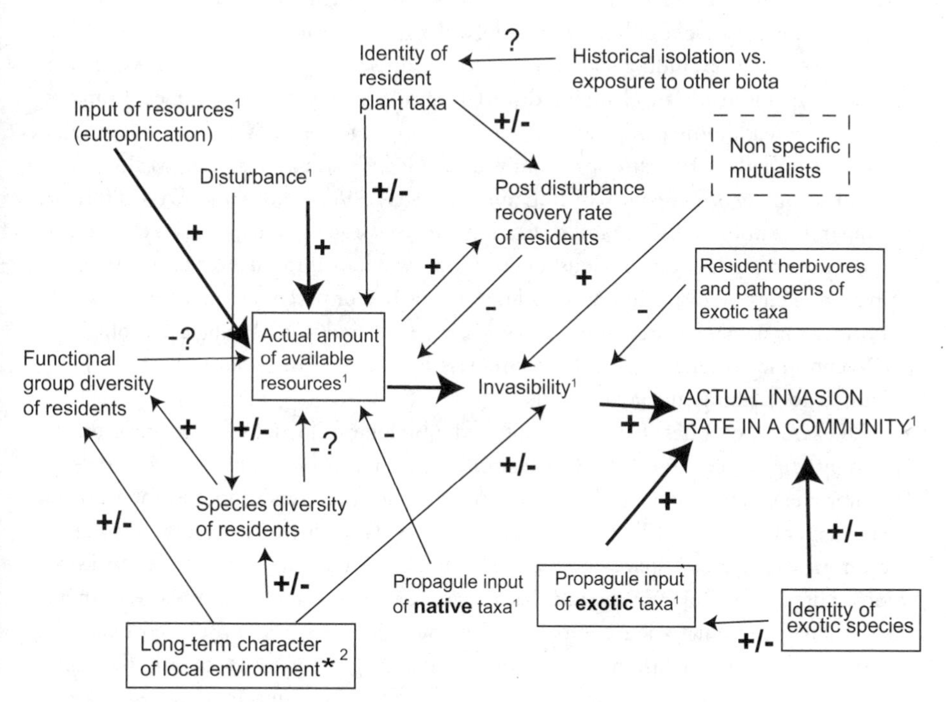

Figure 6.11. Causal relationships between factors and processes that are assumed to be responsible for invasions of exotic species into plant communities. The most important relationships are indicated by thick arrows. *, Spatial heterogeneity, microclimate, and long-term regime of available resources and toxic compounds. 1, time scale: days–years; 2, time scale: years–centuries. The key components are in boxes.

of the "lack of any significant history of coevolution with pests and pathogens" (May 1981: 227)? In other words, actual species richness may not be as important as the richness of assembly history. Mathematical models and computer simulations (Law 1999) certainly are a source of inspiration in this area, but relevant experiments with plant communities still have not been designed. Synthetic experimental plant communities that are so often used for invasibility experiments have a clear advantage of homogeneous substrata and microclimates. However, assembly processes are very short or artificially directed via arbitrary species pool selection, weeding, and reseeding. A stabilized number of Braun–Blanquet vegetation associations described from intensively studied countries (Pot 1995; Schaminée and Stortelder 1996; Rodwell 2000) and the fact that even alien plant species tend to reassemble communities similar to those of their native countries (Lord et al. 2000) indicate that historical development of plant communities cannot be fully substituted by arbitrary decisions.

What Is the Impact?

Numerous studies have documented the wide range of impacts caused by invasive plants. Many invasive taxa have transformed the structure and function of ecosystems by, for example, excluding native taxa, either directly by outcompeting them for resources or indirectly by changing disturbance or nutrient cycling regimes. In many parts of the world, impacts have clear economic implications for humans, for example, as a result of reduced stream flow from watersheds in South African fynbos after alien tree invasions (van Wilgen, Cowling, and Burgers 1996; Calder and Dye 2000) or through disruption to fishing and navigation after invasion by water weeds such as *Eichhornia crassipes.* Managers obviously need to know about impacts because some combination of current distribution, spread rate, and actual or potential impact must define priorities for management action. Impacts can be assessed using biological, ecological, and economic currencies, and much progress has been made recently in developing methods for making such assessments.

An example of a detailed assessment of the biological impact of plant invasions is given by Higgins et al. (1999), who examined how many native taxa would be threatened by alien plants and used this information to prioritize invasive species and sites for management (Figure 6.12). In attempting to assess the value of ecosystem services of South African fynbos systems and the extent to which these values are reduced by invasions, Higgins et al. (1997) showed that the cost of clearing alien plants was very small (less than 5 percent) when compared with the value of services provided by these ecosystems. Their conclusion was that proactive management could increase the value of these ecosystem services by at least 138 percent. The most important ecosystem service was water, and much work has been done on developing models for assessing the value (in monetary terms) of allocating management resources to clearing invasive plants from fynbos watersheds (van Wilgen, Cowling, and Burgers 1996; Hosking and

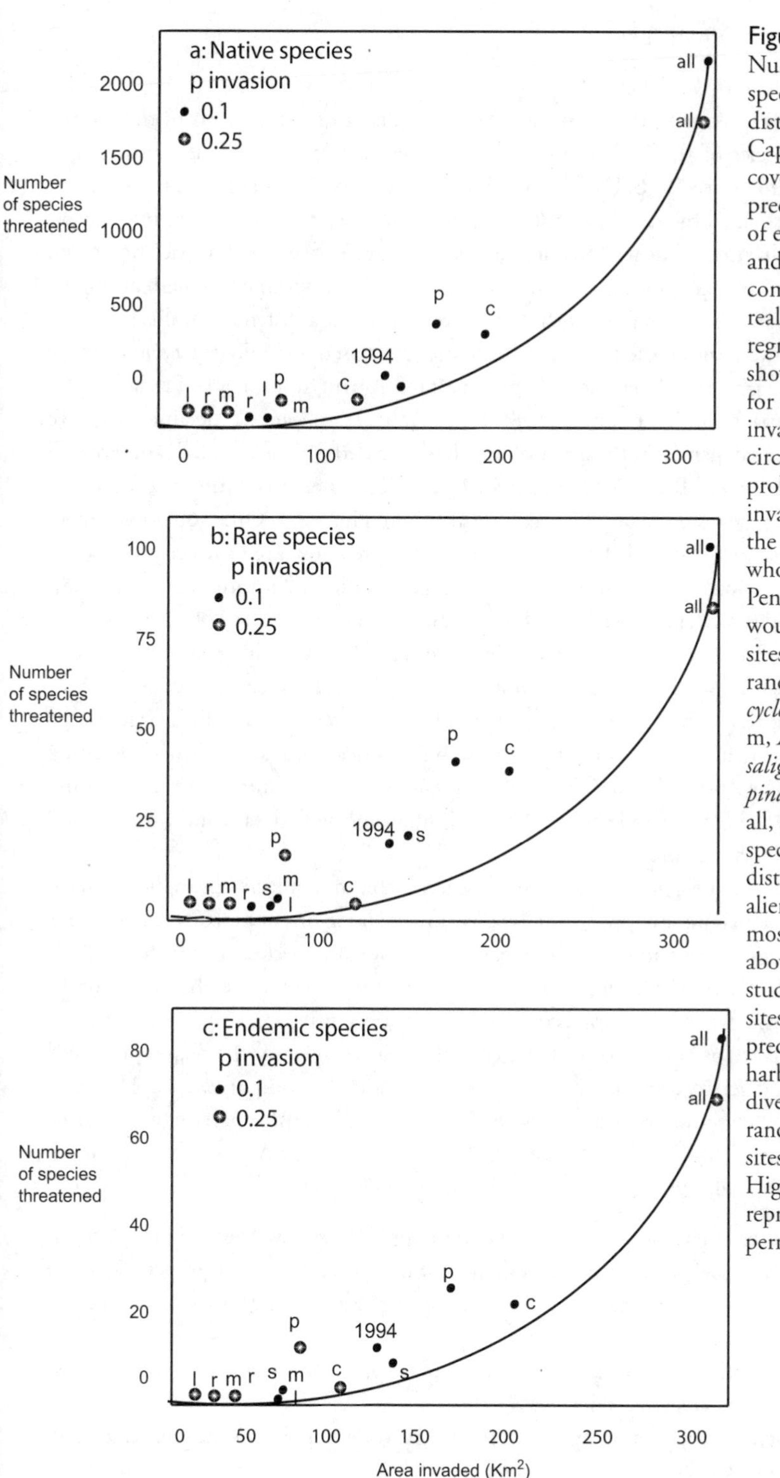

Figure 6.12. Number of native species whose entire distribution on the Cape Peninsula is covered by the predicted distribution of each alien species and for all 6 species combined. Two realizations of the regression model are shown: solid circles are for a .1 probability of invasion, and open circles are for a .25 probability of invasion. The curve is the number of species whose entire Cape Peninsula distribution would be covered if sites were selected randomly. c, *Acacia cyclops;* l, *A. longifolia;* m, *A. mearnsii;* s, *A. saligna;* p, *Pinus pinaster;* r, *P. radiata;* all, all 6 alien plant species; 1994, 1994 distribution of all 6 alien species. Because most alien species fall above the curve, this study shows that the sites that the aliens are predicted to invade harbor more plant diversity than a random selection of sites. Figure from Higgins et al. (1999), reproduced with permission.

Du Preez 1999). Among the most dangerous invaders in riparian areas in the United States are species of the Old World genus *Tamarix* (saltcedar; Tomanek and Ziegler 1961; Sala and Smith 1996; DiTomaso 1998). An economic evaluation of the saltcedar impacts is provided by Zavaleta (2000). Such approaches are very useful, but can we make generalizations that will enable managers to predict the likely impact of a given invasive taxon in a given locality in such a way that the importance of such an impact can be compared objectively with that of another taxon at a different locality?

As we showed earlier, the most reliable predictions based on biological characters are limited to invasiveness (likelihood of species establishment and spread). Predictions of potential impacts probably always will be less reliable. Because a decline in native species richness depends on the cover of invaders (Richardson, Macdonald, and Forsyth 1989; Rejmánek and Rosén 1992; Jäger 1999), indices based on a ratio of cover to frequency should be tested as impact predictors for individual taxa. Other obvious impact indicators may be biological characters of plants that are known to have ecosystem consequences (e.g., high transpiration rates or nitrogen fixation). The phytometer, or indicator, method (Gaudet and Keddy 1988; Keddy, Gaudet, and Fraser 2000) can be used to assess the relative competitiveness of alien plant taxa. However, it is important to realize that invasiveness and impact are not necessarily positively correlated. Some fast-spreading species, such as *Aira caryophyllea* and *Cakile edentula,* exhibit little or no measurable environmental or economic impact. On the other hand, some slowly spreading species, such as *Ammophila arenaria* and *Robinia pseudoacacia,* may have far-reaching environmental effects (stabilization of coastal dunes in the first case and nitrogen soil enrichment in the second).

There may well be a need for a universally acceptable and objectively applicable term for the most damaging invasive plant taxa within given regions or globally. In our view, a potentially useful term to use in this regard is "transformer species," proposed by Wells et al. (1986), referring to a subset of invasive plants that "change the character, condition, form or nature of a natural ecosystem over a substantial area." It is these species, comprising perhaps only about 10 percent of invasive species, that have profound effects on biodiversity and that clearly demand a major allocation of resources for containment, control, and eradication. Several categories of transformers may be distinguished (Richardson et al. 2000c; see also Vitousek 1990; D'Antonio and Vitousek 1992; Chapin et al. 1996; Mack and D'Antonio 1998):

- Excessive users of resources (water, *Tamarix* spp., *Acacia mearnsii;* light, *Pueraria lobata, Rubus armeniacus;* water and light, *Arundo donax;* light and oxygen, *Salvinia molesta, Eichhornia crassipes*); high LAR of many invasive plants (discussed earlier) is an important prerequisite for excessive transpiration.
- Donors of limiting resources (nitrogen, *Acacia* spp., *Lupinus arboreus, Myrica faya, Robinia pseudoacacia, Salvinia molesta*).
- Fire promoters or suppressors (promoters, *Bromus tectorum, Melaleuca quinquenervia, Melinis minutiflora;* suppressors, *Mimosa pigra*).

- Sand stabilizers (*Ammophila* spp., *Elymus* spp.).
- Erosion promoters (*Andropogon virginicus* in Hawaii, *Impatiens glandulifera* in Europe).
- Colonizers of intertidal mudflats and sediment stabilizers (*Spartina* spp., *Rhizophora* spp.).
- Litter accumulators (*Centaurea solstitialis, Eucalyptus* spp., *Lepidium latifolium, Pinus strobus, Taeniatherum caput-medusae*).
- Soil carbon storage modifiers (promoters, *Andropogon gayanus;* suppressors, *Agropyron cristatum;* see Fisher et al. 1994; Christian and Wilson 1999).
- Salt accumulators and redistributors (*Mesembryanthemum crystallinum, Tamarix* spp.).

The potentially most important transformers are taxa that add a new function, such as nitrogen fixation, to the invaded ecosystem (Vitousek and Walker 1989). These can be called discrete trait invaders (Chapin et al. 1996). However, many impacts are not so obvious. For example, invasive *Lonicera* and *Rhamnus* change vegetation structure of the forest, affecting nest predation of birds (Schmidt and Whelan 1999), and *Lythrum salicaria* can have negative impacts on pollination and reproductive success of co-flowering native plants (Grabas and Laverty 1999). It is not yet clear whether an integrated measure of impact for invasive plants is a realistic goal (see Ruiz et al. 1999; Williamson 2001). Nonetheless, we agree with Parker et al. (1999) that it is useful to consider various indices of impact that combine (in some way) the total area occupied, the abundance, and some measure of the impact per individual invading organism.

Control, Contain, or Eradicate?

Many combinations of control methods, usually involving mechanical, biological, and chemical means, are available to managers for containing, controlling, or eradicating alien plants. Wittenberg and Cock (Chapter 9, this volume) discuss the many factors that must be considered in formulating and implementing effective management strategies. Here we discuss only a few supplementary aspects that have emerged from recent studies.

The efficiency of a control operation depends on the currency used to evaluate the operation and the tactics of the control operation. Valid currencies for evaluating control operations include financial cost and ecological impact (e.g., minimizing the threat to native taxa). Sound management strategies demand an objective means for setting priorities. One decision that managers often need to make is whether to prioritize outlying populations of the invader or focus on sites with high conservation value. Using a simple model, Moody and Mack (1988) showed that clearing small, outlying stands first was more efficient than clearing large source stands. Mechanistic models of alien spread also show that clearing the sparsest stands first is most cost-effective and also best for mitigating threats to native biodiversity (Figure 6.13; Higgins, Richardson, and

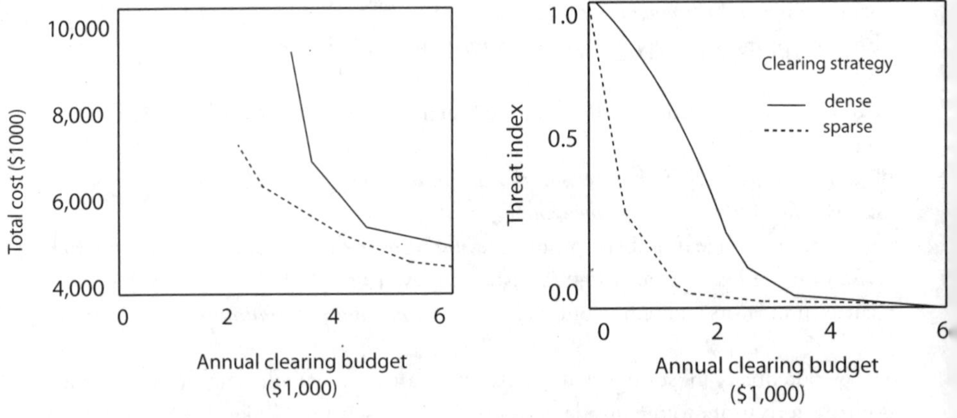

Figure 6.13. Cost of eradicating alien plants from a landscape and the change in threat to the native flora as a function of the annual clearing effort. Results are shown for 2 different alien clearing strategies: the first clears the densest stands first, and the second clears the lowest-density stands first. The results are derived from an alien plant management model parameterized for the Cape Peninsula, South Africa (Higgins, Richardson, and Cowling 2000).

Cowling 2000). Van Wilgen et al. (2000a) give some examples of how spatially explicit models can be used to explore the implications of applying different treatments.

There are several encouraging examples in which widespread alien animals have been eradicated (Dahlsten and Garcia 1989; Myers et al. 2000). Can we eradicate equally widespread and difficult alien plants? For both plants and animals, the extent of an infestation, its configuration in the landscape, and the amount of resources available to managers are key issues. When is complete eradication a realistic goal for invasive plants? Using a unique data set on eradication attempts by the California Department of Food and Agriculture (Table 6.3), we can show that professional eradication of alien weed infestations smaller than 1 hectare is almost always possible. Also, about one-third of all infestations between 1 and 100 hectares have been eradicated. However, costs increase dramatically (Table 6.3). With a realistic amount of resources, it is very unlikely that infestations larger than 100 hectares can be eradicated (Figure 6.14). Early detection of the presence of an invasive taxon can make the difference between being able to use feasible offensive strategies (eradication) and the need to retreat to a defensive strategy that usually means an infinite financial commitment. Obviously, a substantial increase in resources for exclusion and early detection of exotic weeds would be the most profitable investment (see also Cook, Setterfield, and Maddison 1996; Meinesz 1999; Smith, Johnston, and Swanson 1999).

Nevertheless, depending on the potential impact of individual weedy species, even infestations larger than 100 hectares should be targeted for eradication effort or, at least,

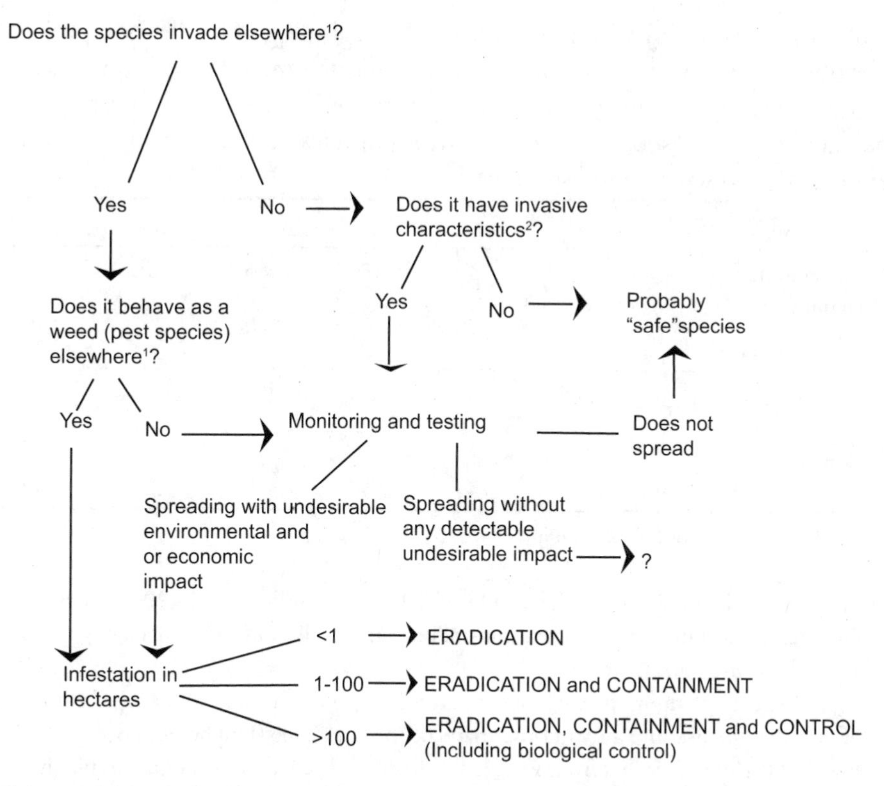

Figure 6.14. Realistic management options for recently introduced plants. 1, Consult literature, experts, and local and global databases on invasive species; 2, See this chapter and relevant references.

substantial reduction and containment. A notable example of a successful quarantine is one established for containment of *Striga asiatica* in parts of North and South Carolina (Ross and Lembi 1985; Eplee 1992; Kaiser 1999; Eplee and Norris 2000). In California, some stands of noxious weeds larger than 100 hectares have been reduced to just a few plants. One large stand of camelthorn (*Alhagi pseudalhagi,* 259 hectares in 1971) probably has been eradicated; only two plants were found in 2000. An exceptional success story is the practically complete eradication (98 percent of properties on which it is known to occur) of *Bassia* (*Kochia*) *scoparia* over the past 8 years in Australia (3,277 hectares, 15,536 work hours; R. P. Randall, pers. comm., 2002). However, if an exotic weed is already widespread, then species-specific biological control (if feasible; Follett and Boston 2000; Pemberton 2000) usually is the only practical way to switch between defensive and offensive strategies. Several successful weed biocontrol projects have already been accomplished in California, South Africa, and Australia (Julien and Grif-

Table 6.3. Areas of initial infestations (at the beginning of eradication projects) of exotic weeds in California, numbers of eradicated infestations, numbers of ongoing projects, and mean eradication effort for 5 infestation area categories; the data include 16 species of noxious weedy species (2 aquatic and 14 terrestrial) representing 50 separate infestations

Initial infestation (ha)		*<0.1*	*0.1–1*	*1.1–100*	*101–1,000*	*>1,000*
No. of eradicated infestations		13	3	5	0	0
No. of ongoing projects		2	4	9	10	4
Mean eradication effort per infestation (work hours)	Eradicated	63	180	1,292	—	—
	Ongoing	174	277	1,577	17,188	42,751

Source: M. Pitcairn and M. Rejmánek, unpublished.

fiths 1998; Villegas 1998; Olckers and Hill 1999). Van Wilgen et al. (2000b) describe the careful integration of biological control into an overall strategy for managing woody alien plants in South Africa.

Of increasing importance in many parts of the world is the need to manage invasions of taxa that are useful plants in other parts of the landscape (Hughes and Styles 1987; Ewel et al. 1999). Insights from invasion ecology are needed, for example, in planning commercial forestry operations using alien conifers in New Zealand to reduce the negative impacts of "wilding spread" from new plantings. Guidelines based on factors affecting spread are now available; these include recommendations on species choice, the siting and design of new plantations (e.g., slope, aspect, exposure relative to prevailing wind), and management requirements in surrounding vegetation (Ledgard and Langer 1999).

Conclusion

Invasion ecology struggles with five basic questions: Which taxa invade? How fast? What makes ecosystems invasible? What is the impact? How can we contain, control, or eradicate harmful invaders? Each of these queries may be accompanied by an equally important subquestion: "In which habitats?" Therefore, these five questions can be dealt with in general terms or, in a more operational way, with respect to particular types of environment. To answer the first two questions, invasion ecology offers five groups of complementary approaches.

- Stochastic approaches allow probabilistic predictions about potential invaders based on initial population size, residence time, and number of introduction events.

- Empirical taxon-specific approaches are based on previously documented invasions of particular taxa.
- Evaluations of the biological characters of invasive and noninvasive taxa give rise to both general and habitat-specific screening procedures.
- Evaluation of environmental compatibility helps to predict whether a particular plant taxon can invade specific habitats.
- Experimental approaches attempt to tease apart intrinsic and extrinsic factors underlying invasion success of particular taxa and factors making ecosystems more or less invasible.

An emerging theory of plant invasiveness based on biological characters has produced several robust predictions that are reviewed in this chapter. Empirical taxon-specific approaches are also most powerful for assessing impacts. To contain, control, or eradicate harmful invaders, invasion ecology provides only strategic foundations. Specific answers emerge on the interface of invasion ecology, pest management and weed science, and restoration ecology.

Despite pessimistic prognoses in recent years, plant invasion ecology has come up with a reasonably robust set of tools that facilitate fairly good predictions in many cases. The problem is that the available tools reviewed in this chapter are not (and perhaps cannot be) integrated into a user-friendly package that could be used by, for example, a customs inspector or even most ecologists. For this reason, we suggest that an urgent priority is the establishment of an "International Invasive Plant Species Data Center" tasked with compiling and maintaining a database with all the information discussed in this chapter. The absence of such an authoritative data source is causing confusion in the literature and resulting in erroneous decisions (e.g., when generalizations or assessments are based on incorrect information regarding which taxa are and which are not invasive).

As for basic/applied research, we suggest six areas on which plant invasion ecology should concentrate:

- We need a better understanding of the relationships between detectable genetic changes and adaptations in invading populations. How important is genetic polymorphism as opposed to phenotypic plasticity?
- Improved understanding is needed on the roles of breeding systems, dispersal modes, and the importance of unusual events in the dynamics of plant invasions. We need to know the causes of lag phases. Such advances are fundamental prerequisites for better screening procedures.
- The causal relationships responsible for actual invasion rates in particular ecosystems (see Figure 6.11) must be quantified.
- Standardized protocols for assessing impact (including microclimatic, biogeochemical, and biotic impacts) of invasive taxa are needed to facilitate the objective prioritization of management actions.

- Strategies for integrating biological control with other forms of invasive species management must be developed (Wajnberg, Scott, and Quimby 2001; Chapter 9, this volume).
- Practical guidelines are needed for environmentally sensible restoration of habitats after removal of invasive aliens (Zavaleta, Hobbs, and Mooney 2001).

Postscript

The final version of this chapter was provided in August 2001. In the meantime, substantial progress in many areas of plant invasion ecology has occurred. Therefore, we are recommending 12 recent publications as supplementary reading: Booth, Murphy, and Swaton (2003, chapter 13), Colautti et al. (2004), Dahler (2003), Dietz and Steinlein (2004), Duncan and Williams (2002), Ehrenfeld (2003), Levine, Adler, and Yelenik (2004), Meyers and Bazely (2003), Rejmánek and Pitcairn (2002), Rejmánek, Richardson, and Pysek (2005), Richardson (2004), Shea and Chesson (2002). Also, recent volumes of *Biological Invasions* and *Diversity and Distributions* should be consulted.

Booth, B. D. , S. D. Murphy, and C. J. Swanton. 2003. *Weed ecology in natural and agricultural systems*, CABI Publishing, Wallingford, United Kingdom.

Colautti, R. I., A. Ricciardi, I. A. Grigorovich, and H. J. Maclsaac. 2004. "Is invasion success explained by the enemy release hypothesis?" *Ecology Letters* 7: 721–733.

Dahler, C. 2003. "Performance comparisons of co-occuring native and alien invasive plants: implications for conservation and restoration," *Annual Review of Ecology, Evolution and Systematics* 34: 183–211.

Dietz, H., and T. Steinlein. 2004. "Recent advances in understanding plant invasions," *Progress in Botany* 65: 539–573.

Duncan, R. P., and P. A. Williams. 2002. "Darwin's naturalization hypothesis challenged," *Nature* 417: 606–609.

Ehrenfeld, J. G. 2003. "Effects of exotic plant invasions on soil nutrient cycling processes," *Ecosystems* 6: 503–523.

Levine, J. M., P. B. Adler, and S. G. Yelenik. 2004. "A meta-analysis of biotic resistance to exotic plant invasions," *Ecology Letters* 7: 975–979.

Myers, J. H., and D. R. Bazely. 2003. *Ecology and control of introdced plants*, Cambridge University Press, Cambridge, United Kingdom.

Rejmánek, M., and M. J. Pitcairn. 2002. "When is eradication of exotic plant pests a realistic goal?" in *Turning the Tide: The Eradication of Invasive Species*, edited by C. R. Weith and M. N. Clout, 249–253. IUCN, Gland, Switzerland and Cambridge, United Kingdom.

Rejmánek, M., D. M. Richardson, and P. Pysek. 2005. "Plant invasions and invasibility of plant communities," in *Vegetation Ecology*, edited by E. van der Maarel, pp. 332–355. Blackwell Science, Oxford.

Richardson, D. M. 2004. "Plant invasion ecology—dispatches from the front line," *Diversity and Distributions*, 10:315–319.

Shea, K., and P. Chesson. 2002. "Community ecology theory as a framework for biological invasions," *Trends in Ecology and Evolution* 17: 170–176.

Acknowledgments

We thank Mark Williamson, Rod Randall, Petr Pysek, Peter Vitousek, Fred Hrusa, Nate Dechorez, and Debra Ayres for valuable discussions and comments on the manuscript. David Richardson acknowledges financial support from the National Research Foundation (South Africa), the University of Cape Town, and the DST-NRF Centre for Excellence for Invasion Biology for work on this contribution.

References

Abbott, R. J. 1992. "Plant invasions, interspecific hybridization and the evolution of new plant taxa," *Trends in Ecology and Evolution* 7:401–405.

Allan, H. H. 1936. "Indigene versus alien in the New Zealand plant world," *Ecology* 17:187–193.

———. 1937. "The origin and distribution of the naturalized plants of New Zealand," *Proceedings of the Linnean Society of London* 1:25–46.

Alpert, P., E. Bone, and C. Holzapfel. 2000. "Invasiveness, invasibility and the role of environmental stress in the spread of non-native plants," *Perspectives in Plant Ecology, Evolution and Systematics* 3:52–66.

Andersen, E. 2000. "Ecological roles of mammals: the case of seed dispersal," in *Priorities for the Conservation of Mammalian Diversity,* edited by A. Entwistle and N. Dunstone, 11–25. Cambridge University Press, Cambridge, UK.

Aptekar, R., and M. Rejmánek. 2000. "The effect of seawater submergence on rhizome bud viability of introduced and native dune grasses (*Ammophila arenaria* and *Leymus mollis*) in California," *Journal of Coastal Conservation* 6:107–111.

Ashton, P. J., and D. S. Mitchell. 1989. "Aquatic plants: Patterns and modes of invasions, attributes of invading species and assessment of control programmes," in *Biological invasions. A global perspective,* edited by J. A. Drake, H. A. Mooney, F. di Castri, R. H. Groves, F. J. Kruger, M. Rejmánek, and M. Williamson, 111–154. Wiley, Chichester, UK.

Auld, B. A., J. Hosking, and R. E. McFadyen. 1983. "Analysis of the spread of tiger pear and parthenium weed in Australia," *Australian Weeds* 2:56–60.

Austin, M. P. 1982. "Use of a relative physiological performance value in the prediction of performance in multispecies mixtures from monoculture performance," *Journal of Ecology* 70:559–570.

Austin, M. P., R. H. Groves, M. F. Fresco, and P. E. Kaye. 1985. "Relative growth of six thistle species along a nutrient gradient with multispecies competition," *Journal of Ecology* 73:667–684.

Ayres, D. R., D. Garcia-Rossi, H. G. Davis, and D. R. Strong. 1999. "Extent and degree of hybridization between exotic (*Spartina alterniflora*) and native (*S. foliosa*) cordgrass (Poaceae) in California, USA determined by random amplified polymorphic DNA (RAPDs)," *Molecular Ecology* 8:1179–1186.

Bachman, K., K. L. Chambers, and H. J. Price. 1985. "Genome size and natural selection: Observations and experiments in plants," in *The Evolution of Genome Size,* edited by T. Cavalier-Smith, 267–276. Wiley, Chichester, UK.

Baker, H. G. 1965. "Characteristics and modes of origin of weeds," in *The Genetics of Colonizing Species,* edited by H. G. Baker and G. L. Stebbins, 147–172. Academic Press, New York.

———. 1974. "The evolution of weeds," *Annual Review of Ecology and Systematics* 5:1–24.

———. 1995. "Aspects of the genecology of weeds," in *Genecology and Ecogeographic Races,* edited by A. R. Kruckeberg, R. B. Walker, and A. E. Leviton, 189–224. Pacific Division AAA, San Francisco, CA.

Baker, H. G., and G. L. Stebbins, eds. 1965. *The Genetics of Colonizing Species.* Academic Press, New York.

Barrett, S. C. H. 1992. "Genetics of weed invasions," in *Applied Population Biology,* edited by S. K. Jain and L. W. Botsford, 91–119. Kluwer Academic Publishers, Dordrecht, The Netherlands.

———. 2000. "Microevolutionary influences of global changes on plant invasions," in *Invasive Species in a Changing World,* edited by H. A. Mooney and R. J. Hobbs, 115–139. Island Press, Washington, DC.

Baruch, Z. 1996. "Ecophysiological aspects of the invasion by African grasses and their impact on biodiversity and function of Neotropical savannas," in *Biodiversity and Savanna Ecosystem Processes,* edited by O. T. Solbrig, E. Medina, and J. F. Silva, 79–93. Springer-Verlag, Berlin.

Baruch, Z., and G. Goldstein. 1999. "Leaf construction cost, nutrient concentration, and net CO_2 assimilation of native and invasive species in Hawaii," *Oecologia* 121:183–192.

Baruch, Z., A. B. Hernández, and M. G. Montila. 1989. "Dinámica del crecimiento, fenología y repartición de biomasa en gramíneas nativas e introducidas de una sabana Neotropical," *Ecotropicos* 2:1–13.

Batianoff, G. N., and A. J. Franks. 1997. "Invasion of sandy beachfronts by ornamental plant species in Queensland," *Plant Protection Quarterly* 12:180–186.

Bayer, R. J. 1999. "New perspectives into the evolution of polyploid complexes," in *Plant Evolution in Man-Made Habitats,* edited by L. W. D. van Raamsdonk and J. C. M. den Nijs, 359–373. Hugo de Vries Laboratory, Amsterdam.

Bazzaz, F. A. 1986. "Life history of colonizing plants: Some demographic, genetic, and physiological features," in *Ecology of Biological Invasions of North America and Hawaii,* edited by H. A. Mooney and J. A. Drake, 96–110. Springer-Verlag, New York.

Becerra, J. X. 1997. "Insects on plants: Macroevolutionary chemical trends in host use," *Science* 276:253–256.

Beerling, D. J., B. Huntley, and J. P. Bailey. 1995. "Climate and the distribution of *Fallopia japonica:* Use of an introduced species to test the predictive capacity of response surfaces," *Journal of Vegetation Science* 6:269–282.

Bennett, M. D., I. J. Leitch, and L. Hanson. 1998. "DNA amounts in two samples of angiosperm weeds," *Annals of Botany* 82(Suppl. A):121–134.

Bierzychudek, P. 1999. "Looking backwards: Assessing the projections of a transition matrix model," *Ecological Applications* 9:1278–1287.

Binggeli, P. 1996. "A taxonomic, biogeographical and ecological overview of invasive woody plants," *Journal of Vegetation Science* 7:121–124.

———. 1998. *An Overview of Invasive Woody Plants in the Tropics.* Online: http://members.tripod.co.uk/WoodyPlantEcology/invasive/index.html.

Bolòs, O. de, and J. Vigo. 1995. *Flora dels Països Catalans.* Volume III. Editorial Barcino, Barcelona.

Bossard, C., J. Randall, and M. Hoshovsky, eds. 2000. *Invasive Plants of California's Wildlands.* University of California Press, Berkeley.

Breiman, L., J. H. Friedman, R. A. Olshen, and C. J. Stone. 1984. *Classification and Regression Trees.* Wadsworth, Belmont, CA.

Brewer, S. W., and M. Rejmánek. 1999. "Small rodents as significant dispersers of tree seeds in a neotropical forest," *Journal of Vegetation Science* 10:165–174.

Bridge, P., P. Jeffries, D. R. Morse, and P. R. Scott, eds. 1998. *Information Technology, Plant Pathology and Biodiversity.* CABI Publishing, Wallingford, UK.

Brochmann, C., and R. Elven. 1992. "Ecological and genetic consequences of polyploidy in arctic *Draba* (Brasicaceae)," *Evolutionary Trends in Plants* 6:111–124.

Bromilow, C. 1995. *Problem Plants of South Africa.* Briza Press, Arcadia, South Africa.

Brown, A. H. D., and J. J. Burdon. 1987. "Mating systems and colonizing success in plants," in *Colonization, Succession and Stability,* edited by A. J. Gray, M. J. Crawley, and P. J. Edwards, 115–131. Blackwell Science, Oxford, UK.

Brzeziecki, B., and F. Kienast. 1994. "Classifying the life-history strategies of trees on the basis of the Grimian model," *Forest Ecology and Management* 69:167–187.

Burdon, J. J., and A. H. D. Brown. 1986. "Population genetics of *Echium plantagineum* L.: Target weed for biological control," *Australian Journal of Biological Science* 30:369–378.

Burke, M. J. W., and J. P. Grime. 1996. "An experimental study of plant community invasibility," *Ecology* 77:776–790.

Cadenasso, M. L., and S. T. A. Pickett. 2001. "Effect of edge structure on the flux of species into forest interiors," *Conservation Biology* 15:91–97.

Cadotte, M. W., and J. Lovett-Doust. 2001. "Ecological and taxonomic differences between native and introduced plants of southwestern Ontario," *EcoScience* 8:230–238.

Calder, I., and P. Dye. 2000. "Hydrological impacts of invasive alien plants," in *Best Management Practices for Preventing and Controlling Invasive Alien Species,* edited by G. Preston, G. Brown, and E. van Wyk, 160–170. Working for Water Programme, Cape Town, South Africa.

Campbell, C. S. 1999. "The evolutionary role of hybridisation in angiosperm agamic complexes, with special emphasis on *Amelanchier* (Rosaceae)," in *Plant Evolution in Man-Made Habitats,* edited by L. W. D. van Raamsdonk and J. C. M. den Nijs, 341–357. Hugo de Vries Laboratory, Amsterdam.

Ceccherelli, G., and F. Cinelli. 1999. "The role of vegetative fragmentation in dispersal of the invasive alga *Caulerpa taxifolia* in the Mediterranean," *Marine Ecology Progress Series (MEPS)* 182:299–303.

Chapin, F. S. III, H. L. Reynolds, C. M. D'Antonio, and V. K. Eckhart. 1996. "The functional role of species in terrestrial ecosystems," in *Global Change and Terrestrial Ecosystems,* edited by B. Walker and W. Steffen, 403–428. Cambridge University Press, Cambridge, UK.

Chicoine, T. K., P. K. Fay, and G. A. Nielsen. 1985. "Predicting weed migration from soil and climate maps," *Weed Science* 34:57–61.

Christensen, R. 1997. *Log-Linear Models and Logistic Regression.* Springer, New York.

Christian, J. M., and S. D. Wilson. 1999. "Long-term impacts of an introduced grass in the northern Great Plains," *Ecology* 80:2397–2407.

Clark, J. S., C. Fastie, G. Hurtt, S. T. Jackson, C. Johnson, G. A. King, M. Lewis, J. Lynch, S. Pacala, C. Prentice, E. W. Schupp, T. Webb, and P. Wyckoff. 1998. "Reid's paradox of rapid plant migration," *BioScience* 48:13–24.

Clark, J. S., M. Silman, R. Kern, E. Macklin, and J. Hille Ris Lambers. 1999. "Seed dispersal near and far: Patterns across temperate and tropical forests," *Ecology* 80:1475–1494.

Cook, C. D. K. 1985. "Range extension of aquatic vascular plant species," *Journal of Aquatic Plant Management* 23:1–6.

Cook, G. D., S. A. Setterfield, and J. P. Maddison. 1996. "Shrub invasion of a tropical wetland: Implications for weed management," *Ecological Applications* 6:531–537.

Cox, G. W. 1999. *Alien Species in North America and Hawaii.* Island Press, Washington, DC.

Cox, J. R., M. H. Martin-R., F. A. Ibarra-F., J. H. Fourie, N. F. G. Rethman, and D. G. Wilcox. 1988. "The influence of climate and soils on the distribution of four African grasses," *Journal of Range Management* 41:127–139.

Crawley, M. J., S. L. Brown, M. S. Heard, and G. G. Edwards. 1999. "Invasion-resistance in experimental grassland communities: Species richness or species identity?" *Ecology Letters* 2:140–148.

Crawley, M. J., P. H. Harvey, and A. Purvis. 1996. "Comparative ecology of the native and alien floras of the British Isles," *Philosophical Transactions of the Royal Society of London* B 351:1251–1259.

Crooks, J. A., and M. E. Soulé. 1999. "Lag times in population explosions of invasive species: Causes and implications," in *Invasive Species and Biodiversity Management,* edited by O. T. Sandlund, P. J. Schei, and A. Viken, 103–125. Kluwer Academic Publishers, Dordrecht, The Netherlands.

Daehler, C. 1998. "The taxonomic distribution of invasive angiosperm plants: Ecological insights and comparison to agricultural weeds," *Biological Conservation* 84:167–180.

Daehler, C., and D. A. Carino. 2000. "Predicting invasive plants: Prospects for a general screening system based on current regional models," *Biological Invasions* 2:93–102.

———. 2001. "Hybridization between native and alien plants and its consequences," in *Biotic Homogenization,* edited by J. L. Lockwood and M. L. McKinney, 81–102. Kluwer Academic/Plenum Publishers, New York.

Dahlsten, D. L., and R. Garcia, eds. 1989. *Eradication of Exotic Pests.* Yale University Press, New Haven, CT.

D'Antonio, C. M., J. Levine, and M. Thomsen. 2001. "Ecosystem resistance to invasion and the role of propagule supply: A California perspective," *Journal of Mediterranean Ecology* 27: 233–245.

D'Antonio, C. M., and P. M. Vitousek. 1992. "Biological invasions by exotic grasses, the grass/fire cycle, and global change," *Annual Review of Ecology and Systematics* 23:63–87.

Darwin, C. 1859. *The Origin of Species by Means of Natural Selection.* Murray, London.

Davis, M. A., J. P. Grime, and K. Thompson. 2000. "Fluctuating resources in plant communities: A general theory of invasibility," *Journal of Ecology* 88:528–534.

Davis, M. A., and M. Pelsor. 2001. "Experimental support for a resource-based mechanistic model of invasibility," *Ecology Letters* 4:421–428.

De'ath, G., and K. E. Fabricius. 2000. "Classification and regression trees: A powerful yet simple technique for ecological data analysis," *Ecology* 81:3178–3192.

Debach, P. 1974. *Biological Control by Natural Enemies.* Cambridge University Press, Cambridge, UK.

De Candolle, A. P. 1855. *Géographie Botanique Raisonné,* Vol. 2. V. Masson, Paris.

Diamond, J. 1986. "Overview: Laboratory experiments, field experiments, and natural experiments," in *Community Ecology,* edited by J. Diamond and T. J. Case, 3–22. Harper & Row, New York.

Diekmann, M., and J. E. Lawesson. 1999. "Shifts in ecological behaviour of herbaceous forest species along a transect from northern central to north Europe," *Folia Geobotanica* 34:127–141.

Dietz, H., M. Fischer, and B. Schmid. 1999. "Demographic and genetic invasion history of a 9-year-old roadside population of *Bunias orientalis* L. (Brasicaceae)," *Oecologia* 120:225–234.

DiTomaso, J. M. 1998. "Impact, biology, and ecology of saltcedar (*Tamarix* spp.) in the southwestern United States," *Weed Technology* 12:326–336.

Drake, J. A., H. A. Mooney, F. di Castri, R. H. Groves, F. J. Kruger, M. Rejmánek, and M. Williamson, eds. 1989. *Biological Invasions. A Global Perspective.* Wiley, Chichester, UK.

Dukes, J. S., and H. A. Mooney. 1999. "Does global change increase the success of biological invaders?" *Trends in Ecology and Evolution* 14:135–139.

Duncan, R. P., M. Bomford, D. M. Forsyth, and L. Conibear. 2001. "High predictability in introduction outcomes and the geographical range size of introduced Australian birds: A role for climate," *Journal of Animal Ecology* 70:621–632.

Egler, F. E. 1983. *The Nature of Naturalization II. The Introduced Flora of Aton Forest, Connecticut.* Claude E. Phillips Herbarium Publication No. 6. Delaware State College, Dover.

Ellenberg, H. 1974. "Zeigerwerte der Gefässpflanzen Mitteleuropas," *Scripta Geobotanica* 9:1–97.

Ellenberg, H., H. E. Weber, R. Düll, V. Wirth, W. Werner, and D. Paulissen. 1992. "Zeigerwerte von Pflanzen in Mitteleuropa," *Scripta Geobotanica* 18:1–248.

Ellstrand, N. C., and K. A. Schierenbeck. 2000. "Hybridization as a stimulus for the evolution of invasiveness in plants?" *PNAS* 97:7043–7050.

Elton, C. S. 1958. *The Ecology of Invasions by Animals and Plants.* Methuen, London.

Eplee, R. E. 1992. "Witchweed (*Striga asiatica*): An overview of management strategies in the USA," *Crop Protection* 11:3–7.

Eplee, R. E., and R. Norris. 2000. "Eradication of *Striga asiatica* from the United States," in *Abstracts of the III International Weed Science Congress,* 212. International Weed Science Society, Oregon State University, Corvallis.

Eriksson, O., and A. Jacobsson. 1999. "Recruitment trade-offs and the evolution of dispersal mechanisms in plants," *Evolutionary Ecology* 13:411–423.

Espinosa, G. F. J., and J. Sarukhán. 1997. *Manual de Malezas del Valle de México.* Universidad Nacional Autonoma de México, México, D.F.

Ewel, J. J., D. J. O'Dowd, J. Bergelson, C. C. Daehler, C. M. D'Antonio, L. D. Gomez, D. R. Gordon, R. J. Hobbs, A. Holt, K. R. Hopper, C. E. Hughes, M. LaHart, R. R. B. Leakey, W. G. Lee, L. L. Loope, D. H. Lorence, S. M. Louda, A. E. Lugo, P. B. McEvoy, D. M. Richardson, and P. M. Vitousek. 1999. "Deliberate introductions of species: Research needs," *BioScience* 49:619–630.

Fenner, M., and W. G. Lee. 2001. "Lack of pre-dispersal predators in introduced Asteraceae in New Zealand," *New Zealand Journal of Ecology* 25:95–100.

Fisher, M. J., I. M. Rao, M. A. Ayrza, C. E. Lascano, J. I. Sanz, R. J. Thomas, and R. R. Vera. 1994. "Carbon storage by introduced deep-rooted grasses in the South American savannas," *Nature* 371:236–238.

Follett, P. A., and J. J. D. Boston, eds. 2000. *Nontarget Effects of Biological Control.* Kluwer Academic Publishers, Dordrecht, The Netherlands.

Forcella, F., and J. T. Wood. 1984. "Colonization potentials of alien weeds are related to their 'native' distributions: Implications for plant quarantine," *Journal of the Australian Institute of Agricultural Science* 50:36–40.

Forcella, F., J. T. Wood, and S. P. Dillon. 1986. "Characteristics distinguishing invasive weeds within *Echium*," *Weed Research* 26:351–364.

Forno, I. W. 1983. "Native distribution of the *Salvinia auriculata* complex and keys to species identification," *Aquatic Botany* 17:71–83.

Fragoso, J. M. V., and J. M. Huffman. 2000. "Seed-dispersal and seedling recruitment patterns by the last neotropical megafaunal element in Amazonia, the tapir," *Journal of Tropical Ecology* 16:369–386.

Franchet, A. 1872. "Sur une florule adventice observée dans le département du Loir-et-Cher," *Bulletin of Société Botanique de France* 19:195–202.

Freckleton, R. P. 2000. "Phylogenetic tests of ecological and evolutionary hypothesis: Checking for phylogenetic independence," *Functional Ecology* 14:129–134.

Futuyma, D. J., M. C. Keese, and D. J. Funk. 1995. "Genetic constrains on macroevolution: The evolution of host affiliation in the leaf beetle genus *Ophraella*," *Evolution* 49:797–809.

Gaudet, C. L., and P. A. Keddy. 1988. "A comparative approach to predicting competitive ability from plant traits," *Nature* 334:242–243.

Gerlach, J. D. 2000. "Using plant functional groups to predict the impact of invasive species on ecosystem functions," PhD dissertation, University of California, Davis.

Goeze, E. 1882. *Pflanzengeographie für Gärtner und Freude des Gartenbaues.* Ulmer, Stuttgart.

Goodin, B. J., A. J. McAllister, and L. Fahrig. 1998. "Predicting invasiveness of plant species based on biological information," *Conservation Biology* 13:422–426.

Govindaraju, D., P. Lewis, and C. Cullis. 1992. "Phylogenetic analysis of pines using ribosomal DNA restriction fragment length polymorphisms," *Plant Systematics and Evolution* 179:141–153.

Grabas, G. P., and M. Laverty. 1999. "The effect of purple loosestrife (*Lythrum salicaria* L.; Lythraceae) on the pollination and reproductive success of sympatric co-flowering wetland plants," *EcoScience* 6:230–242.

Green, R. E. 1997. "The influence of numbers released on the outcome of attempts to introduce exotic bird species to New Zealand," *Journal of Animal Ecology* 66:25–35.

Greene, D. F., and E. A. Johnson. 1989. "A model of wind dispersal of winged or plumed seeds," *Ecology* 70:339–347.

Grime, J. P., J. G. Hodgson, and R. Hunt. 1988. *Comparative Plant Ecology.* Unwin Hyman, London.

Grime, J. P., and R. Hunt. 1975. "Relative growth rate: Its range and adaptive significance in a local flora," *Journal of Ecology* 63:393–422.

Grotkopp, E., M. Rejmánek, and T. L. Rost. 2002. "Toward a causal explanation of plant invasiveness: Seedling growth and life-history strategies of 29 pine (*Pinus*) species," *The American Naturalist* 159(4): 396–419.

Grotkopp, E., R. Stoltenberg, M. Rejmánek, and T. Rost. 1998. "The effect of genome size on invasiveness," *American Journal of Botany* 85(Suppl.):34.

Groves, R. H., F. D. Panetta, and J. G. Virtue, eds. 2001. *Weed Risk Assessment.* CSIRO Publishing, Collingwood, Victoria, Australia.

Groves, R. H., R. C. H. Shepherd, and R. G. Richardson. 1995. *The Biology of Australian Weeds,* Vol. 1. R.G. and F.J. Richardson, Melbourne.

Grubb, P. J. 1998. "A reassessment of the strategies of plants which cope with shortages of resources," *Perspectives in Plant Ecology and Evolutionary Systematics* 1:3–31.

Guo, Q., J. H. Brown, T. J. Valone, and S. D. Kachman. 2000. "Constraints of seed size on plant distribution and abundance," *Ecology* 81:2149–2155.

Hamilton, C. W. 1990. "Variations on a distylous theme in Mesoamerican *Psychotria* subgenus *Psychotria* (Rubiaceae)," *Memoirs of the New York Botanical Garden* 55:62–75.

Haselwood, E. L., and G. G. Motter. 1983. *Handbook of Hawaiian Weeds,* 2nd ed. University of Hawaii Press, Honolulu.

Henderson, L. 1991. "Alien invasive *Salix* spp. (willows) in the grassland biome of South Africa," *South African Forestry Journal* 157:91–95.

Hermanutz, L. A., and S. E. Weaver. 1996. "Agroecotypes or phenotypic plasticity? Comparison of agrestal and ruderal populations of the weed *Solanum ptycanthum,*" *Oecologia* 105:271–280.

Hickman, J. C., ed. 1993. *The Jepson Manual. Higher Plants of California.* University of California Press, Berkeley.

Higgins, S. I., and D. M. Richardson. 1998. "Pine invasions in the Southern Hemisphere: Modelling interactions between organism and environment," *Plant Ecology* 135:79–93.

———. 1999. "Predicting plant migration in a changing world: The role of long-distance dispersal," *American Naturalist* 153:464–475.

Higgins, S. I., D. M. Richardson, and R. M. Cowling. 1996. "The role of plant–environment interactions and model structure on the predicted rate and pattern of invasive plant spread," *Ecology* 77:2043–2054.

———. 2000. "Using a dynamic landscape model for planning the management of alien plant invasions," *Ecological Applications* 10:1833–1848.

Higgins, S. I., D. M. Richardson, R. M. Cowling, and T. H. Trinder-Smith. 1999. "Predicting the landscape distribution of invasive alien plants and their threat to native plant diversity," *Conservation Biology* 13:303–313.

Higgins, S. I., J. K. Turpie, R. Costanza, R. M. Cowling, D. C. Le Maitre, C. Marais, and G. F. Midgley. 1997. "An ecologically-economic simulation model of mountain fynbos ecosystems: Dynamics, valuation and management," *Ecological Economics* 22:155–169.

Hill, M. O., D. B. Roy, and J. O. Mountford. 2000. "Extending Ellenberg's indicator values to a new area: An algorithmic approach," *Journal of Applied Ecology* 37:3–15.

Hiscock, S. J. 2000. "Self-incompatibility in *Senecio squalidus* L. (Asteraceae)," *Annals of Botany* 85(Suppl. A):181–190.

Hnatiuk, R. J. 1990. *Census of Australian Vascular Plants.* Australian Government Publishing Service, Canberra.

Hobbs, R. J. 1989. "The nature and effects of disturbance relative to invasions," in *Biological Invasions: A Global Perspective,* edited by J. A. Drake, H. A. Mooney, F. di Castri, R. H. Groves, F. J. Kruger, M. Rejmánek, and M. Williamson, 389–405. Wiley, Chichester, UK.

Hobbs, R. J., and L. F. Huenneke. 1992. "Disturbance, diversity and invasion: Implications for conservation," *Conservation Biology* 6:324–337.

Hodgson, J. G., P. J. Wilson, R. Hunt, J. P. Grime, and K. Thompson. 1999. "Allocation C-S-R plant functional types: A soft approach to a hard problem," *Oikos* 85:282–294.

Hodkinson, D. J., A. P. Askew, K. Thompson, J. G. Hodgson, J. P. Bakker, and R. M. Bekker. 1998. "Ecological correlates of seed size in British flora," *Functional Ecology* 12:762–766.

Hodkinson, D. J., and K. Thompson. 1997. "Plant dispersal: The role of man," *Journal of Applied Ecology* 34:1484–1496.

Hoffman, A. A., and P. A. Parsons. 1991. *Evolutionary Genetics and Environmental Stress.* Oxford University Press, Oxford, UK.

Hooker, J. D. 1864. "Notes on the replacement of species in the colonies and elsewhere," *Natural History Review* 4:123–127.

Horak, M. J., J. S. Holt, and N. C. Ellstrand. 1987. "Genetic variation in yellow nutsedge (*Cyperus esculentus*)," *Weed Science* 35:506–512.

Hosking, S. G., and M. Du Preez. 1999. "A cost–benefit analysis of removing alien trees in the Tsitsikamma mountain catchment," *South African Journal of Science* 95:442–448.

Hsu, C.-C. 1975. *Illustrations of Common Plants of Taiwan,* Vol. I: *Weeds.* Taiwan Provincial Education Association, Taipei, Taiwan.

Huberty, C. J. 1994. *Applied Discriminant Analysis.* Wiley, New York.

Huffaker, C. B., and C. E. Kennett. 1959. "A ten-year study of vegetational changes associated with biological control of Klamath weed," *Journal of Range Management* 12:69–82.

Hughes, C. E., and B. T. Styles. 1987. "The benefits and potential risks of woody legume introductions," *International Tree Crops Journal* 4:209–248.

Hügin, G. 1995. "Höhengrenzen von Rureral- und Segetalpflanzen in der Alpen," *Flora* 190:169–188.

Hulbert, L. C. 1955. "Ecological studies of *Bromus tectorum* and other annual bromegrasses," *Ecological Monographs* 25:181–213.

Hultén, E., and M. Fries. 1986. *Atlas of Northern European Vascular Plants North of the Tropic of Cancer,* Vols. 1–3. Koeltz Scientific Books, Königstein.

Huston, M. A. 1994. *Biological Diversity.* Cambridge University Press, Cambridge, UK.

Jäger, H. 1999. "Impact of the introduced tree *Cinchona pubescens* Vahl. On the native flora of the highlands of Santa Cruz Island (Galápagos Islands)," dissertation, Carl von Ossietzky Universität, Oldenburg, Germany.

Jolliffe, P. A. 2000. "The replacement series," *Journal of Ecology* 88:371–385.

Jones, P. G., and A. Gladkov. 1999. *FloraMap: A Computer Tool for Predicting the Distribution of Plants and Other Organisms in the Wild.* CIAT CD-ROM Series. International Center for Tropical Agriculture, Cali, Colombia.

Julien, M. H., and M. W. Griffiths, eds. 1998. *Biological Control of Weeds,* 4th ed. CABI Publishing, Wallingford, UK.

Kaiser, J. 1999. "Stemming the tide of invading species," *Science* 285:1836–1841.

Kareiva, P., I. M. Parker, and M. Pascual. 1996. "Can we use experiments and models in predicting the invasiveness of genetically engineered organisms?" *Ecology* 77:1670–1675.

Kartesz, J. T., and C. A. Meacham. 1999. *Synthesis of the North American Flora.* CD-ROM Version 1.0. North Carolina Botanical Garden, Chapel Hill.

Keddy, P., C. Gaudet, and L. H. Fraser. 2000. "Effects of low and high nutrients on the competitive hierarchy of 26 shoreline plants," *Journal of Ecology* 88:413–423.

Knops, J. M. H., D. Tilman, N. M. Haddad, S. Naeem, C. E. Mitchell, J. Haarstad, M. E. Ritchie, K. M. Howe, P. B. Reich, E. Siemann, and J. Groth. 1999. "Effects of plant species richness on invasion dynamics, disease outbreaks, insect abundance and diversity," *Ecology Letters* 2:286–293.

Knops, J. M. H., D. Tilman, S. Naeem, and K. M. Howe. 1997. "Biodiversity and plant invasions in experimental grassland plots," *Bulletin of the Ecological Society of America* Supplement 71:125.

Kolar, C. S., and D. M. Lodge. 2001. "Progress in invasion biology: Predicting invaders," *Trends in Ecology and Evolution* 16:199–204.

Kollmann, J. 2000. "Dispersal of fleshy-fruited species: A matter of spatial scale?" *Perspectives in Plant Ecology, Evolution and Systematics* 3:29–51.

Kotanen, P. M., J. Bergelson, and D. L. Hazlett. 1998. "Habitats of native and exotic plants in Colorado shortgrass steppe: A comparative approach," *Canadian Journal of Botany* 76:664–672.

Kowarik, I. 1995. "Time lags in biological invasions with regard to the success and failure of alien species," in *Plant Invasions,* edited by P. Pysek, K. Prach, M. Rejmánek, and P. M. Wade, 15–38. SPB Academic Publishing, The Hague.

Kriticos, D. J., and R. P. Randall. 2001. "A comparison of systems to analyse potential weed distributions," in *Weed Risk Assessment,* edited by R. H. Groves, F. D. Panetta, and J. G. Virtue, 61–79. CSIRO Publishing, Collingwood, Victoria, Australia.

Krugman, S. L., and J. L. Jenkinson. 1974. "*Pinus* L.," in *Seeds of Woody Plants in the United States,* edited by C. S. Schopmeyer, 598–638. USDA Forest Service Agriculture Handbook 450. U.S. Department of Agriculture, Washington, DC.

Larson, B., and S. C. H. Barrett. 2000. "A comparative analysis of pollen limitation in angiosperms," *Biological Journal of the Linnean Society* 69:503–520.

Lavorel, S., A.-H. Prieur-Richard, and K. Grigulis. 1999. "Invasibility and diversity of plant communities: From patterns to processes," *Diversity and Distributions* 5:41–50.

Law, R. 1999. "Theoretical aspects of community assembly," in *Advanced Ecological Theory,* edited by J. McGlade, 143–171. Blackwell Science, Oxford, UK.

Ledgard, N. J., and E. R. Langer. 1999. *Wildling Prevention. Guidelines for Minimizing the Risk of Unwanted Wildling Spread from New Plantings of Introduced Conifers.* New Zealand Forest Research Institute, Christchurch.

Leishman, M. R. 2001. "Does the seed size/number trade-off model determine plant community structure? An assessment of the model mechanisms and their generality," *Oikos* 93:294–302.

Leitch, I. J., M. W. Chase, and M. D. Bennett. 1998. "Phylogenetic analysis of DNA C-values provides evidence for a small ancestral genome size in flowering plants," *Annals of Botany* 82(Suppl. A):85–94.

Levin, D. A. 2000. *The Origin, Expansion and Demise of Plant Species.* Oxford University Press, Oxford, UK.

Levine, J. M. 2000. "Species diversity and biological invasions: Relating local process to community pattern," *Science* 288:852–854.

Levine, J. M., and C. M. D'Antonio. 1999. "Elton revisited: A review of evidence linking diversity and invasibility," *Oikos* 87:15–26.

Lindacher, R. 1995. "PHANART. Datebank der Gefässpflanzen Mitteleuropas. Erklärung der Kennzahlen, Aufbau und Inhalt." *Veröffentlichungen des Geobotanischen Institutes der ETH, Stiftung Rübel* 125:1–436.

Lindsay, D. R. 1953. "Climate as a factor influencing the mass ranges of weeds," *Ecology* 34:308–321.

Liston, A., W. A. Robinson, D. Pinero, and E. R. Alvarez-Buylla. 1999. "Phylogenetics of *Pinus* (Pinaceae) based on nuclear ribosomal DNA internal transcribed spacer region sequences," *Molecular Phylogenetics and Evolution* 11:95–109.

Lockwood, J. L., D. Simberloff, M. L. McKinney, and B. Von Holle. 2001. "How many, and which, plants will invade natural areas?" *Biological Invasions* 3:1–8.

Lohmeyer, W., and H. Sukopp. 1992. "Agriophyten in der Vegetation Mitteleuropas," *Schriftenreihe für Vegetationskunde* 25:1–185.

Lonsdale, W. M. 1994. "Inviting trouble: Introduced pasture species in northern Australia," *Australian Journal of Ecology* 19:345–354.

———. 1999. "Global patterns of plant invasions and the concept of invasibility," *Ecology* 80:1522–1536.

Lord, J. M., J. B. Wilson, J. B. Steel, and B. J. Anderson. 2000. "Community reassembly: A test using limestone grassland in New Zealand," *Ecology Letters* 3:213–218.

Lorenzi, H. J., and L. S. Jeffery. 1987. *Weeds of the United States and Their Control.* Van Nostrand Reinhold, New York.

Lyons, K. G., and M. W. Schwartz. 2001. "Rare species loss alters ecosystem function: Invasion resistance," *Ecology Letters* 4:358–365.

Mack, M. C., and C. M. D'Antonio. 1998. "Impact of biological invasions on disturbance regimes," *Trends in Ecology and Evolution* 13:195–198.

Mack, R. N. 1996. "Predicting the identity and fate of plant invaders: Emergent and emerging approaches," *Biological Conservation* 78:107–124.

Maillet, J., and C. Lopez-Garcia. 2000. "What criteria are relevant for predicting the invasive capacity of a new agricultural weed? The case of invasive American species in France," *Weed Research* 40:11–26.

Mal, K., J. Lovett-Doust, and L. Lovett-Doust. 1997. "Time-dependent displacement of *Typha angustifolia* by *Lythrum salicaria*," *Oikos* 79:26–33.

Matthei, O. 1995. *Manual de las Malezas que Crecen en Chile.* Alfabeta Impresores, Santiago, Chile.

May, R. N. 1981. "Patterns in multi-species communities," in *Theoretical Ecology. Principles and Applications,* edited by R. N. May, 197–227. Blackwell Scientific Publications, Oxford, UK.

McEvoy, P. B., C. S. Cox, and E. Coombs. 1991. "Successful biological control of ragwort," *Ecological Applications* 1:430–442.

Meinesz, A. 1999. *Killer Algae.* University of Chicago Press, Chicago, IL.

Metz, J. A. J., D. Mollison, and F. van den Bosch. 2000. "The dynamics of invasion waves," in *The Geometry of Ecological Interactions: Simplifying Spatial Complexity,* edited by U. Dickmann, R. Law, and J. A. J. Metz, 482–512. Cambridge University Press, Cambridge, UK.

Michael, P. W. 1981. "Alien plants," in *Australian Vegetation,* edited by R. H. Groves, 44–64. Cambridge University Press, Cambridge, UK.

Milne, R. I., and R. J. Abbott. 2000. "Origin and evolution of invasive naturalized material of *Rhododendron ponticum* L. in the British Isles," *Molecular Ecology* 9:541–556.

Moody, M. E., and R. N. Mack. 1988. "Controlling the spread of plant invasions: The importance of nascent foci," *Journal of Applied Ecology* 25:1009–1021.

Mooney, H. A., and R. J. Hobbs. 2000. *Invasive Species in a Changing World.* Island Press, Washington, DC.

Morgan, K. J. R., and T. R. Seastedt. 1999. "Effects of soil nitrogen reduction on nonnative plants in restored grasslands," *Restoration Ecology* 7:51–55.

Mulvaney, M. 2001. "The effect of introduction pressure on the naturalization of ornamental woody plants on south-eastern Australia," in *Weed Risk Assessment,* edited by R. H. Groves, F. D. Panetta, and J. G. Virtue, 186–193. CSIRO Publishing, Collingwood, Victoria, Australia.

Myers, J. H., D. Simberloff, A. M. Kuris, and J. R. Carey. 2000. "Eradication revisited: Dealing with exotic species," *Trends in Ecology and Evolution* 15:316–320.

Naeem, S., J. M. H. Knops, D. Tilman, K. M. Howe, T. Kennedy, and S. Gale. 2000. "Plant diversity increases resistance to invasion in the absence of covarying extrinsic factors," *Oikos* 91:97–108.

Nathan, R., and H. C. Muller-Landau. 2000. "Spatial patterns of seed dispersal, their determinants and consequences for recruitment," *Trends in Ecology and Evolution* 15:278–285.

Newsome, A. E., and I. R. Noble. 1986. "Ecological and physiological characters of invading species," in *Ecology of Biological Invasions: An Australian Perspective,* edited by R. H. Groves and J. J. Burdon, 1–20. Australian Academy of Science, Canberra.

Novak, S. J., and R. N. Mack. 1993. "Genetic variation in *Bromus tectorum* (Poaceae): Comparison between native and introduced populations," *Heredity* 71:167–176.

———. 2001. "Tracing plant introduction and spread: Genetic evidence from *Bromus tectorum* (cheatgrass)," *BioScience* 51:114–122.

Oberdorfer, E. 1994. *Pflanzensoziologische Exkursionsflora für Süddeutschland und die angrenzenden Gebiete,* 7th ed. Ulmer, Stuttgart.

Olckers, T., and M. P. Hill, eds. 1999. "Biological control of weeds in South Africa," *African Entomology Memoir* 1:1–182.

Oliver, I., A. Pil, D. Britton, M. Dangerfield, R. K. Colwell, and A. J. Beattie. 2000. "Virtual biodiversity assessment systems," *BioScience* 50:441–450.

Owen, S. J. 1997. *Ecological Weeds on Conservation Land in New Zealand: A Database.* Department of Conservation, Wellington.

Panetsos, C., and H. G. Baker. 1968. "The origin of variation in 'wild' *Raphanus sativus* (Cruciferae) in California," *Genetica* 38:243–274.

Panetta, F. D. 1993. "A system of assessing proposed plant introductions for weed potential," *Plant Protection Quarterly* 8:10–14.

Panetta, F. D., and J. Dodd. 1987. "Bioclimatic prediction of the potential distribution of skeleton weed (*Chondrilla juncea* L.) in Western Australia," *Journal of the Australian Institute of Agricultural Science* 53:11–16.

Panetta, F. D., and J. C. Scanlan. 1995. "Human involvement in the spread of noxious weeds: What plants should be declared and when should control be enforced?" *Plant Protection Quarterly* 10:69–74.

Pannel, J. R., and S. C. H. Barrett. 1998. "Baker's Law revisited: Reproductive assurance in a metapopulation," *Evolution* 52:657–668.

Pantone, D. J., B. M. Pavlik, and R. B. Kelley. 1995. "The reproductive attributes of an endangered plant as compared to a weedy congener," *Biological Conservation* 71:305–311.

Parker, I. M., and S. H. Reichard. 1998. "Critical issues in invasion biology for conservation science," in *Conservation Biology,* edited by P. L. Fiedler and P. M. Kareiva, 283–305. Chapman & Hall, New York.

Parker, I. M., D. Simberloff, W. M. Lonsdale, K. Goodell, M. Wonham, P. M. Kareiva, M. Williamson, B. Von Holle, P. B. Moyle, J. E. Byers, and L. Goldwasser. 1999. "Impact: Towards a framework for understanding the ecological effects of invaders," *Biological Invasions* 1:3–19.

Parker, M. A. 2001. "Mutualism as a constraint on invasion success for legumes," *Diversity and Distributions* 7:125–136.

Pattison, R. R., G. Goldstein, and A. Ares. 1998. "Growth, biomass allocation and photosynthesis of invasive and native Hawaiian rainforest species," *Oecologia* 117:449–459.

Pemberton, R. W. 2000. "Predictable risk to native plants in weed biological control," *Oecologia* 125:489–494.

Pheloung, P. C., P. A. Williams, and S. R. Halloy. 1999. "A weed risk assessment model for use as a biosecurity tool evaluating plant introductions," *Journal of Environmental Management* 57:239–251.

Pieterse, A. H., and K. J. Murphy, eds. 1990. *Aquatic Weeds.* Oxford University Press, Oxford, UK.

Poschlod, P., D. Matthies, S. Jordan, and C. Mengel. 1996. "The biological flora of central Europe: An ecological bibliography," *Bulletin of the Geobotanical Institute ETH, Zürich* 62:89–108.

Pot, R. 1995. *Die Pflanzengesellschaften Deutschlands,* 2nd ed. Ulmer, Stuttgart.

Poynton, R. J. 1979. *Tree Planting in Southern Africa,* Vol. 2: *The Eucalypts.* Department of Forestry, Pretoria, South Africa.

Protopopova, V. V. 1991. *Sinantropnaia Flora Ukrainy i Puti eio Razvitia.* Naukova Dumka, Kiev.

Purps, D. M. L., and J. W. Kadereit. 1999. "The evolution of invasive species in *Senecio* (Asteraceae)," in *Plant Evolution in Man-Made Habitats,* edited by L. W. D. van Raamsdonk and J. C. M. den Nijs, 73–89. Hugo de Vries Laboratory, Amsterdam.

Pysek, P. 1997. "Clonality and plant invasions: Can a trait make a difference?" in *The Ecology and Evolution of Clonal Plants,* edited by H. de Kroon and J. van Groenendael, 405–427. Backhuys Publishers, Leiden, The Netherlands.

———. 1998. "Is there a taxonomic pattern to plant invasions?" *Oikos* 92:282–294.

Pysek, P., and K. Prach. 1995. "Invasion dynamics of *Impatiens glandulifera:* A century of spreading reconstructed," *Biological Conservation* 74:41–48.

Pysek, P., K. Prach, and P. Smilauer. 1995. "Relating invasion success to plant traits: An analysis of the Czech alien flora," in *Plant Invasions,* edited by P. Pysek, K. Prach, M. Rejmánek, and P. M. Wade, 237–247. SPB Academic Publishing, The Hague.

Rachich, J., and R. J. Reader. 1999. "An experimental study of wetland invasibility by purple loosestrife (*Lythrum salicaria*)," *Canadian Journal of Botany* 77:1499–1503.

Randall, R. P. 2000. *Weed List* [Australia]. Online: http://www.agric.wa.gov.au/progserv/plants/weeds/weedlist.htm.

Rapoport, E. H. 1991. "Tropical versus temperate weeds: A glance into the present and future," in *Ecology of Biological Invasion in the Tropics,* edited by P. S. Ramakrishnan, 441–451. International Scientific Publications, New Delhi.

———. 1992. "Las implicaciones ecológicas y económicas de la introducción de especies," *Ciéncia & Ambiente* 3(4):69–81.

Ray, P. M., and H. F. Chisaki. 1957. "Studies on *Amsinckia,* I. A synopsis of the genus, with a study of heterostyly in it," *American Journal of Botany* 44:529–536.

Reed, C. F. 1977. *Economically Important Foreign Weeds. Potential Problems in the United States.* Agriculture Handbook No. 498. U.S. Department of Agriculture, Washington, DC.

Reichard, S. H. 1994. "Assessing the potential of invasiveness in woody plants introduced in North America," PhD dissertation, University of Washington, Seattle.

Reichard, S. H., and C. W. Hamilton. 1997. "Predicting invasions of woody plants introduced into North America," *Conservation Biology* 11:193–203.

Rejmánek, M. 1989. "Invasibility of plant communities," in *Biological Invasions. A Global Perspective,* edited by J. A. Drake, H. A. Mooney, F. di Castri, R. H. Groves, F. J. Kruger, M. Rejmánek, and M. Williamson, 369–388. Wiley, Chichester, UK.

———. 1995. "What makes a species invasive?" in *Plant Invasions,* edited by P. Pysek, K. Prach, M. Rejmánek, and P. M. Wade, 3–13. SPB Academic Publishing, The Hague.

———. 1996a. "Species richness and resistance to invasions," in *Diversity and Processes in Tropical Forest Ecosystems,* edited by G. H. Orians, R. Dirzo, and J. H. Cushman, 153–172. Springer-Verlag, Berlin.

———. 1996b. "A theory of seed plant invasiveness: The first sketch," *Biological Conservation* 78:171–181.

———. 1999. "Invasive plant species and invasible ecosystems," in *Invasive Species and Biodiversity Management,* edited by O. T. Sandlund, P. J. Schei, and A. Viken, 79–102. Kluwer Academic Publishers, Dordrecht, The Netherlands.

———. 2000a. "Invasive plants: Approaches and predictions," *Austral Ecology* 25:497–506.

———. 2000b. "On the use and misuse of transition matrices in plant population biology," *Biological Invasions* 2:315–317.

Rejmánek, M., and S. Reichard. 2001. "Predicting invasiveness," *Trends in Ecology and Evolution* 16:545.

Rejmánek, M., and D. M. Richardson. 1996. "What attributes make some plant species more invasive?" *Ecology* 77:1655–1661.

———. 2003. "What makes some conifers more invasive?" Proceedings of the Fourth International Conifer Conference. 2003. *Acta Horticulturae* 615:375–380.

Rejmánek, M., G. R. Robinson, and E. Rejmánková. 1989. "Weed–crop competition: Experimental design and models for data analysis," *Weed Science* 37:276–284.

Rejmánek, M., and E. Rosén. 1992. "Influence of colonizing shrubs on species–area relationships in alvar plant communities," *Journal of Vegetation Science* 3:625–630.

Ricciardi, A., W. W. M. Steiner, R. N. Mack, and D. Simberloff. 2000. "Toward a global information system for invasive species," *BioScience* 50:2139–2244.

Rice, P. M. 1998. *INVADERS Database,* Release 6.5. Division of Biological Sciences, University of Montana, Missoula. Online: http://invader.dbs.umt.edu.

Richardson, D. M. 1989. "The ecology of *Pinus* and *Hakea* invasions," PhD dissertation, Botany Department, University of Cape Town, South Africa.

———. 2001. "Plant invasions," in *Encyclopedia of Biodiversity,* Vol. 4, edited by S. Levin, 677–688. Academic Press, San Diego, CA.

Richardson, D. M., N. Allsopp, C. M. D'Antonio, S. J. Milton, and M. Rejmánek. 2000a. "Plant invasions: The role of mutualisms," *Biological Reviews* 75:65–93.

Richardson, D. M., W. J. Bond, W. R. J. Dean, S. I. Higgins, G. F. Midgley, S. J. Milton, L. Powrie, M. C. Rutherford, M. J. Samways, and R. E. Schulze. 2000b. "Invasive alien organisms and global change: A South African perspective," in *The Impact of Global Change on Alien Species,* edited by H. A. Mooney and R. J. Hobbs, 303–349. Island Press, Washington, DC.

Richardson, D. M., R. M. Cowling, and D. C. Le Maitre. 1990. "Assessing the risk of invasive success in *Pinus* and *Banksia* in South African mountain fynbos," *Journal of Vegetation Science* 1:629–642.

Richardson, D. M., I. A. W. Macdonald, and G. G. Forsyth. 1989. "Reductions in plant species richness under stands of alien trees and shrubs in the fynbos biome," *South African Forestry Journal* 149:1–8.

Richardson, D. M., P. Pysek, M. Rejmánek, M. G. Barbour, F. D. Panetta, and C. J. West. 2000c. "Naturalization and invasion of alien plants: Concepts and definitions," *Diversity and Distributions* 6:93–107.

Richardson, D. M., and M. Rejmánek. 1998. "*Metrosideros excelsa* takes off in the fynbos," *Veld & Flora* 85:14–16.

Richardson, D. M., and M. Rejmánek. 2004. "Conifers as invasive aliens: a global survey and predictive framework." *Diversity and Distributions* 10: 321–331.

Richardson, D. M., M. Rouget, and M. Rejmánek. 2004. "Using natural experiments in the study of alien tree invasions: Opportunities and limitations," in *Experimental Approaches in Conservation Biology,* edited by M. Gordon and S. Bartol, 180–201. University of California Press, Berkeley.

Ridgway, R. L., W. P. Gregg, R. E. Stinner, and A. G. Brown, eds. 1999. *Invasive Species Databases.* U.S. Department of the Interior, Washington, DC.

Roché, B. F., C. T. Roché, and R. C. Chapman. 1994. "Impacts of grassland habitat on yellow starthistle (*Centaurea solstitialis* L.) invasion," *Northwest Science* 68:86–96.

Rodwell, J. S., ed. 2000. *British Plant Communities,* Vol. 5. Cambridge University Press, Cambridge, UK.

Rodwell, J. S., S. Pignatti, L. Mucina, and J. H. J. Schaminée. 1995. "European vegetation survey: Update on progress," *Journal of Vegetation Science* 6:759–762.

Room, P. M., K. L. S. Harley, I. W. Forno, and D. P. A. Sands. 1981. "Successful control of the floating weed *Salvinia,*" *Nature* 294:78–80.

Room, P. M., and M. H. Julien. 1995. "*Salvinia molesta* D. S. Mitchell," in *The Biology of Australian Weeds,* Vol. 1., edited by R. H. Groves, R. C. H. Shepherd, and R. G. Richardson, 217–230. R.G. & F.J. Richardson, Melbourne.

Ross, M. A., and C. A. Lembi. 1985. *Applied Weed Science.* Burgess, Minneapolis.

Rothera, S. L., and A. J. Davy. 1986. "Polyploidy and habitat differentiation in *Deschampsia caespitosa,*" *New Phytologist* 102:449–467.

Rouget, M., and D. M. Richardson. 2002. "Understanding patterns of plant invasion at different spatial scales: Quantifying the roles of environment and propagule pressure" in *In Plant Invasions: Ecological Threats Management and Solutions*, edited by L. E. Child, J. H. Brook, G. Brundu et al., 3–15. Backhuys Publishers, Leiden, The Netherlands.

Rozefelds, A. C. F., and R. Mackenzie. 1999. "The weed invasion in Tasmania in the 1870s: Knowing the past to predict the future," in *12th Australian Weed Conference Papers & Proceedings,* edited by A. C. Bishop, M. Boersma, and C. D. Barnes, 581–583. Tasmanian Weed Society Inc., Hobart, Australia.

Ruiz, G. M., P. Fofonoff, A. N. Hines, and E. D. Grosholz. 1999. "Non-indigenous species as stressors in estuarine and marine communities: Assessing invasion impacts and interactions," *Limnology and Oceanography* 44:950–972.

Rumney, G. R. 1968. *Climatology and the World's Climates.* Macmillan, New York.

Sala, A., and S. D. Smith. 1996. "Water use by *Tamarix ramosissima* and associated phreatophytes in a Mojave Desert floodplain," *Ecological Applications* 6:888–898.

Salisbury, E. 1961. *Weeds & Aliens.* Collins, London.

Sanders, R. W. 1976. "Distributional history and probable ultimate range of *Galium pedemontanum* (Rubiaceae) in North America," *Castanea* 41:73–80.

Sandlund, O. T., P. J. Schei, and A. Viken, eds. 1999. *Invasive Species and Biodiversity Management.* Kluwer Academic Publishers, Dordrecht, The Netherlands.

Schaffers, A. P., and K. V. Sykora. 2000. "Reliability of Ellenberg indicator values for moisture, nitrogen and soil reaction: A comparison with field measurements," *Journal of Vegetation Science* 11:225–244.

Schaminée, J. H. J., and A. H. F. Stortelder. 1996. "Recent developments in phytosociology," *Acta Botanica Neerlandia* 45:443–459.

Schmidt, K. A., and C. J. Whelan. 1999. "Effects of exotic *Lonicera* and *Rhamnus* on songbird nest predation," *Conservation Biology* 13:1502–1506.

Schmidt, W. 1981. "Uber das Konkurrenzverhalten von *Solidago canadensis* und *Urtica dioica,*" *Verhandlugen der Gesellschaft für Ökologie* 11:173–188.

Schoen, D. J., M. O. Johnston, A.-M. L'Hereux, and J. V. Marsolais. 1997. "Evolutionary history of the mating system in *Amsinckia* (Boraginaceae)," *Evolution* 51:1097–1099.

Scott, J. K., and F. D. Panetta. 1993. "Predicting the Australian weed status of southern African plants," *Journal of Biogeography* 20:87–93.

Sharma, B. D., and D. S. Pandey. 1984. *Exotic Flora of Allahabad District.* Flora of India, Series IV. Botanical Survey of India, Delhi.

Shigesada, N., and K. Kawasaki. 1997. *Biological Invasions: Theory and Practice.* Oxford University Press, Oxford, UK.

Shigesada, N., K. Kawasaki, and Y. Takeda. 1995. "Modeling stratified diffusion in biological invasions," *American Naturalist* 146:229–251.

Siggins, H. 1933. "Distribution and rate of fall of conifer seeds," *Journal of Agricultural Research* 47:119–128.

Silvertown, J., and J. B. Wilson. 2000. "Spatial interactions among grassland plant populations," in *The Geometry of Ecological Interactions: Simplifying Spatial Complexity,* edited by U. Dickmann, R. Law, and J. A. J. Metz, 28–47. Cambridge University Press, Cambridge, UK.

Simberloff, D. 1995. "Why do introduced species appear to devastate islands more than mainland areas?" *Pacific Science* 49:87–97.

Simberloff, D., and B. Von Holle. 1999. "Positive interactions of nonindigenous species: Invasional meltdown?" *Biological Invasions* 1:21–32.

Smith, H. A., W. S. Johnston, and S. R. Swanson. 1999. "The implications of variable or constant expansion rates in invasive weed infestations," *Weed Science* 47:62–66.

Smith, M. D., and A. K. Knapp. 1999. "Exotic plant species in a C-4 dominated grassland: Invasibility, disturbance, and community structure," *Oecologia* 120:605–612.

Smith, S. D., T. E. Huxman, S. F. Zitzer, T. N. Charlet, D. C. Housman, J. S. Coleman, L. K. Fenstermaker, J. R. Seeman, and R. S. Nowak. 2000. "Elevated CO_2 increases productivity and invasive species success in an arid ecosystem," *Nature* 408:79–82.

Space, J. 2000. *Pacific Island Ecosystems at Risk,* PIER-CD, Version 2 (CD-ROM). U.S. Forest Service Institute of Pacific Islands Forestry, Honolulu, Hawaii.

Specht, R. L., and A. Specht. 1999. *Australian Plant Communities.* Oxford University Press, Oxford.

Stadler, J., A. Trefflich, S. Kloz, and R. Brandl. 2000. "Exotic plant species invade diversity hot spots: The alien flora of northwestern Kenya," *Ecography* 23:169–176.

Stohlgren, T. J., D. Binkley, G. W. Chong, M. A. Kalkham, L. D. Schell, K. A. Bull, Y. Otsuki, G. Newman, M. Bashkin, and Y. Son. 1999. "Exotic plant species invade hot spots of native plant diversity," *Ecological Monographs* 69:25–46.

Strasberg, D. 1995. "Processus d'invasion par les plantes introduites à La Réunion et dynamique de la végétation sur les coulées volcaniques," *Ecologie* 26:169–180.

Strong, D. R., J. H. Lawton, and T. R. E. Southwood. 1984. *Insects on Plants: Community Pattern.* Harvard University Press, Cambridge, MA.

Sultan, S. E. 1987. "Evolutionary implications of phenotypic plasticity in plants," *Evolutionary Biology* 21:127–178.

Sultan, S. E., and F. A. Bazzaz. 1993. "Phenotypic plasticity in *Polygonum persicaria.* III. The evolution of ecological breadth for nutrient environment," *Evolution* 47:1050–1071.

Sutherst, R. W., G. F. Maywald, T. Yonow, and P. M. Stevens. 1999. *CLIMEX: Predicting the Effects of Climate on Plants and Animals.* CSIRO Publishing, Collingwood, Victoria, Australia.

Swarbrick, J. T., and D. B. Skarratt. 1994. *The Bushweed 2 Database on Environmental Weeds in Australia.* Gatton College, University of Queensland, Gatton.

Symstad, A. J. 2000. "A test of the effects of functional group richness and composition on grassland invasibility," *Ecology* 81:99–109.

Thébaud, C., and M. Debussche. 1991. "Rapid invasion of *Fraxinus ornus* L. along the Herault River system in southern France: The importance of seed dispersal by water," *Journal of Biogeography* 18:7–12.

Thébaud, C., A. C. Finzi, L. Affre, M. Debussche, and J. Escarre. 1996. "Assessing why two introduced *Conyza* species differ in their ability to invade Mediterranean old fields," *Ecology* 77:791–804.

Thébaud, C., and D. Simberloff. 2001. "Are plants really larger in their introduced ranges?" *American Naturalist* 157:231–236.

Thompson, K., J. G. Hodgson, and C. G. Rich. 1995. "Native and alien invasive plants: More of the same?" *Ecography* 18:390–402.

Tilman, D. 1997. "Community invasibility, recruitment limitation, and grassland biodiversity," *Ecology* 78:81–92.

———. 1999. "The ecological consequences of changes in biodiversity: A search for general principles," *Ecology* 80:1455–1474.

Tomanek, G. W., and R. L. Ziegler. 1961. *Ecological Studies on Salt Cedar.* Division of Biological Sciences, Fort Hays Kansas State College, Hays, Kansas.

Toney, J. C., P. M. Rice, and F. Forcella. 1998. "Exotic plant records in the northwest United States, 1950–1996: An ecological assessment," *Northwest Science* 72:198–213.

Trepl, L. 1994. "Zur Rolle interspezifischer Konkurrenz bei der Einbürgerung von Pflanzenarten," *Archives of Nature Conservation and Landscape Research* 33:61–84.

Tucker, K. C., and D. M. Richardson. 1995. "An expert system for screening potentially invasive alien plants in South African fynbos," *Journal of Environmental Management* 44:309–338.

Turnbull, L. A., M. J. Crawley, and M. Rees. 2000. "Are plant populations seed-limited? A review of seed sowing experiments," *Oikos* 88:225–238.

Turner, I. M. 1995. "A catalogue of the vascular plants of Malaya," *The Gardens' Bulletin Singapore* 47:1–346.

UCPE. 1996. *Unit of Comparative Plant Ecology Annual Report.* The University, Sheffield, UK.

van den Bosch, F., R. Hengeveld, and J. A. J. Metz. 1992. "Analysing the velocity of animal range expansion," *Journal of Biogeography* 19:135–150.

van Wilgen, B. W., R. M. Cowling, and C. J. Burgers. 1996. "Valuation of ecosystem services. A case study from South African fynbos ecosystems," *BioScience* 46:184–189.

van Wilgen, B. W., and W. R. Siegfried. 1986. "Seed dispersal properties of three pine species as a determinant of invasive potential," *South African Journal of Botany* 52:546–548.

van Wilgen, B. W., D. M. Richardson, and S. Higgins. 2000a. "Integrated control of alien plants in terrestrial ecosystems," in *Best Management Practices for Preventing and Controlling Invasive Alien Species*, edited by G. Preston, G. Brown, and E. van Wyk, 118–128. Working for Water Programme, Cape Town, South Africa.

van Wilgen, B. W., F. van der Heyden, H. G. Zimmermann, D. Magadlela, and T. Willems. 2000b. "Big returns from small organisms: Developing a strategy for the biological control of invasive alien plants in South Africa," *South African Journal of Science* 96:148–152.

Venette, R. C. 1997. "Assessment of the colonization potential of introduced species during biological invasions," PhD dissertation, University of California, Davis.

Vilá, M., E. Weber, and C. M. D'Antonio. 2000. "Conservation implications of invasion by plant hybridization," *Biological Invasions* 2:207–217.

Villasenor, J. L., and F. J. Espinosa. 1998. *Cataloga de Malezas de Mexico.* Universidad Nacional Autonoma de Mexico, Mexico, D.F.

Villegas, B. 1998. "Implementation status of biological control of weeds," in *Biological Control Program Annual Summary, 1997,* edited by D. M. Woods, 35–38. California Department of Food and Agriculture, Sacramento.

Vitousek, P. M. 1990. "Biological invasions and ecosystem processes: Towards an integration of population biology and ecosystem studies," *Oikos* 57:7–13.

Vitousek, P. M., and L. R. Walker. 1987. "Colonization, succession and resource availability: Ecosystem-level interactions," in *Colonization, Succession and Stability,* edited by A. J. Gray, M. J. Crawley, and P. J. Edwards, 207–223. Blackwell Science, Oxford, UK.

———. 1989. "Biological invasion by *Myrica faya* in Hawaii: Plant demography, nitrogen fixation, ecosystem effects," *Ecological Monographs* 59:247–265.

Wajnberg, E., J. K. Scott, and P. C. Quimby, eds. 2001. *Evaluating Indirect Ecological Effects of Biological Control.* CABI Publishing, Wallingford, UK.

Walter, H. 1968. *Die Vegetation der Erde in öko-physiologischer Betrachtung.* Fischer, Jena, Germany.

Walton, C., and N. Ellis. 1997. *A Manual for Using the Weed Risk Assessment System (WRA) to Assess New Plants.* Plant Quarantine Policy Branch, Australian Quarantine and Inspection Service, Canberra.

Webb, C. J., W. R. Sykes, and P. J. Garnock-Jones. 1988. *Flora of New Zealand,* Vol. IV: *Naturalised Pteridophytes, Gymnosperms, Dicotyledons.* Botany Division, D.S.I.R., Christchurch.

Wells, M. J., R. J. Poynton, A. A. Balsinhas, C. F. Musil, H. Joffe, E. van Hoepen, and S. K. Abbott. 1986. "The history of introduction of invasive alien plants to southern Africa," in *The Ecology and Management of Biological Invasions in Southern Africa,* edited by I. A. W. Macdonald, F. J. Kruger, and A. A. Ferrar, 21–35. Oxford University Press, Cape Town, South Africa.

Westoby, M., S. A. Cunningham, C. R. Fonseca, J. M. Overton, and I. J. Wrignt. 1998. "Phylogeny and variation in light capture area deployed per unit investment in leaves: Design for selecting study species with a view to generalizing," in *Inherent Variation in Plant Growth,* edited by H. Lambers, H. Poorter, and M. M. I. Van Vuuren, 539–566. Backhuys Publishers, Leiden, The Netherlands.

White, T. A., B. D. Campbell, and P. D. Kemp. 1997. "Invasion of temperate grassland by a subtropical annual grass across an experimental matrix of water stress and disturbance," *Journal of Vegetation Science* 8:847–854.

Wilen, C. A., J. S. Holt, N. C. Elstrand, and R. G. Dhaw. 1995. "Genotypic diversity of kikuyugrass (*Penisetum clandestinum*) populations in California," *Weed Science* 43:209–214.

Wilkinson, D. M. 1997. "Plant colonisation: Are wind dispersed seeds really dispersed by birds at large spatial and temporal scales?" *Journal of Biogeography* 24:61–65.

Williams, D. G., R. N. Mack, and R. A. Black. 1995. "Ecophysiology of introduced *Pennisetum setaceum* on Hawaii: The role of phenotypic plasticity," *Ecology* 76:1569–1580.

Williams, P. A. 1996. *A Weed Risk Assessment Model for Screening Plant Imports into New Zealand.* MAF Policy, Wellington.

Williams, P. A., E. Nicol, and M. Newfield. 2001. "Assessing the risk to indigenous biota of new plant taxa new to New Zealand," in *Weed Risk Assessment,* edited by R. H. Groves, F. D. Panetta, and J. G. Virtue, 100–116. CSIRO Publishing, Collingwood, Victoria, Australia.

Williamson, M. 1989. "Mathematical models of invasion," in *Biological Invasions: A Global Perspective,* edited by J. A. Drake, H. A. Mooney, F. di Castri, R. H. Groves, F. J. Kruger, M. Rejmánek, and M. Williamson, 329–350. Wiley, Chichester.

———. 1996. *Biological Invasions.* Chapman & Hall, London.

———. 2001. "Can the impacts of invasive plants be predicted?" in *Plant Invasions: Species Ecology and Ecosystem Management,* edited by G. Brundu, J. Brock, I. Camarada, L. Child, and M. Wade, 11–20. Backhuys Publishers, Leiden, The Netherlands.

Williamson, M., and A. Fitter 1996. "The characters of successful invaders," *Biological Conservation* 78:163–170.

Willis, A. J., J. Memmott, and R. I. Forrester. 2000. "Is there evidence for post-invasion evolution of increased size among invasive plant species?" *Ecology Letters* 3:275–283.

Wilson, J. B., G. L. Rapson, M. T. Sykes, A. J. Watkins, and P. A. Williams. 1992. "Distribution and climatic correlations of some exotic species along roadsides in South Island, New Zealand," *Journal of Biogeography* 19:183–194.

Wilson, S. D., and P. A. Keddy. 1986. "Measuring diffuse competition along an environmental gradient: Results from a shoreline plant community," *American Naturalist* 127:862–869.

Wu, S. H., S. M. Chaw, and M. Rejmánek. 2003. Naturalized Fabaceae (Leguminsae) species in Taiwan: the first approximation. *Botanical Bulletin of Academia Sinica* 44: 59–66.

Winkler, E., and M. Fischer. 1999. "Two fitness measures for clonal plants and the importance of spatial aspects," *Plant Ecology* 141:191–199.

Zavaleta, E. 2000. "Valuing ecosystem services lost to *Tamarix* invasion in the United States," in *The Impact of Global Change on Invasive Species,* edited by H. A. Mooney and R. J. Hobbs, 261–300. Island Press, Washington, DC.

Zavaleta, E., R. J. Hobbs, and H. A. Mooney. 2001. "Viewing invasive species removal in a whole-ecosystem context," *Trends in Ecology and Evolution* 16:454–459.

7

Facilitation and Synergistic Interactions between Introduced Aquatic Species

Anthony Ricciardi

Interactions between introduced species have long been ignored or presumed to be competitive and mutually detrimental. However, in recent years studies have suggested that the establishment and persistence of introduced species are commonly facilitated by other introductions; the best examples are plant invasions that have been aided by animal pollinators and seed dispersers (Simberloff and Von Holle 1999; Richardson et al. 2000). Terrestrial studies also reveal that multiple introductions can produce synergistic impacts (in which the joint effect of two or more invasions is greater than the sum of their individual effects), which might accumulate over time, a phenomenon called invasional meltdown (Richardson, Cowling, and Lamont 1996; Simberloff and Von Holle 1999).

By contrast, aquatic invasions are often treated as isolated events whose impacts are independent of one another. Judging from the scientific literature and invasive species conferences, it seems aquatic ecologists and fishery managers have overlooked the potential importance of facilitation between invaders, possibly because the consequences of such interactions are less conspicuous in aquatic systems. This chapter examines evidence that aquatic invasions may produce synergistic impacts that pose a formidable challenge to conservation and resource management. Herein, *invasion* is defined as the establishment of a reproducing population by an introduced species, and *facilitation* is defined as an interaction in which one species has a positive effect on the persistence or population growth of another species.

Case Studies

Facilitation of Introduced Species by Zebra Mussels in the Great Lakes

The effects of several invaders of the North American Great Lakes have been altered by the introduction of the Eurasian zebra mussel (*Dreissena polymorpha*) in the mid-1980s. *Dreissena* provides other benthic invertebrates with nourishment (in the form of fecal deposits) and shelter (interstitial spaces between clumped mussel shells), causing local enhancement of benthic invertebrate abundance and diversity (Ricciardi, Whoriskey, and Rasmussen 1997; Ricciardi 2003). Among the invertebrates responding positively to zebra mussel colonization is a Eurasian amphipod crustacean (*Echinogammarus ischnus*), which is replacing a North American amphipod in the Great Lakes–St. Lawrence River system (Van Overdijk et al. 2003; A. Ricciardi, unpublished data, 1998). By colonizing silty sediments in western Lake Erie, *Dreissena* facilitated the expansion of *Echinogammarus* into habitats that would otherwise be unsuitable (Bially and MacIsaac 2000). Field experiments demonstrate that the presence of *Dreissena* can cause a 20-fold increase in *Echinogammarus* biomass (Stewart, Miner, and Lowe 1998). This produces an abundant prey resource for the round goby (*Neogobius melanostomus*), a Eurasian fish that feeds primarily on amphipods during its juvenile stage and on zebra mussels during its adult life (Shorygin 1952; Diggins et al. 2002). Shortly after the round goby invaded the St. Clair River, populations of native logperch (*Percina caprodes*) and mottled sculpin (*Cottus bairdi*) declined (Jude, Janssen, and Crawford 1995). Mottled sculpin are nearly extirpated from a harbor in southern Lake Michigan by competition with the round goby for shelter and spawning sites (Janssen and Jude 2001). The goby's ability to consume zebra mussels gives it a competitive advantage over sculpin and logperch. Moreover, its feeding activities reduce benthic invertebrate (non-mussel) biomass, thereby affecting the food resources of other benthic fishes (Kuhns and Berg 1999).

Dreissena is involved in at least two mutualistic interactions in the Great Lakes. The first case, demonstrated experimentally, is the European faucet snail (*Bithynia tentaculata*), whose abundance is several times higher in dreissenid mussel patches (Ricciardi, Whoriskey, and Rasmussen 1997). Mussel shells provide the snail with grazing area and protection from large predators (Ricciardi, Whoriskey, and Rasmussen 1997; Stewart, Miner, and Lowe 1999) and might also release it from competition with larger native snails, such as pleurocerids, which tend to be excluded from dense mussel patches (Ricciardi, Whoriskey, and Rasmussen 1997; Haynes, Stewart, and Cook 1999). In exchange, *Bithynia*'s grazing activities prevent mussels from becoming excessively fouled by sponges and other attached organisms (A. Ricciardi, unpublished data), which can reduce mussel recruitment and survival (Ricciardi et al. 1995; Lauer et al. 1999).

A second mutualistic interaction observed in Lake St. Clair and Saginaw Bay (Lake

Huron) is supported by correlation. In both of these ecosystems, zebra mussel filtration has increased water clarity, thus stimulating prolific growth of exotic weeds such as Eurasian milfoil (*Myriophyllum spicatum*) and curly pondweed (*Potamogeton crispus*) (Skubinna, Coon, and Batterson 1995; MacIsaac 1996). The weeds act as substrates for settling mussel larvae (possibly alleviating intraspecific competition with adult mussels) and also facilitate the dispersal of attached mussels via rafting on fragmented vegetation (Horvath and Lamberti 1997). As a result of the system-wide changes produced by *Dreissena,* the fish community in Lake St. Clair shifted from dominance by commercially important walleye (*Stizostedion vitreum*) to bass (*Micropterus* spp.) and pike (*Esox lucius*) (MacIsaac 1996). Similar cascading impacts have been observed in the Potomac River after invasion by the Asiatic clam (*Corbicula fluminea*), whose intense filtration activity caused increased water clarity and prolific growth of the exotic weeds *Hydrilla verticillata* and *Myriophyllum spicatum.* Habitat provided by weed beds resulted in increased populations of introduced largemouth bass (*Micropterus salmoides*) (Phelps 1994; Serafy, Harrell, and Hurley 1994).

Through their filtration activities in the lower Great Lakes, zebra and quagga mussels probably stimulated the proliferation of botulism bacteria, which thrives in decaying vegetation and possibly accumulations of mussel feces. Outbreaks of Type E botulism have occurred in Lake Erie each summer since 1999 and are responsible for bird and fish die-offs. Carcasses of tens of thousands of waterfowl, particularly fish-eating and scavenging birds such as common loons (*Gavia immer*), red-breasted mergansers (*Mergus serrator*), and ring-billed gulls (*Larus delawarensis*), have been found on the shores of Lake Erie and, more recently, Lake Ontario. There have also been large die-offs of freshwater drum (*Aplodinotus grunniens*), a native fish that feeds on other fishes and mollusks, including dreissenid mussels (Morrison, Lynch, and Dabrowski 1997). The botulin toxin has been found in dreissenid mussels as well as their principal predator, the round goby, which is commonly found in the stomachs of affected birds. Therefore, it is hypothesized that the round goby is transferring toxin from the dreissenid mussels to higher trophic levels (Campbell et al. 2002; McLaughlin 2002; W. Stone, pers. comm., 2002).

Facilitation of Exotic Planktivorous Fishes by the Sea Lamprey in the Great Lakes

Synergistic impacts also resulted from the indirect facilitation of exotic planktivores by the sea lamprey (*Petromyzon marinus*) in the Great Lakes. It is not known when the sea lamprey became established in the basin, but the species was recorded in Lake Ontario as early as the 1830s. It may have entered the lake through the Erie Canal, which opened a passage to the Atlantic Ocean in 1819, and subsequently gained access to the upper Great Lakes by passing Niagara Falls through the Welland Canal (Coon 1999). The sea lamprey invaded Lake Erie by 1921 and had spread to Lake Huron, Lake Michigan, and Lake Superior by 1947 (Lawrie 1970).

Sea lampreys are voracious external parasites of other fish, to which they attach using a rasping suctorial mouth that causes bloody lesions in their victims. Host fish often die from multiple attacks. In each of the upper Great Lakes, invasion by the lamprey was immediately followed by a sharp decline in the resident lake trout (*Salvelinus namaycush*) population, which was already weakened by intense fishing (Lawrie 1970). Consequently, lake trout were extirpated from Lake Michigan and nearly eliminated from lakes Huron and Superior. Stocks of lake whitefish and deepwater ciscoes (*Coregonus* spp.) also collapsed, coincident with the order of establishment of the sea lamprey in each lake (Christie 1974). In combination with overfishing, the sea lamprey contributed to the extinctions of three endemic fishes—the deepwater cisco (*Coregonus johannae*), the shortnose cisco (*Coregonus reighardi*), and the blackfin cisco (*Coregonus nigrippinus*)—as well as the extirpation of populations of shortjaw cisco (*Coregonus zenithicus*) from the Great Lakes (Coon 1999).

The near total extinction of lake trout, the dominant piscivore, from the upper Great Lakes facilitated the explosive proliferation of two invasive planktivores, alewife (*Alosa pseudoharengus*) and rainbow smelt (*Osmerus mordax*). The alewife, like the sea lamprey, probably penetrated the Great Lakes from the Atlantic drainage via shipping canals, whereas rainbow smelt was intentionally stocked in a Michigan lake that became the source for populations in the upper Great Lakes (Smith 1970; Christie 1974). Before lake trout populations collapsed, alewife did not appear in Lake Michigan and were sparse in Lake Huron. In Lake Superior, alewife became common only after lake trout abundance diminished to its lowest level in the early 1960s (Smith 1970). After sea lamprey were controlled by lampricides in the mid-1960s, lake trout began to increase, and alewife subsequently declined (Smith 1970). Alewife populations were further reduced by stocking of nonindigenous strains of lake trout and Pacific salmonids in the 1960s (Stewart, Kitchell, and Crowder 1981). Similarly, rainbow smelt reached peak abundances in Lake Michigan, Lake Huron, and Lake Superior soon after lake trout populations crashed (Christie 1974).

Alewife undergo boom-and-bust cycles, in which die-offs litter beaches and clog water intakes (Kitchell and Crowder 1986). Population explosions of alewife and smelt triggered changes in the composition and abundance of zooplankton in the upper Great Lakes. Size-selective predation by alewife suppressed populations of large-bodied cladocerans and copepods in Lake Michigan (Wells 1970). Moreover, alewife and smelt consumed the pelagic eggs and larvae of several native planktivores, including important forage fishes such as emerald shiner (*Notropis atherinoides*) and lake herring (*Coregonus artedii*), all of which abruptly declined (Christie 1974; Crowder 1980; Stewart, Kitchell, and Crowder 1981). Commercial yields of lake herring crashed in lakes Superior, Michigan, Huron, and Ontario (Christie 1974). Thus, a combination of competition and predation by these exotic planktivores, facilitated by the sea lamprey, reduced native planktivores and overall fishery productivity (Smith 1970; Stewart, Kitchell, and Crowder 1981).

Facilitation of a Japanese Seaweed by Exotic Invertebrates on the Atlantic Coast of North America

Another instructive example of unanticipated synergy involves a recent series of invasions along the Atlantic coast of Nova Scotia, Canada. Species interactions within the rocky subtidal community were studied for several years before and after the establishment of a Japanese green alga (*Codium fragile* ssp. *tomentosoides*) and a European epiphytic bryozoan (*Membranipora mebranacea*) in the early 1990s. Before these invasions, the structure and stability of the community were regulated by sea urchin grazing. In the absence of intense grazing, the subtidal zone normally is dominated by kelp beds (*Laminaria longicruris*), which limit the abundance of understory alga species (Johnson and Mann 1988). Periodic formation of dense feeding aggregations of sea urchins (*Strongylocentrotus droebachiensis*) defoliated kelp beds, leaving open barrens dominated by thin crusts of coralline algae (Scheibling, Hennigar, and Balch 1999).

Recurrent outbreaks of disease caused by the amoeba *Paramoeba invadens* resulted in mass mortalities of sea urchins in the early 1980s, 1993, and 1995 (Scheibling and Hennigar 1997). After each outbreak, the elimination of sea urchins was accompanied by increases in kelp cover and biomass, reaching levels comparable to those of mature kelp beds within a few years (Miller 1985; Scheibling 1986). Thus, a cyclical shift of community states was driven by the population dynamics of sea urchins, which responded to the population dynamics of the amoeba. *Paramoeba invadens* is thought to be a nonindigenous species periodically introduced to the coastal waters of Nova Scotia by the movement of warm water masses, which are also responsible for the high temperatures that stimulate its growth. This hypothesis is favored by experiments demonstrating the inability of *Paramoeba* to survive in laboratory culture below 2°C (the winter minimum temperature along the northwest Atlantic is 0°C to –2°C) (Jellett and Scheibling 1988). Outbreaks occur in late summer during unusually warm years (Scheibling and Hennigar 1997), consistent with laboratory findings that show strong temperature dependence of the amoeba's growth rate (Jellett and Scheibling 1988). Furthermore, *Paramoeba* is waterborne and can be cultured on marine bacteria, indicating that it is not an obligate parasite of urchins (Jones and Scheibling 1985), yet it has been found only in diseased urchin tissues and not in healthy urchins or in the natural environment (Jellett et al. 1989).

In the 1990s, a sequence of events caused an unprecedented transformation of the subtidal ecosystem. In 1995, a mass mortality of sea urchins resulted from a *Paramoeba* outbreak during anomalous warm temperatures and large-scale mixing of ocean currents (Scheibling and Hennigar 1997). This event interrupted the transition to urchin barrens and caused kelp beds to reestablish. Warm water temperatures also stimulated prolific growth of the bryozoan *Membranipora,* which was first recorded in the Gulf of Maine in the late 1980s (Lambert, Levin, and Berman 1992) and has been present on the Atlantic coast of Nova Scotia since at least 1992 (Scheibling, Hennigar, and Balch

1999). The suppression of sea urchins by *Paramoeba* in 1995 created an abundance of substrate for *Membranipora,* which forms calcareous coatings on kelp fronds. Intense bryozoan colonization increases the brittleness of kelp fronds so that they are easily damaged by heavy wave action, which destroys the kelp canopy (Dixon, Schroeter, and Kastendiek 1981; Lambert, Levin, and Berman 1992; Scheibling, Hennigar, and Balch 1999). The loss of the kelp canopy usually is temporary because recruitment from a local spore source can regenerate a kelp bed (Scheibling, Hennigar, and Balch 1999).

After the successive outbreaks of *Paramoeba* and *Membranipora,* the Japanese alga *Codium fragile* became abundant in the shallow subtidal for the first time. *Codium* had already been present on the Nova Scotia coast since the early 1990s (Bird, Dadswell, and Grund 1993) but at low densities because of intense sea urchin grazing in barrens and because of competition in kelp beds (Scheibling 2000). Previously, *Membranipora* outbreaks resulted in temporary kelp defoliation, followed by kelp resurgence and subsequent colonization by the bryozoan. In 1995, *Codium* replaced kelp before it could resurge. *Codium*'s growth was promoted by the reduction in kelp canopy cover, which permitted increased light penetration to the bottom sediments (Scheibling 2000). Because *Membranipora* rarely colonizes *Codium* (R. E. Scheibling, pers. comm., 2000), its population probably will diminish in the absence of suitable macroalgal substrate. Thus, *Membranipora* has acted as a transient facilitator of the *Codium* invasion by releasing it from competition with *Laminaria.* An identical facilitation of *Codium* by *Membranipora* has occurred in the Gulf of Maine (Harris and Tyrrell 2001).

The emergence of *Codium* as the dominant alga has disturbing implications for Atlantic coastal fisheries. Unlike kelps that form a canopy with an understory of algae, *Codium* is a branching alga that forms a short bushy meadow almost impenetrable to large fish and invertebrates such as lobster. There appear to be no herbivores capable of limiting the growth and expansion of these meadows. In mixed-diet feeding experiments involving *Codium,* sea urchins prefer kelp, although they will feed on *Codium* in the absence of kelp (Prince and LeBlanc 1992; Scheibling and Anthony 2001). However, a single diet of *Codium* inhibits urchin gonadal development (Scheibling and Anthony 2001). Therefore, if the shift in algal dominance from kelp to *Codium* is sustained, it should have negative consequences for urchin population growth and thus for an important commercial fishery (Hatcher and Hatcher 1997). Furthermore, if reductions in kelp beds cause lower lobster yields (Wharton and Mann 1981), the *Codium–Membranipora* synergism will affect the Atlantic lobster fishery.

Does Facilitation Increase Rates of Invasion in Aquatic Ecosystems?

The invasional meltdown model predicts that ecosystems subjected to a chronically high frequency of species introductions will become progressively unstable and easier to invade, as each introduced species has the potential to facilitate subsequent invaders

(Simberloff and Von Holle 1999). Indeed, observed rates of invasion are increasing in several aquatic ecosystems, including the Great Lakes, San Francisco Bay, the Baltic Sea, and the Mediterranean Sea (Ribera and Boudouresque 1995; Cohen and Carlton 1998; Leppäkoski and Olenin 2000; Ricciardi 2001). These emergent patterns are thought to reflect temporal variation in dispersal opportunity, sampling bias (search effort), and changes in the resistance of the recipient environment to invasion (Ruiz et al. 2000; MacIsaac, Grigorovich, and Ricciardi 2001).

There are sparse data available to test the effect of facilitation on invasion rates. Facilitation may have lowered environmental resistance to invasion in the Great Lakes, where documented cases of facilitation among introduced species are more common than competition and amensalism (Ricciardi 2001). Moreover, several predator–prey and parasite–host interactions in the Great Lakes are strongly asymmetric in benefiting one invading species at a negligible cost to another, thus acting more as a commensal relationship (Ricciardi 2001). Virtual commensalisms of this kind are common among coevolved relationships.

In the Baltic Sea, nonindigenous species dominate the biomass at all trophic levels and form food web links involving coevolved species and species that share no coevolutionary history (Leppäkoski and Olenin 2000). In the Vistula River delta, the North American crab (*Rhithropanopeus harrisi*) feeds primarily (as juveniles) on *Cordylophora* and (as adults) on *Dreissena polymorpha* (Leppäkoski 1984). *Cordylophora* uses *Dreissena* shells as attachment substrate, and *Dreissena* larvae are its principal food source. Patches of *Dreissena* and of the North American barnacle (*Balanus improvisus*) provide habitat for several Ponto-Caspian amphipod species (Köhn and Waterstraat 1990; Olenin and Leppäkoski 1999). *Balanus* itself uses zebra mussel shells as attachment surfaces (Olenin and Leppäkoski 1999) and probably benefits from direct exposure to mussel filtration currents as it does in its commensal relationship with *Mytilus* (Laihonen and Furman 1986). However, none of these species needed any previous invasion to become established.

In fact, overall there are only a few documented cases in which the presence of an aquatic invader led to the establishment of another invader. There is little evidence linking increased invasion rates to facilitation in aquatic ecosystems, which supports the view that aquatic invasions are governed more by dispersal opportunity and physical habitat conditions than by the composition of the recipient community (Moyle and Light 1996; MacIsaac, Grigorovich, and Ricciardi 2001). However, it is clear that facilitation can enhance the abundance, persistence, and local spread of aquatic invaders. Additional examples include *Alepes djeddaba,* a carangid fish native to the Indian Ocean, which increased in abundance in the Mediterranean Sea after the appearance of swarms of the scyphomedusan jellyfish *Rhopilema nomadica,* a Red Sea migrant introduced in the 1970s; juvenile *Alepes* shelter among the jellyfish tentacles (Galil 2000). The spread and population growth of the Chinese macrofouling mussel *Limnoperna fortunei* in the Parana–Rio de la Plata system were enhanced by the previous invasion of the Asiatic clam (*Corbicula fluminea*), which provided hard substratum for colonization by *Limnoperna* in vast stretches of the silty river bottom (Darrigran and de Drago 2000; D. Boltovskoy, pers. comm., 2000). The variety of examples in the scientific literature demonstrates the ubiquity of this phenomenon.

Are Coadapted Invaders More Likely to Cause Invasional Meltdown?

Invasional meltdown is hypothesized to occur through one of two processes: frequent disturbance through species introductions progressively lowers community resistance to invasion, and increased introductions lead to a higher frequency of potential facilitations and synergies. Evidence for the first process is in computer simulations showing that frequent and simultaneous species introductions increase invasion success (Lockwood et al. 1997). The second process is favored by an influx of coadapted propagules. Certain coadapted species combinations ("invasion cartels") might act as mutual attractors that promote the assembly of foreign food webs in new ecosystems. Highly active invasion corridors may introduce numerous species from the same endemic region or from a region each has invaded previously. Consequently, invasion corridors tend to reunite groups of coadapted species, either in simultaneous introductions (e.g., a host arriving with its parasites) or in successive introductions, thereby assembling contiguous links of a nonindigenous food web.

Over the past two decades, the Great Lakes have been invaded predominantly by species native to the Black and Caspian seas, that is, Ponto-Caspian species (Ricciardi and MacIsaac 2000; Ricciardi 2001). This influx is attributable largely to shipping traffic linking the Great Lakes to western European ports, which have become increasingly invaded by Ponto-Caspian species (MacIsaac, Grigorovich, and Ricciardi 2001). Even though they still make up only a minor proportion (about 11 percent) of all introduced biota, Ponto-Caspian species dominate facilitations in the Great Lakes, and their food webs are being reassembled in the region (Ricciardi 2001). For example, the introductions of dreissenid mussels, amphipod *Echinogammarus ischnus,* and the round goby (*Neogobius melanostomus*) reassembled a tripartite cartel of Ponto-Caspian species in the Great Lakes. In fact, the number of coadapted foreign (predominantly Eurasian) food web links in the Great Lakes has increased exponentially over the past century (Figure 7.1). As discussed previously, the *Dreissena*–round goby interaction is apparently responsible for outbreaks of avian botulism in Lake Erie and Lake Ontario. In western Europe, sequential invasions by Ponto-Caspian species completed the parasitic life cycle of the trematode *Bucephalus polymorphus.* The introductions of the zebra mussel (*Dreissena polymorpha,* the trematode's first intermediate host) and the pike perch (*Stizostedion lucioperca,* its definitive host) allowed *B. polymorphus* to spread into inland waters and cause high mortality in local populations of cyprinid fishes, which act as secondary intermediate hosts (Combes and Le Brun 1990).

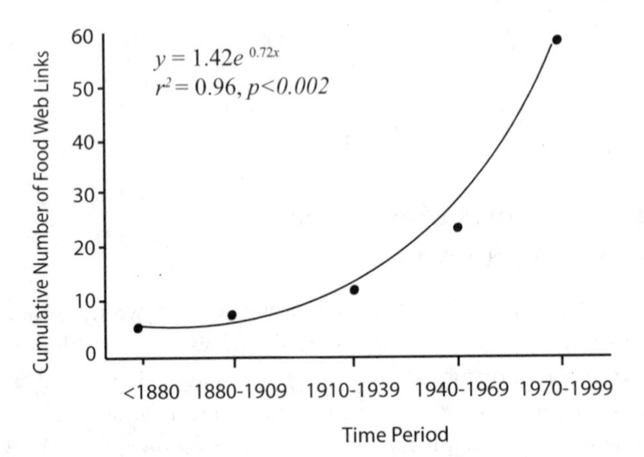

Figure 7.1. Increasing rate of establishment of coadapted foreign links in Great Lakes food webs. Each data point represents the cumulative number of links established at each 30-year interval. Model fitted by least-squares regression. Data are from Ricciardi (2001) and Mills et al. (1993).

If coadaptation reduces the intensity of predation and parasitism (Levin et al. 1982), then positive interactions probably dominate invasion cartels, and successive introductions of coadapted species might produce a higher rate of invasion than would introductions of unacquainted species. This is an alternative to the enemy release hypothesis, which relates the success of an invader to the absence of its natural predators and parasites in the invaded region (Wolfe 2002). Introduced predators and parasites may increase the invasion success of their natural hosts by differentially affecting the host's competitors: naive resident hosts that have not had selection pressure to adapt to their new enemies. While its natural host is present, the parasite is not limited by rapid declines of resident hosts. In this way, a fungal parasite (*Aphanomyces astaci*) transported with shipments of the American crayfish (*Pacifastacus leniusculus*) caused a large-scale plague that wiped out many native crayfish populations in Europe in the 1930s, including commercially important stocks (Reynolds 1988). Although the *Aphanomyces–Pacifastacus* relationship normally would be considered an exploitive interaction, in this case it is virtually mutualistic because the pathogen eliminated the crayfish's competitors. A similar, terrestrial example is the ongoing replacement of the red squirrel (*Sciurus vulgaris*) in the United Kingdom by the North American gray squirrel (*S. carolinensis*), which is promoted by the vulnerability of the native species to a viral disease introduced by the invading species (Tompkins, White, and Boots 2003). In Australia, introduced parasites are thought to have likewise aided the success of their introduced host fishes by reducing native fish populations (Dove 1998).

Recently, Fukami, Simberloff, and Drake (n.d.) found theoretical and experimental evidence that an invading prey species often serves as an additional resource to an invading coadapted predator, and the predator releases its prey from competition with native species. This finding is supported by several terrestrial case studies, including the introduction of the brown tree snake (*Boiga irregularis*) and its coevolved prey to Guam, which caused the extinction of several endemic species of birds, bats, and reptiles through hyperpredation (Fritts and Rodda 1998). Similarly, native fishes have been drastically reduced by the North American predator northern pike (*Esox lucius*), introduced into lakes in Spain, where its populations are sustained by another prey resource, the American crayfish (*Procambarus clarkii*) (Elvira, Nicola, and Almodovar 1996). In western North America, the invasion of bullfrogs (*Rana catesbeiana*) is facilitated by bluegill sunfish (*Lepomis macrochirus*), which increase the survival of their tadpoles by reducing densities of predatory macroinvertebrates (Adams, Pearl, and Bury 2003). Because introduced bullfrogs exacerbate declines in native frog populations (Kats and Ferrer 2003), the sunfish–bullfrog commensalism contributes to the impoverishment of amphibian communities. A coadapted predator–prey or parasite–host cartel thus can function mutualistically to exert a synergistic impact on the recipient community.

Implications for Policy and Management

These examples demonstrate that introduced species interact in unanticipated ways to alter aquatic ecosystems, with potentially serious consequences for biodiversity and fishery management. The aforementioned case studies corroborate terrestrial studies showing that complex combinations of direct and indirect species interactions can structure entire communities and affect ecosystem function (Richardson, Cowling, and Lamont 1996; Richardson et al. 2000; Callaway and Walker 1997; Levine 1999). Facilitation can magnify the ecological impact of an introduced species across multiple trophic levels. If trophic cascades are more common in aquatic ecosystems than in terrestrial ecosystems (Strong 1992), aquatic ecosystems might be particularly susceptible to synergistic impacts of introduced species. Unanticipated synergies reduce our capacity to predict and manage invasion threats. Therefore, we need new theoretical perspectives on the community ecology of invaders, particularly for inland and coastal aquatic ecosystems, which are being disturbed by an increasing number of invasions worldwide (Ruiz et al. 1997; Cohen and Carlton 1998; Ricciardi 2001). In particular, researchers should attempt to incorporate direct and indirect facilitation into impact models.

Risk assessments of aquatic species introductions must consider the presence of potential facilitators in the recipient community and the potential impact of simultaneous or rapidly successive invasions by coadapted species. Databases that are intended to provide managers with information to aid monitoring, risk assessment, and control of invaders should list species known to facilitate the invader's establishment and impact as well as commensal and parasitic organisms known to benefit from the invader's presence. The identification and control of keystone facilitators may help reduce further invasions. Examples of concern include components of parasite life cycles, keystone predators, and ecological engineers (*sensu* Jones, Lawton, and Shachak 1994) likely to produce major ecosystem shifts and thus change the rules of existence for other species.

If the invasional meltdown model is valid, an increased frequency of species introduction will lead to a rapid accumulation of invaders and synergistic impacts (Simberloff and Von Holle 1999), which will cause the ecosystem to become increasingly unstable and difficult to manage. This justifies efforts to reduce inoculation pressure on ecosystems and refutes any argument that strict controls on ballast water discharge (a major vector for aquatic invasions worldwide) are unwarranted if future invasions are inevitable through more subtle vectors. Even a partial reduction of inoculation pressure might slow the buildup of feedback cycles that can destabilize an ecosystem, giving resource managers more time to adjust to changing conditions or to develop better prevention strategies.

Finally, greater effort should be made to control invasion corridors that are linked to centers of endemism because these may deliver large numbers of coadapted organisms and thus assemble synergistic invasion cartels. Although invasion cartels are expected to contribute (at least initially) to an invasional meltdown, a new equilibrium community ultimately could be reached if the dominant invasion corridors do not change; any preexisting equilibrium could be disrupted by a new suite of coadapted species. Therefore, invasion cartels might undermine efforts to restore natural communities, not only by replacing native species but also by shifting the community toward an alternative stable state (Lockwood 1997).

Acknowledgments

I thank Jim Carlton for commenting on the manuscript. This chapter also benefited from stimulating discussions with Bob Scheibling at Dalhousie University.

References

Adams, M. J., C. A. Pearl, and R. B. Bury. 2003. "Indirect facilitation of an anuran invasion by non-native fishes," *Ecology Letters* 6:343–351.

Bially, A., and H. J. MacIsaac. 2000. "Fouling mussels (*Dreissena*) colonize soft sediments in Lake Erie and facilitate benthic invertebrates," *Freshwater Biology* 43:85–98.

Bird, C. J., M. J. Dadswell, and D. W. Grund. 1993. "First record of the potential nuisance algae *Codium fragile* ssp. *tomentosoides* (Chlorophyta, Caulerpales) in Atlantic Canada," *Proceedings of the Nova Scotia Institute of Science* 40:11–17.

Callaway, R. M., and L. R. Walker. 1997. "Competition and facilitation: A synthetic approach to interactions in plant communities," *Ecology* 78:1958–1965.

Campbell, M., L. Gauriloff, H. Domske, and E. Obert. 2002. "Environmental correlates with outbreaks of Type-E avian botulism in the Great Lakes," in *Botulism in Lake Erie Workshop Proceedings,* February 2002, 6–16. New York–Pennsylvania–Ohio Sea Grant, Buffalo, NY.

Christie, W. J. 1974. "Changes in the fish species composition of the Great Lakes," *Journal of the Fisheries Research Board of Canada* 31:827–854.

Cohen, A. N., and J. T. Carlton. 1998. "Accelerating invasion rate in a highly invaded estuary," *Science* 279:555–558.

Combes, C., and N. Le Brun 1990. "Invasions by parasites in continental Europe," in *Biological Invasions in Europe and the Mediterranean Basin,* edited by F. Di Castri, A. J. Hansen, and M. Debussche, 285–296. Kluwer Academic Publishers, Dordrecht, The Netherlands.

Coon, T. G. 1999. "Ichthyofauna of the Great Lakes Basin," in *Great Lakes Fisheries Policy and Management,* edited by W. W. Taylor and C. P. Ferreri, 55–71. Michigan State University Press, East Lansing.

Crowder, L. B. 1980. "Alewife, rainbow smelt and native fishes in Lake Michigan: Competition or predation?" *Environmental Biology of Fishes* 5:225–233.

Darrigran, G., and I. E. de Drago. 2000. "Invasion of the exotic freshwater mussel *Limnoperna fortunei* (Dunker, 1857) (Bivalvia: Mytilidae) in South America," *Nautilus* 114:69–73.

Diggins, T. P., J. Kaur, R. K. Chakraborti, and J. V. DePinto. 2002. "Diet choice by the exotic round goby (*Neogobius melanostomus*) as influenced by prey motility and environmental complexity," *Journal of Great Lakes Research* 28:411–420.

Dixon, J., S. C. Schroeter, and J. Kastendiek. 1981. "Effects of the encrusting bryozoan, *Membranipora membranacea,* on the loss of blades and fronds by the giant kelp, *Macrocystis pyrifera* (Laminariales)," *Journal of Phycology* 133:125–134.

Dove, A. D. M. 1998. "A silent tragedy: Parasites and the exotic fishes of Australia," *Proceedings of the Royal Society of Queensland* 107:109–113.

Elvira, B., G. G. Nicola, and A. Almodovar. 1996. "Pike and red swamp crayfish: A new case on predator–prey relationship between aliens in central Spain," *Journal of Fish Biology* 48:437–446.

Fritts, T. H., and G. H. Rodda. 1998. "The role of introduced species in the degradation of island ecosystems: A case history of Guam," *Annual Review of Ecology and Systematics* 29:113–140.

Fukami, T., D. Simberloff, and J. A. Drake. n.d. "Predator–prey invasional meltdown." Unpublished manuscript.

Galil, B. S. 2000. "A sea under siege: alien species in the Mediterranean," *Biological Invasions* 2:177–186.

Harris, L. G., and M. C. Tyrrell. 2001. "Changing community states in the Gulf of Maine: Synergism between invaders, overfishing and climate change," *Biological Invasions* 3:9–21.

Hatcher, B. G., and A. I. Hatcher. 1997. "Research directions and management options for sea urchin culture in Nova Scotia," *Bulletin of the Aquaculture Association of Canada* 97:62–65.

Haynes, J. M., T. W. Stewart, and G. E. Cook. 1999. "Benthic macroinvertebrate communities in southwestern Lake Ontario following invasion of *Dreissena:* Continuing change," *Journal of Great Lakes Research* 25:828–838.

Horvath, T. G., and G. A. Lamberti. 1997. "Drifting macrophytes as a mechanism for zebra mussel (*Dreissena polymorpha*) invasion of lake-outlet streams," *American Midland Naturalist* 138:29–36.

Janssen, J., and D. J. Jude. 2001. "Recruitment failure of mottled sculpin *Cottus bairdi* in Calumet Harbor, southern Lake Michigan, induced by the newly introduced round goby *Neogobius melanostomus,*" *Journal of Great Lakes Research* 27:319–328.

Jellett, J. F., J. A. Novitsky, J. A. Cantley, and R. E. Scheibling. 1989. "Non-occurrence of *Paramoeba invadens* in the water column and sediments off Halifax, Nova Scotia," *Marine Ecology Progress Series* 56:205–209.

Jellett, J. F., and R. E. Scheibling. 1988. "Effect of temperature and food concentration on the growth of *Paramoeba invadens* (Amoebida: Paramoebidae) in monoxenic culture," *Applied Environmental Microbiology* 54:1848–1854.

Johnson, C. R., and K. H. Mann. 1988. "Diversity, patterns of adaptation, and stability of Nova Scotian kelp beds," *Ecology Monographs* 58:129–154.

Jones, C. G., J. H. Lawton, and M. Shachak. 1994. "Organisms as ecosystem engineers," *Oikos* 69:373–386.

Jones, G. M., and R. E. Scheibling. 1985. "*Paramoeba* sp. (Amoebida, Paramoebidae) as the possible causative agent of sea urchin mass mortality off Nova Scotia," *Journal of Parasitology* 71:559–565.

Jude, D. J., J. Janssen, and G. Crawford. 1995. "Ecology, distribution, and impact of the newly introduced round (and) tubenose gobies on the biota of the St. Clair and Detroit Rivers," in *The Lake Huron Ecosystem: Ecology, Fisheries and Management,* edited by M. Munawar, T. Edsall, and J. Leach, 447–460. SPB Academic Publishing, Amsterdam.

Kats, L. B., and R. P. Ferrer. 2003. "Alien predators and amphibian declines: Review of two decades of science and the transition to conservation," *Diversity and Distributions* 9:99–110.

Kitchell, J. F., and L. B. Crowder. 1986. "Predator–prey interactions in Lake Michigan: Model predictions and recent dynamics," *Environmental Biology of Fishes* 16:205–211.

Köhn, J., and A. Waterstraat. 1990. "The amphipod fauna of Lake Kummerow (Mecklenburg, German Democratic Republic) with reference to *Echinogammarus ischnus* Stebbing, 1899)," *Crustaceana* 58:74–82.

Kuhns, L. A., and M. B. Berg. 1999. "Benthic invertebrate community responses to round goby (*Neogobius melanostomus*) and zebra mussel (*Dreissena polymorpha*) invasion in southern Lake Michigan," *Journal of Great Lakes Research* 25:910–917.

Laihonen, P., and E. R. Furman. 1986. "The site of settlement indicates commensalism between bluemussel and its epibiont," *Oecologia* 71:38–40.

Lambert, W. J., P. S. Levin, and J. Berman. 1992. "Changes in the structure of a New England (USA) kelp bed: The effects of an introduced species?" *Marine Ecology Progress Series* 88:303–307.

Lauer, T. E., D. K. Barnes, A. Ricciardi, and A. Spacie. 1999. "Evidence of recruitment inhibition of zebra mussels (*Dreissena polymorpha*) by a freshwater bryozoan (*Lophopodella carteri*)," *Journal of the North American Benthological Society* 18: 406–413.

Lawrie, A. H. 1970. "The sea lamprey in the Great Lakes," *Transactions of the American Fisheries Society* 99:766–775.

Leppäkoski, E. 1984. "Introduced species in the Baltic Sea and its coastal ecosystems," *Ophelia* (Suppl. 3):123–135.

Leppäkoski, E., and S. Olenin. 2000. "Non-native species and rates of spread: Lessons from the brackish Baltic Sea," *Biological Invasions* 2:151–163.

Levin, B. R., A. C. Allison, H. J. Bremmermann, B. C. Clarke, R. Frentzel-Beyme, W. D. Hamilton, S. A. Levin, R. M. May, and H. R. Thieme. 1982. "Evolution of parasites and hosts," in *Population Biology of Infectious Diseases,* edited by R. M. Anderson and R. M. May, 213–243. Springer-Verlag, New York.

Levine, J. M. 1999. "Indirect facilitation: Evidence and predictions from a riparian community," *Ecology* 80:1762–1769.

Lockwood, J. L. 1997. "An alternative to succession: Assembly rules offer guides to restoration efforts," *Restoration and Management Notes* 15(1):45–50.

Lockwood, J. L., R. D. Powell, M. P. Nott, and S. L. Pimm. 1997. "Assembling ecological communities in time and space," *Oikos* 8:549–553.

MacIsaac, H. J. 1996. "Potential abiotic and biotic impacts of zebra mussels on the inland waters of North America," *American Zoologist* 36:287–299.

MacIsaac, H. J., I. A. Grigorovich, and A. Ricciardi. 2001. "Reassessment of species invasions concepts: The Great Lakes basin as a model," *Biological Invasions* 3:405–416.

McLaughlin, G. S. 2002. "Conceptual model of Type-E botulism in Lake Erie," in *Botulism in Lake Erie Workshop Proceedings,* February 2002, 48–50. New York–Pennsylvania–Ohio Sea Grant, Buffalo, NY.

Miller, R. J. 1985. "Succession in sea urchin and seaweed abundance in Nova Scotia, Canada," *Marine Biology* 84:275–286.

Mills, E. L., J. H. Leach, J. T. Carlton, and C. L. Seacor. 1993. "Exotic species in the Great Lakes: A history of biotic crises and anthropogenic introductions." *Journal of Great Lakes Restoration,* 19(1): 1–54.

Morrison, T. W., W. E. Lynch Jr., and K. Dabrowski. 1997. "Predation on zebra mussels by freshwater drum," *Journal of Great Lakes Research* 23:177–189.

Moyle, P. B., and T. Light. 1996. "Fish invasions in California: Do abiotic factors determine success?" *Ecology* 77:1666–1670.

Olenin, S., and E. Leppäkoski. 1999. "Non-native animals in the Baltic Sea: Alteration of benthic habitats in coastal inlets and lagoons," *Hydrobiologia* 393:233–243.

Phelps, H. L. 1994. "The Asiatic clam (*Corbicula fluminea*) invasion and system-level ecological change in the Potomac River estuary near Washington, D.C.," *Estuaries* 17:614–621.

Prince, J. S., and W. G. LeBlanc. 1992. "Comparative feeding preference of *Strongylocentrotus droebachiensis* (Echinoidea) for the invasive seaweed *Codium fragile* ssp. *tomentosoides* (Chlorophyceae) and four other seaweeds," *Marine Biology* 113:159–163.

Reynolds, J. D. 1988. "Crayfish extinctions and crayfish plague in central Ireland," *Biological Conservation* 45:279–285.

Ribera, M. A., and C. F. Boudouresque. 1995. "Introduced marine plants, with special reference to macroalgae: Mechanisms and impact," *Progress in Phycological Research* 11:187–267.

Ricciardi, A. 2001. "Facilitative interactions among aquatic invaders: Is an 'invasional meltdown' occurring in the Great Lakes?" *Canadian Journal of Fisheries and Aquatic Sciences* 58:2513–2525.

———. 2003. "Predicting the impacts of an introduced species from its invasion history: An empirical approach applied to zebra mussel invasions," *Freshwater Biology* 48:972–981.

Ricciardi, A., and H. J. MacIsaac. 2000. "Recent mass invasion of the North American Great Lakes by Ponto-Caspian species," *Trends in Ecology and Evolution* 15:62–65.

Ricciardi, A., F. L. Snyder, D. O. Kelch, and H. M. Reiswig. 1995. "Lethal and sublethal effects of sponge overgrowth on introduced dreissenid mussels in the Great Lakes–St. Lawrence River system," *Canadian Journal of Fisheries and Aquatic Sciences* 52:2695–2703.

Ricciardi, A., F. G. Whoriskey, and J. B. Rasmussen. 1997. "The role of the zebra mussel (*Dreissena polymorpha*) in structuring macroinvertebrate communities on hard substrata," *Canadian Journal of Fisheries and Aquatic Sciences* 54:2596–2608.

Richardson, D. M., N. Allsopp, C. D'Antonio, S. J. Milton, and M. Rejmánek. 2000. "Plant invasions: The role of mutualisms," *Biological Reviews* 75:65–93.

Richardson, D. M., R. M. Cowling, and B. B. Lamont. 1996. "Non-linearities, synergisms and plant extinctions in South African fynbos and Australian kwongan," *Biodiversity and Conservation* 5:1035–1046.

Ruiz, G. M., J. T. Carlton, E. D. Grosholz, and A. H. Hines. 1997. "Global invasions of marine and estuarine habitats by non-indigenous species: Mechanisms, extent, and consequences," *American Zoologist* 37:621–632.

Ruiz, G. M., P. W. Fofonoff, J. T. Carlton, M. J. Wonham, and A. H. Hines. 2000. "Invasion of coastal marine communities in North America: Apparent patterns, processes, and biases," *Annual Reviews in Ecology and Systematics* 31:481–531.

Scheibling, R. E. 1986. "Increased macroalgal abundance following mass mortalities of sea urchins (*Strongylocentrotus droebachiensis*) along the Atlantic coast of Nova Scotia, Canada," *Oecologia* 68:186–198.

———. 2000. Species invasions and community change threaten the sea urchin fishery in Nova Scotia." Workshop on the Coordination of Sea Urchin Research in Atlantic Canada, June 1–2, Moncton, New Brunswick. Online: http://crdpm.cus.ca/oursin/PDF/SCHEIB.PDF.

Scheibling, R. E., and S. X. Anthony. 2001. "Feeding, growth and reproduction of sea urchins (*Strongylocentrotus droebachiensis*) on single and mixed diets of kelp (*Laminaria* spp.) and the invasive alga *Codium fragile* ssp. *tomentosoides*," *Marine Biology* 139:139–146.

Scheibling, R. E., and A. W. Hennigar. 1997. "Recurrent outbreaks of disease in sea urchins *Strongylocentrotus droebachiensis* in Nova Scotia: Evidence for a link with large-scale meteorologic and oceanographic events," *Marine Ecology Progress Series* 152:155–165.

Scheibling, R. E., A. W. Hennigar, and T. Balch. 1999. "Destructive grazing, epiphytism, and disease: The dynamics of sea urchin–kelp interactions in Nova Scotia," *Canadian Journal of Fisheries and Aquatic Sciences* 56:2300–2314.

Serafy, J. E., R. M. Harrell, and L. M. Hurley. 1994. "Mechanical removal of *Hydrilla* in the Potomac River, Maryland: Local impacts on vegetation and associated fishes," *Journal of Freshwater Ecology* 9:135–143.

Shorygin, A. A. 1952. *The feeding and food relationships of fishes in the Caspian Sea (Acipenseridae, Cyprinidae, Gobiidae, Percidae, and predatory herrings).* (In Russian). Pristchepromisdat Publishers, Moscow.

Simberloff, D., and B. Von Holle. 1999. "Positive interactions of nonindigenous species: Invasional meltdown?" *Biological Invasions* 1:21–32.

Skubinna, J. P., T. G. Coon, and T. R. Batterson. 1995. "Increased abundance and depth of submersed macrophytes in response to decreased turbidity in Saginaw Bay, Lake Huron," *Journal of Great Lakes Research* 21:476–488.

Smith, S. H. 1970. "Species interactions of the alewife in the Great Lakes," *American Fisheries Society* 1970:754–765.

Stewart, D. J., J. F. Kitchell, and L. B. Crowder. 1981. "Forage fishes and their salmonid predators in Lake Michigan," *Transactions of the American Fisheries Society* 110:751–763.

Stewart, T. W., J. G. Miner, and R. L. Lowe. 1998. "Macroinvertebrate communities on hard substrates in western Lake Erie: Structuring effects of *Dreissena*," *Journal of Great Lakes Research* 24:868–879.

———. 1999. "*Dreissena*-shell habitat and anti-predator behavior: Combined effects on survivorship of snails co-occurring with molluscivorous fish," *Journal of the North American Benthological Society* 18:488–498.

Strong, D. R. 1992. "Are trophic cascades all wet? Differentiation and donor-control in speciose ecosystems," *Ecology* 73:747–754.

Tompkins, D. M., A. White, and M. Boots. 2003. "Ecological replacement of native red squirrels by invasive greys driven by disease," *Ecology Letters* 6:189–196.

Van Overdijk, C. D. A., I. A. Grigorovich, T. Mabee, W. J. Ray, J. J. H. Ciborowski, and H. J. MacIsaac. 2003. "Microhabitat selection by the invasive amphipod *Echinogammarus ischnus* and native *Gammarus fasciatus* in laboratory experiments and in Lake Erie," *Freshwater Biology* 48:567–578.

Wells, L. 1970. "Effects of alewife predation on zooplankton populations in Lake Michigan," *Limnology and Oceanography* 15:556–565.

Wharton, W. G., and K. H. Mann. 1981. "Relationship between destructive grazing by the sea urchin *Strongylocentrotus droebachiensis,* and the abundance of American lobster, *Homarus americanus,* on the Atlantic coast of Nova Scotia," *Canadian Journal of Fisheries and Aquatic Sciences* 38:1339–1349.

Wolfe, L. M. 2002. "Why alien invaders succeed: Support for the escape-from-enemy hypothesis," *American Naturalist* 160:705–711.

8
Assessing Biotic Invasions in Time and Space: The Second Imperative

Richard N. Mack

Assessment, that is, the broad sense determination of the size, importance, or value of a phenomenon, commodity, or population, is a fundamental goal anytime there is need to predict the potential for change over time. Such appraisals occur at almost any level of detail, from so-called rapid assessments to protracted, highly quantitative assessments, as illustrated with the mandatory decennial census of the United States (Forstall 1996). However much assessments or appraisals may differ in scope and detail, they all are conducted to provide answers to basic, recurring questions on changing status, usually in order to predict the future condition. Thus, a nation counts and characterizes its inhabitants to determine their current composition by age, sex, health, occupation, and location. Such information forms a guide, albeit an imperfect one, to the future values in these categories (Champion et al. 1996; Day 1996;).

Assessments over time and space are essential in the study of biotic invasions (*sensu* Mack et al. 2000) because these phenomena involve populations that change rapidly in their spatial distribution, abundance, and demographic composition. Even though all populations change numerically and in their geographic extent, biotic invasions are excellent examples of such change. Thanks to human transport, immigrants may traverse a previously insurmountable physical barrier in days or even hours. Once in a new range, the immigrants and their descendants may disperse widely, swiftly establish new colonies, and undergo rapid demographic and genetic changes as well (Williamson 1996; Mack et al. 2000). Even if the immigrant population persists and occupies the limits of its new range, the range's boundaries continue to be dynamic. Occasionally, the ranges may even contract as the once prominent invader succumbs to forces in the new range (Mack 2000 and references therein).

Given the potential influence by an invader in its new range, there has long been strong incentive to gauge the course of invasions as potential guides to their control if not their eradication (Sindal and Michael 1992; Lonsdale 1993; Pysek and Prach 1995). In a sense, these assessments usually are initiated comparatively late in the invasion; emigration from the native or another donor range and the species' establishment in the new range have already occurred. Control, and even eradication, may have been possible during those earlier stages; control is much more difficult once the species has proliferated and spread (Mack et al. 2000 and references therein). Consequently, I refer here to assessment of biotic invasions as the second imperative, that is, an essential goal after opportunities to achieve the first (eradication of the immigrants upon entry) or even the zeroth (prevention of the initiation of the emigration altogether) goals have been missed. Delineation of assessment to the second imperative is not meant to diminish its importance, however; it usually is the earliest action taken in curbing an invasion (Auld, Hosking, and McFadyen 1983). It is nonetheless essential in providing the basis for rationally marshaling the most effective control measures, implementing these measures, and continuously evaluating the effects of these control measures, the third, fourth, and fifth imperatives, respectively.

As viewed here, assessment has two major components: detection of the organisms in the new range and the tabulation and presentation of this potentially enormous amount of survey information in forms that can be readily understood. Detection tools vary, but most involve visual recognition of organisms in the field. Yet detecting even a representative number of individuals in the field may be daunting, especially if they are small, are cryptic, or, as is usually the case, occur in mixtures with native species. Next, spatial components must be determined, such as the areal extent of the invader, its apparent paths of dispersal, and its current disposition in relation to additional vulnerable habitat (Mack 2000). Thus, effective representation of detection information also forms an important part of assessment. Assessment for almost any invasive species, other than those that produce diseases in humans or their crops and domesticated animals, is sketchy. We certainly lack comprehensive information on the current distribution of all but a few plant and vertebrate invaders (Mooney 1999). Emergence of a global assessment of invaders—even for arguably the best-known group, plant invaders—remains firmly in the future.

Given these limitations in our current knowledge, I have several goals in this chapter. By necessity, this evaluation is neither exhaustive nor comprehensive. First, I evaluate the detection tools for invasive species in three large taxonomic groups: vascular plants and the invasive insects and fungal parasites that attack plants. I also identify some tools or procedures that, although little used, offer the potential for success. Second, I provide examples where the clear intent was to provide visual conclusions on the status of the invasion in space. To establish from the outset a high standard by which modern assessments of invasions may be evaluated graphically and also to emphasize the commonality of questions that prompt assessments of invasions, I outline briefly a famous invasion and the remarkably lucid assessment that it sparked.

An Early Standard for Assessing Invasions: Minard's Map of Napoleon's Invasion of Russia, 1812–1813

Napoleon's disastrous invasion and cataclysmic retreat from Russia have been active subjects for historians for the last 180 years; the Library of Congress alone houses 254 volumes in which this invasion is a main topic. At some point, a visual distillation of the events is needed to understand the dynamics of this or any other invasion. A map that merely shows the general path of invasion is inadequate because there may be multiple paths of entry, and invaders, human or otherwise, neither arrive, disperse, nor perish at regular intervals. The most effective maps of invasions portray the fate of the invaders in stark demographic and environmental terms so that one can instantly understand the size and distribution of the invaders and the essential parameters that determined the outcome.

Minard's 1869 "Carte Figurative" on the campaign of the Grand Army incorporates all these features (Figure 8.1). Because the invaders were an army, detection was not the hurdle that arises in documenting other biotic invaders; the raw data consist of the detailed unit records of the army on the march. These tabular data become immediately graphic and meaningful, thanks to Minard's distillation and portrayal. After crossing the Niemen River at Kowno on the Polish–Russian border with more than 422,000 troops, Napoleon's legions steadily shrank through a combination of pitched battles, skirmishes, and desertions; the time and place of these events are shown clearly on Minard's

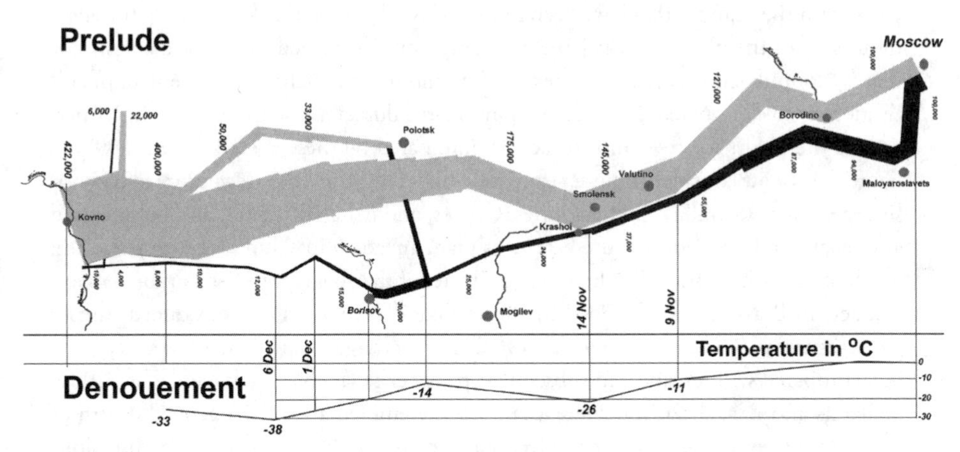

Figure 8.1. Minard's classic 1869 map of the French army's catastrophic campaign in Russia, 1812–1813, illustrates the major features of a superb graphic assessment of an invasion. Instantly discernible are the changes in the location, disposition, and size of Napoleon's army as it snaked its way to Moscow, as well as the army's numerical decline as it retreated. It may be the best graphic of an invasion ever compiled (Tufte 1983).

map. The army had shrunk to 100,000 by the time it reached Moscow, but almost all these survivors died in the retreat. The shrinking width of the black column graphically shows the rapid losses, such as the results of the disastrous winter crossing of the Berezina River, where more than 20,000 perished. Once rejoined with the remnants of units that had been dispatched earlier, barely 10,000 troops recrossed the Niemen River in 1813. Minard effectively portrayed five pertinent statistics on the army all on a single black-and-white map: the size and location of the army on a two-dimensional surface, the direction of its movement, and the frigid temperatures on selected dates. It may well be the best statistical graphic ever drawn (Tufte 1983) and is an appropriate standard by which to judge more recent attempts to report assessments of biotic invasions.

Vascular Plant Invasions

The global growth of agriculture, including agronomy, horticulture, forestry, and agroforestry, has been a product of the accelerating pace of commerce and human transport. Much of modern agriculture is practiced with plants maintained beyond their native ranges: most agricultural economies, certainly those in the temperate biomes, operate with introduced crops (Harlan 1975). Furthermore, an increasing proportion of forestry is practiced in plantations with nonindigenous species (Zobel, van Wyk, and Stahl 1987). Nonindigenous species also have been used as soil stabilizers and are widely used for ornamental purposes (Mack 2001).

The number of plant species that have become invasive as a result of these global range extensions is small, perhaps less than several thousand (Holm et al. 1979), compared with the number that have been immigrants. However, this figure belies the enormous area of the planet that is largely occupied by naturalized or invasive organisms, that is, nonindigenous organisms persistent without the benefit or requirement of cultivation. I once estimated that 4.02 million square kilometers, or approximately 3.1 percent of the earth's ice-free land surface, is dominated by these species (Mack 1997), a figure that I now consider an underestimate. I underevaluated the area occupied by invasive species in Australia (cf. Humphries, Groves, and Mitchell 1991), and I also lacked any reliable information on the extent of plant invaders in China, central Asia, and much of sub-Sahel Africa. Furthermore, estimates of the geographic distribution of plant invaders in Brazil are confounded by the varying degree to which introduced pasture grasses, such as *Brachiaria brizantha,* owe their new ranges to deliberate, repeated sowing, compared with escape and subsequent persistence (R. N. Mack, pers. obs., 2000; Williams and Baruch 2001). These examples are emblematic of the general dearth of information, except in the most anecdotal terms, for the extent of plant invasions across most of the earth's land surface. Even in the United States, we lack firm values for the extent of economically important invasive plants in the midcontinent, such as leafy spurge (*Euphorbia esula*); Spencer (1996) estimates that it already occupied 2 million hectares. Little assessment of plant invaders has been completed in most subtropical and tropical regions, yet consistent reports suggest the presence of major invasions. For example, the forests of Madagascar have been almost completely destroyed in the late twentieth century (Sussman, Green, and Sussman 1996) and replaced with agroforestry species and naturalized herbs (R. N. Mack, pers. obs., 2000). *Lantana camara* is widespread in the northern and eastern section of the island (H. Evans, pers. obs., 1999), and *Opuntia* spp. reportedly are invasive in the south (Decary 1947, as cited in Rauh 1995). But the area now occupied by invasive species in Madagascar is estimated largely from deductions based on assessment of the areas now denuded of native forest. Other invasive species that reportedly occupy large areas include *Chromolaena odorata* in western and southern Africa, India, and Oceania (McFadyen and Skarratt 1996) and *Lantana camara* in India (S. T. Murphy, pers. obs., 2004). Given the reports of severe economic damage attributed to these invaders across a broad geographic range, their areal extent probably is on the order of 100,000 square kilometers or higher.

Clearly, no comprehensive global assessment of the areal spread of plant invaders is yet possible. Before such a global effort can be launched for this or any other major taxonomic group, we need standardization of both detection methods and reporting schemes. I have reviewed elsewhere the major tools that have been used in detecting and evaluating plant invasions (Mack 2000). My purpose here is to provide a synthetic overview and evaluation of the most promising of these tools and protocols, especially in comparison to tools used for other major taxonomic groups.

Detecting Plant Invaders: Traditional Tools

Vascular plants present clear advantages for assessment: they are generally recognizable and have a consistent habit. Moreover, invasive terrestrial plants as vegetative organisms remain immobile. As Harper (1977, p. 515) observed, terrestrial "plants stand still and wait to be counted." But to be counted, they must first be detected. Although the dispersal agents themselves, most notably seeds, can be exceedingly difficult to detect, most invasive plants are large enough to be seen and identified with the unaided eye. Fortunately, such minute angiosperms as *Wolffia columbiana* (1–1.5 mm) have not yet proved invasive anywhere. The lack of minute higher plants as invaders is little comfort, however, given the difficulty of detecting individual plants in a landscape, especially if they are widely separated.

Detection of terrestrial plants is conceptually straightforward, no matter how difficult in practice; it takes only the time, initiative, and thoroughness to examine plots so small that the likelihood of plant detection is high. Such detection and mapping are carried to their extreme in demographic studies in which each plant, upon its emergence, is recorded in a mapped census (Sarukhan and Harper 1973; Mack and Pyke 1983). Even when the demography of trees is followed (Condit et al. 1996, 1999), the land area searched is small in comparison to the potential new range of a plant invader. Clearly, demographic census protocols cannot be used to assess a plant invader throughout its new range. A conceptually similar approach, a grid-based floristic survey, is perhaps the most detailed practical scheme for detecting nonindigenous species. The plant search is conducted within regularly shaped parcels of land that have been demarcated into a grid system of cells. In Western Europe, where this system has been most widely and consistently practiced, the cells are several kilometers across. Searchers carefully walk across each grid cell. Although this lacks the thoroughness of a plot census, the results are nonetheless impressive. In the Netherlands, for example, comparison of the results between surveys conducted about 50 years apart has revealed numerous new (or at least previously undetected) introductions and radical changes in ranges among the nonindigenous species. Accuracy of grid-based surveys is a function of the diligence of the searchers (Mack 2000 and references therein). Even though the area in Western Europe that has been examined in this manner is truly impressive (more than the combined area of the Netherlands, Germany, Great Britain, and Ireland), it is dwarfed by the earth's land surface area (130 million square kilometers). As conceptually attractive as grid-based surveys may appear, global assessment of plant invasions cannot rely on this approach.

Herbarium specimens offer irrefutable evidence of the location and date of collection (i.e., detection) for any plant, including an invasive organism. Any ambiguity or error is limited to the accuracy of the collection data. Unlike other raw data collected for later assessment, a herbarium specimen's identification can be verified through annotation. But herbarium records present severe limitations in their use in assessments of an invasion (Mack 2000). Among these problems is the collection bias that underlies all but the most range-restricted species. Collection locales arise usually by chance detection by multiple collectors over time. Collection intensity is highest near herbaria themselves and not evenly distributed across a landscape (Soberon, Llorente, and Benitez 1996). Most important, the lack of records of a species at a locale, whatever the area, cannot be validly interpreted to indicate absence, only "no detection." Even evidence of a species' presence at a site in the past does not guarantee that it will remain present. In fact, some invasive species have actually contracted in their new ranges over time, as in the Netherlands (Mack 2000). The habitat information provided on herbarium labels usually is insufficient from which to draw valid conclusions on any link between a species' presence and its environment. And the portrayal of species' locations in maps prepared for floras often reports the species only at the county level. For example, assembly of an atlas of the flora of New England is under way; when finished it will report each species' occurrence at the county level (Figure 8.2) (Angelo and Boufford 1998). Although useful in forming a coarse-grain impression of a species' range, such maps do not reveal the number of collections, their exact location, and the collection dates—all necessary information if any assessment were to be made of the species' regional status and role.

Unlike the status of plant detection, accurate determination of the location of terrestrial plants once detected has improved enormously with development of global

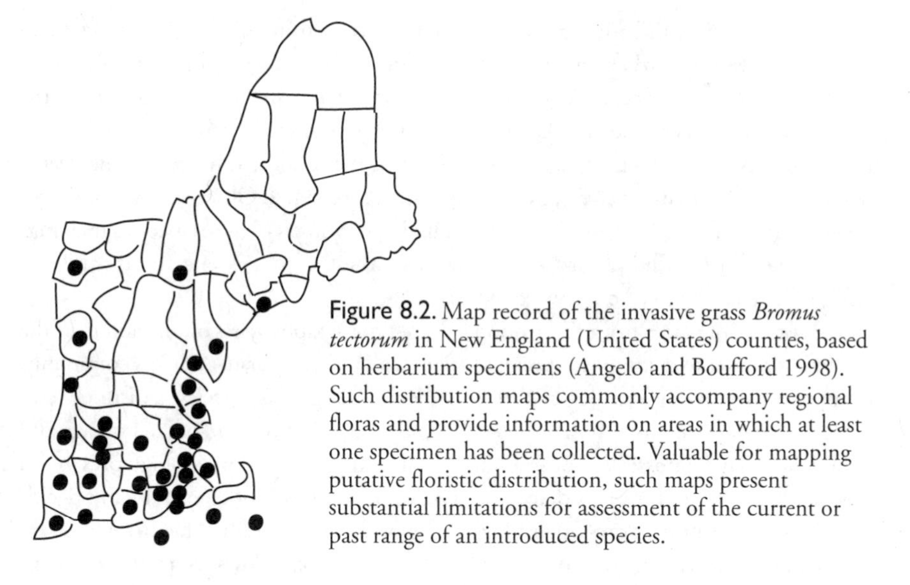

Figure 8.2. Map record of the invasive grass *Bromus tectorum* in New England (United States) counties, based on herbarium specimens (Angelo and Boufford 1998). Such distribution maps commonly accompany regional floras and provide information on areas in which at least one specimen has been collected. Valuable for mapping putative floristic distribution, such maps present substantial limitations for assessment of the current or past range of an introduced species.

positioning systems (GPSs) (Lass and Callihan 1993; Akers 2000). A ground observer can form a two-dimensional map of an invasion by recording locations along a traverse of the invasion's outer boundaries or form a dot map through the accumulation of the coordinates for individual plants. Either approach would prove effective where the infestation remains localized, as in multiple small foci, or where the invader displays a sprawling phalanx-like spread (e.g., kudzu [*Pueraria montana*] in the southeastern United States; bridal creeper [*Asparagus asparagoides*] and African carrion flower [*Orbea variegata*] in South Australia). The versatility of GPSs in mapping plant invasions has yet to be fully exploited.

Detecting Invasive Plants from Aloft: Aerial Photography

Aerial detection of plants has long attracted ecologists and land managers as a means to resolve some of the problems outlined earlier (Krumpe 1972); there are indisputable advantages in being able to view a plant-covered landscape from above. Unfortunately, any such remote sensing of a plant species is frustrated by plants' universal possession of chlorophyll. So abundant is chlorophyll in most healthy leaves that it hampers detection of other plant pigments, such as anthocyanins and xanthophylls, that could provide a diagnostic spectral signature. Successes in identifying unambiguously the spectral signatures of plant species (i.e., an array of unique reflectances) has been limited to distinguishing major plant groups, such as conifers, dicotyledonous species, and grasses, from each other (Avery and Berlin 1992). Few species have so distinctive a spectral signature as to be identifiable against a background composed of other species. These problems plague the identification of nonindigenous species from any height, especially in the preinvasive stages when the plants are rare and scattered. Attempts to resolve these limitations have followed at least two major lines of investigation: detection of the reflectance of plants in the visible and infrared range through photography and detection of plants by their reflectance in the electromagnetic spectrum beyond the visible (e.g., Landsat Thematic imagery, SPOT "Système Probatoíre d'Observation de la Terre" imagery, and AVIRIS "Airborne Visual-Infrared Spectrometry" hyperspectral imaging) (Rock, Skole, and Choudhury 1993). Each of these topics is enormous, and my remarks here are limited to a few examples.

Assessment of plants from conventional aerial photography relies on the ability of the photo interpreter to distinguish species based on color, shape, form, and, less commonly, canopy texture (Avery and Berlin 1992). If the nonindigenous species is distinctive in one or more of these properties, its detection and even quantification of its coverage and abundance may be possible. For example, color changes tied to phenology present an obvious opportunity for plant detection. The wholesale death of chlorophyll in deciduous leaves in autumn allows other pigments to be detected readily. The distinctive yellow-orange to orange-brown color of *Tamarix ramossissima* allows its areal assessment along riparian areas in the southwestern United States where it is invasive (Everitt and Deloach 1990). Thus, any invasive species in temperate and higher latitudes with the potential to undergo distinct autumnal color change is a candidate for such photographic aerial assessment (e.g., *Sapium sebiferum* in southeastern United States and *Larix decidua* in New Zealand). Unfortunately, few invaders display such radical differences in leaf color and reflectance in their new ranges compared with the native species. Other species are detectable during flowering and fruit maturation, when flowers, floral bracts, or fruits may be prominently colored. For example, the yellow to yellow-green leafy bracts of leafy spurge (*Euphorbia esula*) are detected in color aerial photography in June in their introduced range in the northern Great Plains of the United States (Everitt et al. 1995). Flower color has been used extensively to detect invasive species; even white flowers can be diagnostic if the plant populations are sufficiently large. Of course, detection of invasive species based on floral or fruit color depends on the absence of native species displaying simultaneously the same color.

Distinctive morphology or stand physiognomy is potentially useful in detecting invasive plants from aloft. Isolated plants or plants that readily form a virtual monoculture can be detected and correctly identified. These cases usually involve the invasive species displaying a plant habit that is not represented in the native flora (e.g., invasive trees among treeless native vegetation). *Salix fragilis* is invasive along rivers and streams in New Zealand and has often supplanted the low-stature native riparian species (Wardle 1991). Thus, an arboreal margin has emerged along New Zealand rivers that can be readily detected. The drought-tolerant African tree *Acacia nilotica* is invading grasslands in central Queensland (Australia), where no native trees occur, and represents a similar opportunity for aerial detection (Humphries, Groves, and Mitchell 1991).

Spatial assessments of invasive species in aquatic environments are perhaps the most common cases in which unique plant morphology has aided detection. Repeatedly, invasive macrophytes such as *Salvinia molesta, Pistia stratoides, Hydrilla verticillata,* and the infamous *Eichhornia crassipes* have invaded rivers, lakes, and ponds in many new ranges (Sculthorpe 1967). These species often invade environments in which there are few or no native macrophytes. Against a water background, unimpaired by the confounding cover of native macrophytes, these species are readily detected from the air. Repeated aerial assessments have been particularly informative in these cases, given the extraordinary rapidity with which many invasive macrophytes occupy their new ranges. In 6 years (1975–1981), *Hydrilla verticillata* spread across about 44 percent of Lake Conroe (8,100 hectares) in east Texas, a chronology that is effectively documented through aerial color infrared (CIR) photographic surveys that were conducted annually. The biocontrol of *Hydrilla* with introduction of green carp in 1981 was also recorded (Figure 8.3). Amazingly, the carp reduced the *Hydrilla* below the level of aerial detection within 2 years. The chronicle of *Hydrilla* at Lake Conroe is strengthened substantially by the field surveys across the lake several weeks after each annual aerial photographic survey

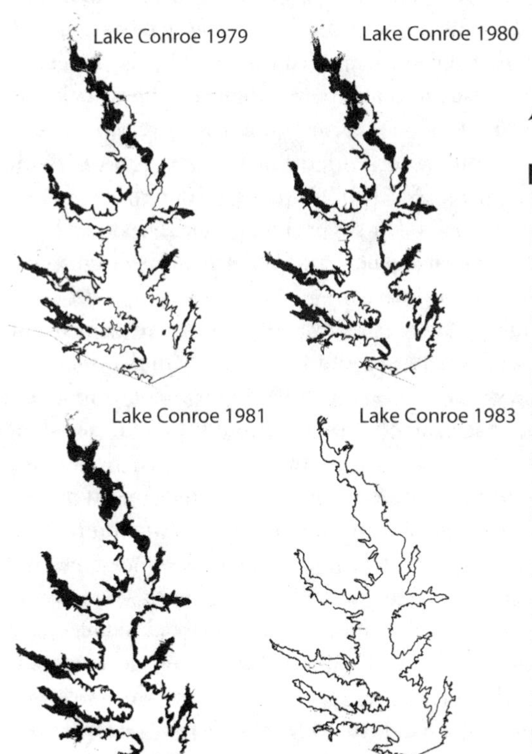

Figure 8.3. Color infrared aerial photography can effectively detect the growth of an introduced aquatic macrophyte and its eventual control; even a black-and-white representation from the photographic survey effectively reveals the spatial dynamics of plant spread. *Hydrilla verticillata* expanded rapidly in Lake Conroe (Montgomery County, Texas) after its first detection in 1975; by 1979, 1980, and 1981 it covered 29, 41, and 44 percent, respectively, of the lake surface. Introduction of green carp as a biocontrol agent in late 1981 proved swiftly effective. By 1983, no *Hydrilla* could be detected in aerial photographs or field surveys. Figure after Martyn et al. (1986).

(Martyn et al. 1986). The Lake Conroe record illustrates how following a distinctive invasive species with aerial photography verified with routine field surveys can produce a detailed record of an invasion's rise and decline.

One of the most intensive aerial assessments of invasive plants was conducted recently in the Everglades. The ill-conceived water diversion that largely destroyed the natural hydrology of the Everglades and facilitated the proliferation of invasive species, principally *Melaleuca quinquenervia* and *Schinus terebinthifolius,* has been extensively reported (Davis and Ogden 1994). Recently, a water management plan has been readied for implementation that would reverse the region's hydrology to its pre-1917 condition (i.e., before diversion). As a result, it became imperative to map the plant communities so as to gauge changes produced through direct and indirect control efforts on these invasive species (Doren, Rutchey, and Welch 1999). This ambitious undertaking has included not only preparation of a 1:15,000 scale vegetation map but also other high-resolution maps that highlight specific tracts in the Everglades National Park, Big Cypress National Preserve, Biscayne National Park, Florida Panther Wildlife Refuge, and Water Conservation Area 3 (Doren, Rutchey, and Welch 1999). Before aerial mapping was conducted, a detailed and comprehensive vegetation classification system was developed that distinguished eight major vegetation types that were further subdivided, based on species composition (Madden, Jones, and Vilchek 1999). Each type (e.g., mangrove forest, oak–sabal forest) could be reliably identified from air photos. Incredibly, the aerial mapping history of the Everglades region is poor, although the first black-and-white photographs were prepared in 1940. But the next comprehensive aerial photographs that could serve as a basis for comparison with the new survey apparently were not taken until 1985 (Doren, Rutchey, and Welch 1999). Thus, no retrospective assessment can be reconstructed on the invasions that swept through the Everglades in the twentieth century—all the more reason that any new survey be given maximum opportunity to establish a baseline for future comparisons.

Mapping the extent of the Everglades' native and invasive species relied primarily on 1:7,000 scale CIR aerial photographs. Exhaustive efforts were made so that these primary data were registered accurately against geodetic survey points. By incorporating data from GPS and SPOT satellite images, air photo surveys could be conducted along reproducible flight paths, ensuring adequate photo overlap (Madden, Jones, and Vilchek 1999). The photographs themselves have a ground resolution of about 30 centimeters, a scale that greatly facilitates species identification and quantification of vegetative cover (McCormick 1999). Photo interpretation of the CIR photographs was made from print format enlargements (1 x 1 meter). The photographs were overlain with polyester film on which the vegetation boundaries could be marked and registered with land features (Welch, Madden, and Doren 1999). The resulting images eventually were reduced to computer-generated, color-coded maps of vegetation, which were verified by field surveys. Identification of the highly invasive *M. quinquenervia* was aided by the tree's characteristic cylindrical crown, distinct shadow, and tendency to occur in monospecific stands that appear reddish-orange in CIR (McCormick 1999). As control efforts continue (Doren and Jones 1997), this detailed vegetation survey of the Everglades will allow repeated assessments of the major invasive species.

Detecting Invasive Plants from Aloft: Satellite Imagery

For the last 40 years, images of the earth's surface have been available from satellites, as both traditional photographs and images constructed from other segments of the electromagnetic spectrum. An enormous amount of geographic information has been derived from this imagery, including the gross-scale distribution of vegetation (Alexander and Millington 2000). The incentive to detect, compile, and report the areal extent of plant invasions from satellite imagery is apparent, and many attempts have been made in this regard (e.g., Ullah et al. 1989; Dewey, Price, and Ramsey 1991; Lass and Callihan 1997). The limitations encountered in tracing plant invasions from aerial photography are compounded in satellite imagery, even though photographic images from space may have remarkable clarity. Using other segments of the spectrum, as in Landsat Thematic or SPOT imagery, as the primary plant detection tools has been even more problematic (Rutchey and Vilchek 1999; Bulman 2000).

Nevertheless, the vastness of the earth's terrestrial and freshwater surfaces that would need to be mapped if a truly global assessment of plant invasions were to be assembled suggests strongly that any such assessment will rely largely on satellite imagery. In this regard, the recent mapping of almost 88,000 square kilometers in the Cape Floristic Region (CFR) in South Africa with Landsat Thematic imagery proves instructive (Lloyd, van den Berg, and van Wyk 1999). In this ambitious survey the goal was to prepare a 1:250,000 scale map of the vegetation across the CFR as differentiated into eight broad categories, three of which include different coverage levels of nonindigenous vegetation (less than 20, 20–75, and more than 75 percent cover). Eight Landsat Thematic scenes (185 x 180 kilometers each), derived from overflights from December 1997 to February 1998 were the primary source of data. Subsequent vegetation classification was constructed through an unsupervised (i.e., automatic pattern recognition) procedure (ISODATA "Iteraline Self-Organizing Data Analysis Technique") with successive iterations and revisions with supervised classification protocols (Lloyd, van den Berg, and van Wyk 1999). The comparatively small percentage of the region that supports medium to high coverage of nonindigenous species (less than 25 percent) belies the influence these species are collectively exerting on the natural environments. For example, much of the area now dominated by alien species is along the coastline, which is the native range of hundreds of endemic species (Richardson et al. 1996; Higgins et al. 1999). The vegetation maps produced in this study provide powerful visual evidence of the extent of these threats to biodiversity and the areas that need immediate protection.

As implied earlier, results of assessments of vegetation from satellite imagery have varied and often substantial limitations, which are illustrated in the CFR study. The ability to distinguish between major vegetation groups varied; dense stands of invasive pines initially could not be distinguished from pine plantations. Some native species (e.g., *Leucadendron* spp.) could not be distinguished reliably from invasive vegetation. The dissimilarity in morphology between the woody invasive species and the native coastal species probably did facilitate their accurate identification, however (D. M. Richardson, pers. comm., 2000). By definition, the widespread cover category (less than 20 percent nonindigenous vegetation) includes sites that range from undisturbed native vegetation to those in which nonindigenous species have already gained a substantial foothold. Furthermore, sites smaller than 25 hectares were not included in the final map. But immediate control measures often prove most effective in such small sites. Accuracy of the basic interpretations among the eight vegetation types is also problematic. Independent interpretation of the imagery data differed by as much as 53 percent from interpretations formed in the original study. Field surveys (i.e., ground-truthing) to resolve these discrepancies were limited (Lloyd, van den Berg, and van Wyk 1999).

Despite the limitations in using satellite imagery to assess plant invasions, this general class of tools probably will become a standard for global assessment, if only by default. It is extremely unlikely that much of the earth's surface, even in regions where vegetation has already been much altered by plant invasions, will be recorded through low-altitude aerial photography with the detail of the Everglades Mapping Project (Doren, Rutchey, and Welch 1999). Instead, greater reliance probably will be placed on distinguishing land features, including the extent of invasive plants, by detection from satellite-based remote sensing. Which, if any, of the emerging technologies in this field will prove equal to the challenge of constructing a global assessment of plant invasions remains undecided (Mack 2000).

Monoclonal Antibody Screening: A Possible New Tool for Plant Detection

Improved detection of the spatial distribution of wind-pollinated plant invaders may prove possible through innovations in molecular biology. The need to identify the source of pollen that produces allergic reactions in humans has sparked rapid development of monoclonal antibodies that are diagnostic for pollen protein allergens (Yli-Panula 1997; Schappi et al. 1999). The technology underpinning this approach is straightforward and well established: the basic goal is to identify an epitope, the point on the antigen to which an antibody attaches, unique to the target species. Spectrophotometric techniques, such as enzyme-linked immunosorbent assay (ELISA), are the most sensitive (Goding 1983), but even "dot blots" of pollen samples can be diagnostic (V. R. Franceschi, pers. comm., 2002).

It may prove feasible to collect airborne pollen of invasive species and then perform antibody screening of the pollen across a known array of species' antibodies. Samplers for airborne pollen have long been used and can sample volumetrically in prescribed intervals, thus providing quantification of the pollen collected per unit time and air speed (Ogden et al. 1974; Lewis, Vinay, and Zenger 1983). A network of samplers in a landscape could simultaneously screen for an array of species suspected to be in the area, and the data could be used to map their distribution and perhaps their abundance. Among the species for which antibodies have already been developed are some troublesome nonindigenous species, including *Cynodon dactylon* (Lovborg, Baker, and Tovey 1998), *Imperata cylindrica* (Kumar et al. 1998), and *Ambrosia artemisiifolia* (Rafnar et al. 1998). Many other wind-pollinated invasive species could be detected in this manner, even before they were otherwise detected in a new range (e.g., *Parthenium hysterophorus,* almost all grasses, and pines). Antibody screening for the presence of specific invaders would be particularly useful for grasses. Grass invaders worldwide form a long list (Holm et al. 1979) and often are difficult to detect in native grasslands. Monoclonal antibodies apparently are unexplored as tools in the assessment of plant invasions but may well deserve future evaluation.

Invasive Insects

Invasive insects provide enormous economic incentive for their spatial assessment; Pimentel et al. (2000) estimate that in the United States alone invasive insects cause about $20 billion in damage each year. Compared with vascular plants, insects are small, often are highly mobile, and may be exceedingly cryptic. As a result, the occurrence and unexploited new ranges of invasive insects often are assessed by mapping of their plant hosts. Damaged plants often can be detected more readily than the insect grazers themselves (Southwood 1978) and may serve as first evidence of insects in a new range. The mobility of winged insects also frustrates attempts to provide estimates of their spatial distribution because their new range may change daily. Taken together, these features of invasive insects make them challenging organisms to assess accurately.

As discussed earlier in connection with vascular plants, reliance on museum collections as the raw data from which to assess invasions presents substantial limitations. This information is nonetheless useful in forming testable hypotheses about past, current, and even future invasions. Heretofore, a common difficulty in their greater application has been simply logistical: plotting the raw collection data onto maps is laborious and entails replotting for each purpose (Jalas and Suominen 1972). Recently developed software, such as the Mexican Comision Nacional de Biodiversidad (CONABIO) project for Mexican flora (Soberon, Llorente, and Benitez 1996) and BioLink for insects in Australia (Shattuck 2000), compile collection records so that dot maps can be instantly prepared from location data. Current application of BioLink for insects is particularly informative because these mapped records can be compared and displayed simultaneously with environmental, including climatic, records (Shattuck 2000).

Detecting Invasive Insects

Many highly mobile insects can be detected directly with a network of traps and lures (Ridgway, Silverstein, and Inscoe 1990; Price 1997). Here the detail and predictive value of the resulting maps are directly proportional to the density and geographic distribution of the detection stations. Maps prepared from trap data are used extensively in monitoring the growth of insect populations that may reach plague proportions (e.g., gypsy moths in the United States; Liebhold et al. 1989). Dense networks of traps prove invaluable even as tripwires in first detection of nonindigenous insects, long before they become invasive or even established (Hadwen et al. 1998). Early detection has been essential for assessment and any effective control of nonindigenous dipterans, such as the highly destructive Mediterranean fruit fly in the United States (Carey 1996a, 1996b), the Asian papaya fruit fly in northern Queensland (Hadwen et al. 1998), and *Aedes albopictus,* the Asian tiger mosquito. *A. albopictus* is a vector of pathogenic arboviruses, including eastern equine encephalomyelitis (Moore and Mitchell 1997).

This chapter focuses on invasive plants and the invasive insect and fungal pests of plants. Public interest in the assessment of these species can be uncertain or temporary. However, assessments of invasive organisms that either cause disease or serve as vectors of parasites in humans and domesticated animals command enormous resources. In the United States, the Centers for Disease Control is chiefly responsible for the detection, tabulation, and assessment of these organisms, including the nonindigenous insect vectors of these parasites. The ongoing assessment of *Aedes albopictus* is typical of these efforts. Since its first detection in the United States in 1975, the mosquito has been the subject of an extensive assessment. By 1997 it had spread to 678 counties in 25 states (Moore and Mitchell 1997); it had entered a twenty-sixth state by 1998 (C. G. Moore, pers. comm., 1999). The chronological map of the spread of this mosquito (Figure 8.4) illustrates the scope and detail that are needed when the invasive species may spread disease in humans or domestic animals. National maps of county records are instantly informative for tracking the spatial spread of an invasive species. For instance, information gleaned from the multicolored map for *A. albopictus* raises intriguing questions and hypotheses that focus subsequent investigation.

- Although *A. albopictus* was first detected along coastal Texas and Louisiana, the mosquito was not detected in upstate Louisiana until more than a decade later.
- The mosquito was detected in much of Kentucky before its detection in Tennessee. But once it arrived in Tennessee, it apparently spread rapidly.
- Distant, isolated occurrences of *A. albopictus* occurred early in the invasion; it clearly did not move in a wavelike manner across its new range. New foci of the mosquito were established by long-distance transport.
- The general pattern of spread from the mid-1980s onward was simultaneously east and west from the Gulf Coast.
- Closest approximation to a wavelike spread occurred from north to south in Florida.

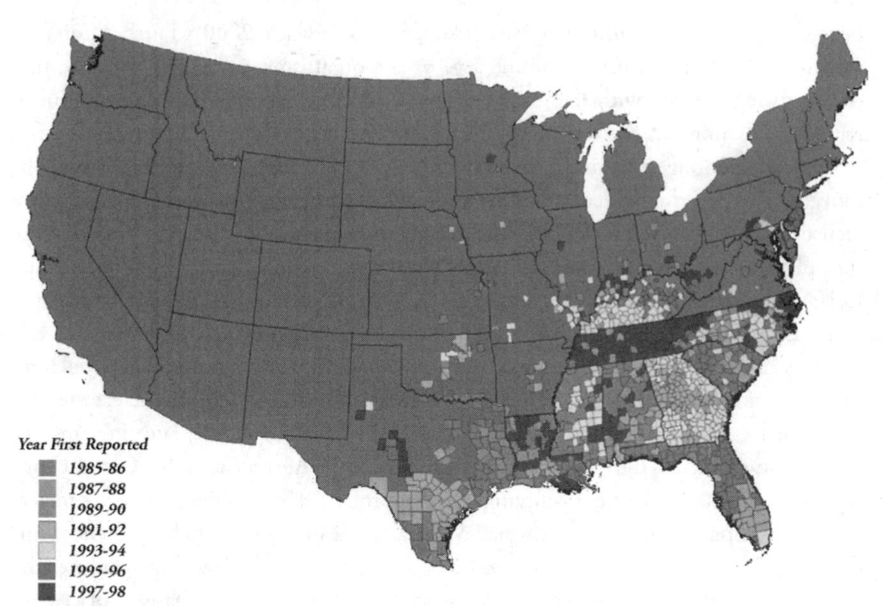

Figure 8.4. Chronology of the invasion of *Aedes albopictus* in the United States, 1985–1998, based on the date of first detection of the mosquito in each U.S. county. Displaying the county records in color-coded 2-year intervals allows instant recognition of major aspects of the spread of this insect vector of eastern equine encephalomyelitis (Moore 1999).

It is curious that the mosquito did not reach subtropical Florida locales, such as metropolitan Miami, before sites further north in Florida.

- There is no evidence that the invasion is slackening; new range continued to be occupied through 1998.

Observations gleaned even from a nonquantitative map of the mosquito's spread can prove valuable in predicting the future course of the invasion and identifying where resources could be marshaled to contain its spread. For instance, early recognition that the mosquito occurred in areas surrounding Tennessee should have signaled that the state was an immediate target for the mosquito's entry. The apparent saltatorial movements of the mosquito to new foci far removed from the Gulf Coast support the contention that its spread across the United States has been primarily in scrap tires or other commerce in which water collects (Moore and Mitchell 1997).

As outlined earlier in connection with similar maps for plant invaders, maps based on first detection have substantial limitations. Most important, the contrast reported in these data is between "detection" and "no detection," compared with the more definitive categories of "present" and "absent." But there is no assurance that the mosquito was indeed absent as opposed to simply undetected. In addition, without repeated surveys in all counties there is no assurance that such a nonindigenous species may have been extirpated in a county since its first detection (Mack 2000). Furthermore, no attempt is made to report the abundance of the organisms; a single occurrence in a county is represented with the same weight as multiple occurrences. Map location is recorded no more precisely than by county, and counties differ substantially in area.

One of the longest-running programs for detecting, mapping, and predicting the movements of grazing insects is not for an invasive insect at all but for voracious native locusts in Australia (Magor 1970). Three native orthopterans reach extraordinarily large (i.e., plague) proportions across New South Wales, Queensland, South Australia, and Victoria: the migratory locust (*Locusta migratoria*), spur-throated locust (*Austracris guttulosa*), and Australian plague locust (*Chortoicetes terminifera*). Although each of these species can be a serious threat to rangeland grasses, most concern is directed at the annual population dynamics of the Australian plague locust. The goal in assessment here is to detect the timing of the locust population buildups and predict the direction and the rate of the insects' spread eastward across the continent. For this purpose the Australian Plague Locust Commission conducts regular field sampling in the approximately 1 million square kilometers of suitable habitat in which locusts may occur across 2 million square kilometers of this four-state area. The ability of locusts to move as much as 500 km overnight dictates the sampling protocol. In sampling other, less mobile populations emphasis is placed on detecting all populations, including those comprising isolated nascent foci (Mack 2000). For plague locusts emphasis is directed instead to early detection of the large, fast-moving, potentially devastating swarms (Devenson and Hunter 2000). The primary sampling procedure is conducted on foot along 250-meter transects at approximately 10-kilometer intervals along roadways in suitable locust habitat. Estimates of adult and nymph numbers from these surveys are rapidly tallied and reported. Additional information on swarm movements and abundance is gathered from a network of eight light traps at widely separated sites in the region (Devenson and Hunter 2000). If detection for rapidly moving swarms is needed across a large area, surveys may be conducted from helicopters flying at extraordinarily low (3- to 5-meter) elevations (Anonymous 1997).

Data from these surveys are compiled monthly into maps comprising grid cells that display both the current geographic distribution and the estimated density of the locust populations (Anonymous 2000; Figure 8.5). Density estimates range in categories from isolated (scattered) to concentrated or swarm numbers. The product is a semiquantitative yet effective tool for prediction. For example, arranging maps in a time series from September to the following June can reveal the growth of the populations of *C. terminifera.* The locusts have reached plague proportions in 12 episodes between 1933 and 1990 alone (Devenson and Hunter 2000). Clearly, the locusts do not reach the same population sizes each year, and this knowledge, combined with detailed regional meteorologic information, further refines the predictive value of the surveys. The sheer magnitude of these locust swarms facilitates their detection; populations smaller than swarms would be much more difficult to detect and follow. The ultimate goal of the Australian Plague Locust Commission, as with curbing the growth of invasive species populations elsewhere, is to maximize control efforts by accurate assessment of the

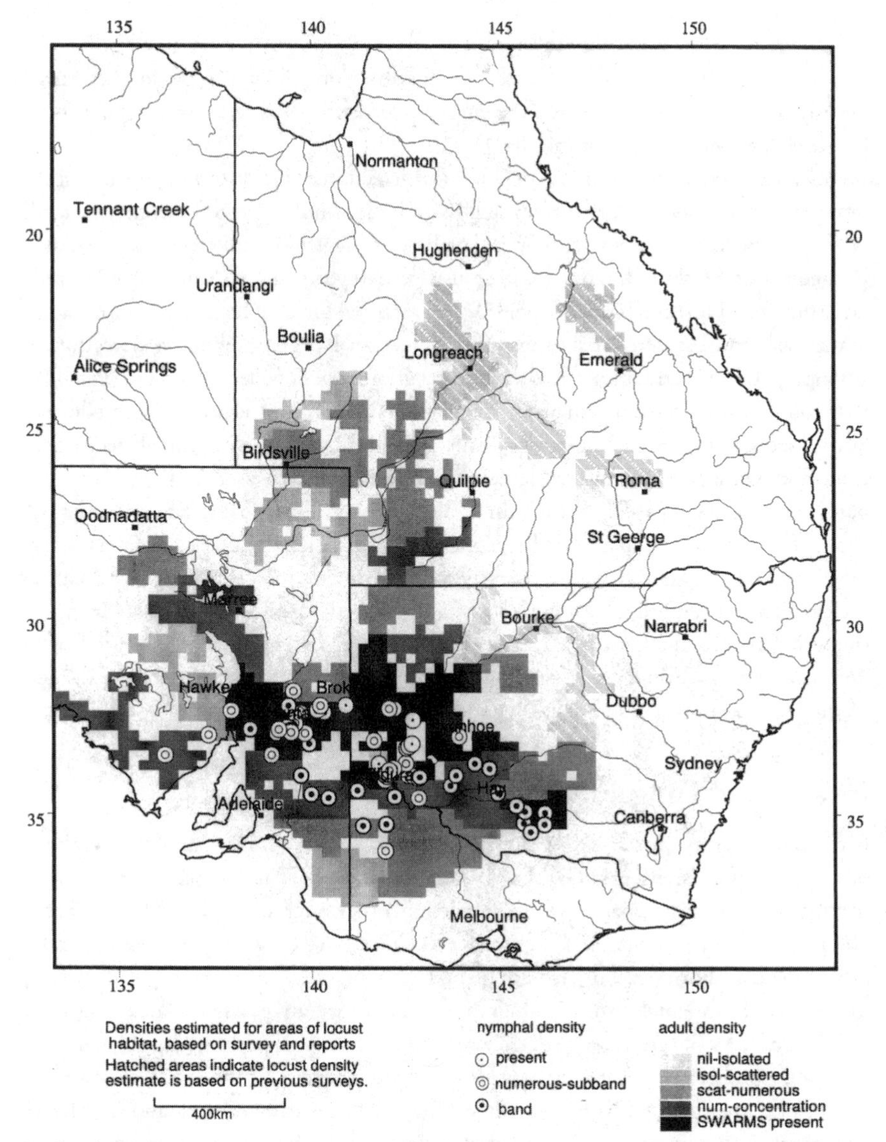

Figure 8.5. Native migratory insects that reach plague proportions, such as the Australian plague locust (*Chortoicetes terminifera*), display the same rapid changes in range and abundance that exemplify invasive winged insects. Assessing these recurring incursions across Australia involves a combination of monthly ground sampling, light trap collections, and aerial surveys. Populations from small populations of nymphs to swarms of adults are monitored in both time and space. Data compiled for the period November 2–December 6, 2000, illustrate the frequent sampling and assessment needed to predict the insects' flight directions and future densities (Anonymous 2000).

organisms' spread in time and space. Successful control of plague locusts entails destroying at least 30 percent of the migratory swarms before they consume forage grasses (D. Hunter, pers. comm., 2001).

Advances in radar entomology, which is the detection of the direction and abundance of winged insects as they are intercepted by radar (Reynolds 1988), may greatly facilitate the comprehensiveness and speed of detection of both migratory and invasive winged insects. Although radar has been used for 50 years in detecting massed populations of winged insects (Reynolds 1988), the potential for establishing networks of permanent ground-based radar stations offers a new level of comprehensive detection. As with arrays of traps and lures, these radar installations would collectively form "tripwires" that could constantly monitor insect movements year-round, even in remote locales. Such insect-monitoring radar searches only a narrow (5- to 50-meter) arc directly overhead (Beerwinkle, Witz, and Schleider 1993; Beerwinkle et al. 1995). Yet a comprehensive and dense network of such units could conceivably provide an extraordinarily detailed picture of the spatial and temporal changes in winged insect invaders. Obvious limitations include the radar's ability to discriminate insects from ground clutter, such as adjacent vegetation and other low-flying winged organisms, and its ability to distinguish between insect taxa. Nevertheless, the approach offers the potential of an objective, semiquantitative tool for continuous monitoring and subsequent assessment of winged insect invaders (Smith, Reynolds, and Riley 2000).

Invasions of Fungal Plant Parasites

Economic incentive for assessing the new range of plants' fungal parasites has always been strong (Christensen 1965). Fungi rank collectively as major parasites of plants, rivaling and perhaps surpassing the toll taken by other microorganisms (Pimentel et al. 2000). The speed with which most fungi can be transported by air or water currents has placed great importance on predicting their current range as a means of identifying source areas from which future immigrant spores or vegetative cells will arise. Although the disseminules of fungi can be detected through air and water sampling (Ogden et al. 1974; Edmonds 1979), detection of both the occupied and as yet unoccupied new range of parasitic fungi often involves the detection of proxies: the infected and uninfected host plants in the new range. In fact, useful plant proxies for detecting fungal parasites may not be the host of concern. Roses have long been planted in vineyards because they are likely to display disease symptoms caused by powdery mildew before symptoms of the parasite appear on the neighboring grape plants. Detection, tabulation, and portrayal of the dynamism of plants' fungal parasites share many of the components found in the assessment method for invasive vascular plants and insects. Unlike that of these other groups of invaders, the native range of fungal parasites often is uncertain, however. This limitation does not diminish the need to tabulate and represent the dynamic spread of fungi; however, it does leave equivocal whether the spread being monitored represents an invasion or reentry to part of the native range.

With thousands of plant fungal parasites in the United States (USDA 1960), whether comprehensive assessment of the ranges of invasive fungal parasites can be achieved is problematic. In this chapter I have chosen only three examples of invasive parasites: one has a strong historic component, another has a recent but devastating history, and the third has both a historic and a more recent record of devastation. Each example illustrates protocols used to assess this enormously important group of invaders.

Historic Assessment of the North American Range of Puccinia gracilis

Introduction of Eurasian agriculture into North America often caused not only the immigration of Eurasian crops but also entry of their parasites and the parasites' other hosts. For example, introduction of wheat was followed by the accidental introduction of the stem rust (*Puccinia gracilis*) and the equally unfortunate but deliberate intro-

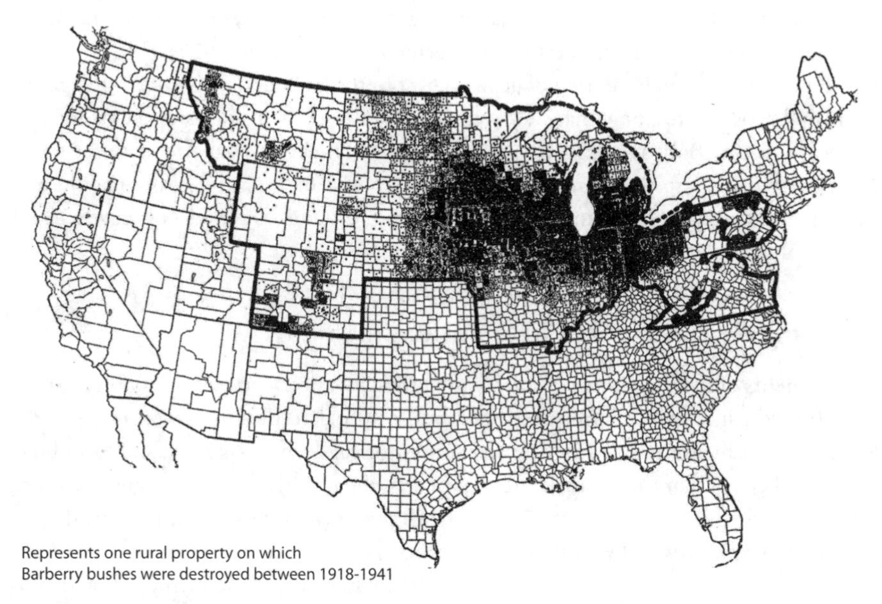

Figure 8.6. Assessment of *Berberis vulgaris* across much of the Great Plains in the United States in the 1920s probably is the largest-scale detailed assessment of a plant invasion ever attempted. Because the ultimate object of this detection effort was to reduce the alternative host for stem rust, *Puccinia graminis,* the assessment simultaneously became a very large (and exceptionally detailed) assessment of a fungal parasite as well (Meier 1933).

duction of the stem rust host, common barberry (*Berberis vulgaris*), as early as the seventeenth century. Recognition of the causal link between common barberry and rust infection in wheat was firmly established in the late nineteenth century, even though some astute farmers understood the correlation much earlier (Stevens 1933).

Despite this hard-won knowledge, little control was practiced on the spread of this invasive shrub until stem rust threatened wheat production throughout the Upper Great Plains early in the twentieth century. Eventually a truly enormous effort was organized to detect and immediately destroy every common barberry (Hutton 1927; Stakman and Fletcher 1930). In an era before aerial detection of any kind, all the detection and subsequent assessment of the range of barberry were performed through laborious ground surveys. Elsewhere I outline the scale of this vast undertaking (Mack 2000). Important here is that in the course of destroying more than 14 million *B. vulgaris* plants across a 13-state region (Stakman and Fletcher 1930), a county-by-county map was prepared of common barberry's new range (i.e., before its wholesale destruction; Figure 8.6); mapping the distribution of *B. vulgaris* in effect mapped simultaneously the distribution of the rust. If no additional barberry could be detected on foot in an area, the investigators used any disease in wheat as an indication that common barberry and therefore fruiting fungi remained nearby. These plants and their parasitic fungi were then hunted down and destroyed (Freeman and Melander 1924).

It probably would have been of little interest to the thousands involved in this monumental program to know that they were simultaneously forming assessments for two invasive species. Both assessments were byproducts of their real goal: to reduce the range of *P. graminis*. Despite its limitations, this byproduct of a control effort may be the largest assessment for biotic invaders yet attempted. It stands as a model of diligence and geographic scope.

Assessing the Invasion of Phytophthora infestans

Assessments of the dynamism within many biotic invasions are by necessity retrospective (i.e., the invasion occurred long ago, even before detection tools were developed), and the assessment must be based entirely on historical documents (Mack 2000 and references therein). For the parasitic fungi of plants, the records nevertheless may be quite detailed if the host is an important crop. Few crops have been so central to the diet of so many people as was the white potato in the nineteenth century. Thus, its attack by the late potato blight (*Phytophthora infestans*) in Europe and soon thereafter in the United States and Canada in the 1840s caused enormous crop loss and human suffering. The character and rate of spread of this invasive fungal parasite across its new ranges were documented in detail, despite the lack of formal detection procedures or even universal knowledge that a fungus caused the blight (Bourke 1964). The timing and location of detailed accounts of destroyed potato fields form the raw data on which an assessment can be assembled of the nineteenth-century spread of *P. infestans*. Local newspapers and officials heralded first detection of the disease in rural America. An annual account of the devastation was also dutifully chronicled in the annual reports of the Commissioner of Patents, who was then responsible for agriculture within the federal bureaucracy (Stevens 1933). Damage by the fungus was first detected in 1843 in a five-state region surrounding Philadelphia and New York City, major U.S. seaports. The implication, difficult to test, is that the parasite entered the United States at one or both ports. Within a year, it had spread to 11 states and had reached as far inland as Illinois by 1845 (Stevens 1933). A similar epidemiology (i.e., rapid spread from a putative point of entry at a seaport) occurred a few years later in Western Europe (Bourke 1964).

Stevens's (1933) account of the spread of the late potato blight owes its detail to these anonymous reporters. But biologists today rarely consider dreary government statistics of the nineteenth century as sources of information on biotic invasions. Nevertheless, a surprisingly extensive amount of detailed information was being assembled in the United States before the mid-nineteenth century, in the annual reports of government offices (U.S. Congress 1840) or other contemporary sources. Even for the spread of *P. infestans,* more could be gleaned from these records. For example, a detailed examination of all accounts in newspapers in the paths of the invasion and other contemporaneous sources could clarify whether the spread of *P. infestans* moved as a wave or phalanx or contained outliers that become established through the long-distance transport of infected potatoes. Such accounts remain a largely unevaluated, if not unappreciated, source of otherwise irretrievable information.

Invasion by *P. infestans* has had a sequel in the late twentieth century that probably will continue to unfold. In 1984, sexually reproducing genotypes of the fungus were first detected in Western Europe and soon thereafter in many other locations worldwide (Fry et al. 1993; Fry and Goodwin 1997). The speed of modern commerce, including the international trade in seed potatoes, probably sped the movement of infected potatoes with the virulent A2 mating types. In the past decade, these newly arrived genotypes have displaced the descendants of the original immigrant stock, the asexual A1 genotype. Few potato-producing regions remain free of the A2 genotypes, and all are vulnerable to invasion of the *P. infestans* A2 mating types (Fry et al. 1993). In such cases, rapid assessment of the invasion is needed to identify vulnerable new ranges and potential routes for further spread. Diseased host plants provide much of the fundamental evidence, even for this still contemporary invasion.

Invasion of the Generalist Plant Parasite Phytophthora cinnamomi *in Australia*

The need for new tools for rapid assessment of a fungal plant parasite is exemplified by the invasion of the virulent soilborne oomycete *Phytophthora cinnamomi* throughout much of forested Australia in the last 80 years (Weste and Marks 1987). It appears the rust is native to soils in eastern Australia but was accidentally carried to western Australia and elsewhere through road construction, thus greatly expanding its geographic range and the floristic range of hosts (Podger and Ashton 1970). So far, the fungus attacks at least 1,000 plant species in an array of plant families, but dicotyledonous species are primarily susceptible (Zentmyer 1980). Sensitivity to the parasite varies between hosts, although many species experience high mortality, such as the Australian natives *Xanthorrhoea australis, X. platyphylla,* and most members of the Magnoliidae and Proteaceae (Wills 1993).

Detection and subsequent mapping of the *P. cinnamomi* invasion have been performed with traditional tools; dead and dying trees have created dynamic openings in infected forests that can be detected in retrospective analysis of aerial photographs and field surveys (Hill, Tippett, and Shearer 1994; Peters and Weste 1997). But early detection and rapid assessment of the fungus cannot rely completely on visual detection of infected hosts; hosts often appear healthy in natural communities, only to develop the disease and die a few weeks later (Cahill and Hardham 1994). Effective control measures depend greatly on detection of contaminated soil before the disease is expressed and identifying of parasite-free soil in the potential geographic range (Cahill 1999). Direct detection of soil pathogens is complicated by the physical complexity and microbial diversity of soils, although soil microbiological sampling techniques have long provided satisfactory results, particularly the soil dilution plate method and the serial dilution endpoint method (Johnson and Curl 1972). However, spread of *P. cinnamomi* across such vast areas in Australia has been so swift and often so unpredictable that much more rapid tests have been needed (Wilson, Aberton, and Cahill 2000). Baiting the parasite with cotyledonary tissue of *Eucalyptus sieberi* or other species known to express rapidly the disease symptoms of *P. cinnamomi* has been used extensively (Cahill and Hardham 1994). Even though results of these tests may be available in a few days, these techniques still take too much time to assess whether the soil fungus is present. Furthermore, all the aforementioned techniques entail visual identification of the species' hyphae, a skill that may be in short supply.

Development of immunoassays specific for *P. cinnamomi* in the last 10 years has revolutionized the speed and potential geographic scope of detection for this invasive fungus. Soil samples may be collected rapidly in the field for analysis, using the so-called dipstick immunoassay (Cahill and Hardham 1994). Results usually are unambiguous and can be produced without knowledge of fungal taxonomy. Repeatedly, sites in which no disease symptoms had yet appeared among the native plants were found by immunoassay to already harbor the fungal parasite (Wilson, Aberton, and Cahill 2000). Rapid immunoassay also has the ability to detect quantitative changes in fungal populations: *P. cinnamomi* populations often decline locally in abundance over time (Weste 1997).

Rapid identification tools based on the specificity of immunoassay may revolutionize what has been until recently an exceedingly laborious and inaccurate arena of assessment. With widespread implementation of such molecular tools, the detail of any assessment of invasion by a fungal parasite would be limited only by the number of soil samples. The glowing predictions for these procedures (Dewey, Thornton, and Gilligan 1997; Cahill 1999) may well prove justified.

Conclusion

Assessment of invaders in time and space is a vital component of our understanding of these enormously important biological phenomena (Mooney 1999), even when the invaders are armies and the invaded space is Russia in the nineteenth century (Tufte 1983). Such appraisals form the basis for prediction about the course of future invasions. As with any vibrant science, any list of conclusions on this topic is dynamic and likely to change substantially in the future. Nevertheless, the following conclusions seem warranted.

- Assessment of a biotic invasion has at least two essential components: comprehensive detection of invasive organisms in time and space and the effective representation of this voluminous information in a form that instantly conveys the essential dynamism of this phenomenon.
- Invasive insects and parasitic fungi are small, often cryptic, and rapidly dispersed. Both their present and potential new ranges are commonly detected by assessment of the location and extent of their plant hosts. Thus, infected and uninfected hosts serve as proxies for these invaders' current and potential ranges.
- Despite current limitations in accuracy and detail, satellite imagery by default will become the only practical tool by which any assessment of plant invasions will approach global coverage; other, more refined tools are simply impractical at the scale of 10,000 square kilometers and above.
- When control is the goal, assessment of terrestrial plant invaders places emphasis on detecting even the remotest foci because they may give rise to major infestations if left unattacked. In contrast, flying, congregating insect invaders (as exemplified even by native migratory locusts) can move en masse very rapidly. The first goal of assessment here is rapid detection of the swarms rather than the small foci.
- Retrospective assessment of invasions by plant fungal parasites may allow a level of detail not possible with other invasive organisms because the rapid, devastating economic impact of these parasites on important crops prompts intense public interest and recordkeeping. Archival resources remain largely untapped as a resource with which to follow the dynamism in past invasions.
- Development of rapid, specific immunoassays for species has already revolutionized the detection of plant fungal parasites and even offers the potential for detection and quantification of wind-pollinated plant invaders. Conceivably, these tools could supplant many current detection tools for assessing the spatial dynamism of some major taxonomic groups of invaders.
- Although we are probably several decades from completing a truly global assessment

of even a representative taxonomic grouping of biotic invasions, wide implementation of tools outlined in this chapter will contribute to this goal.

Acknowledgments

I thank E. D. Deveson, R. F. Doren, H. Evans, V. R. Franceschi, W. E. Fry, D. M. Hunter, J. W. Lloyd, P. Marshall, W. MacDonald, C. G. Moore, L. Neville, S. T. Murphy, and D. M. Richardson for assistance and advice in the preparation of this manuscript. I thank J. Cullen, W. M. Lonsdale, S. Shattuck, K. Smith, and other staff of CSIRO Entomology (Canberra) for their invaluable assistance in the preparation of this chapter.

References

Akers, P. 2000. "GPS and surveying of weed populations. Equipment and costs," *Wildland Weeds* 3(2):10, 12–15.

Alexander, R., and A. C. Millington, eds. 2000. *Vegetation Mapping: From Patch to Planet.* Wiley, Chichester, UK.

Angelo, R., and D. E. Boufford. 1998. "Atlas of the flora of New England: Poaceae," *Rhodora* 100:101–233.

Anonymous. 1997. *Field Operations Manual.* Department of Primary Industry and Energy, Australian Plague Locust Commission, Canberra.

———. 2000. "General situation and forecast to mid-January 2001." *Locust Bulletin* 11. Department of Primary Industry and Energy, Australian Plague Locust Commission, Canberra.

Auld, B. A., J. Hosking, and R. E. McFadyen. 1983. "Analysis of the spread of tiger pear and parthenium weed in Australia," *Australian Weeds* 2:56–60.

Avery, T. E., and G. L. Berlin. 1992. *Fundamentals of Remote Sensing and Airphoto Interpretation,* 5th ed. Macmillan, New York.

Beerwinkle, K. R., J. D. Lopez, P. G. Schleider, and P. D. Lingren. 1995. "Annual patterns of aerial insect densities at altitudes from 500 to 2400 meters in east-central Texas indicated by continuously-operating vertically-oriented radar," *Southwestern Entomologist* Suppl. 18:63–79.

Beerwinkle, K. R., J. A. Witz, and P. G. Schleider. 1993. "An automated, vertical looking, X-band radar system for continuously monitoring aerial insect activity," *Transactions of the American Society of Agricultural Engineers* 36:965–970.

Bourke, P. M. A. 1964. "Emergence of Potato Blight, 1843–46," *Nature* 203:805–808.

Bulman, D. 2000. "Is the application of remote sensing to weed mapping just 'S-pie in the sky'?" *Plant Protection Quarterly* 15:127–131.

Cahill, D. M. 1999. "Detection, identification and disease diagnosis of soilborne pathogens," *Australasian Plant Pathology* 28:34–44.

Cahill, D. M., and A. R. Hardham. 1994. "A dipstick immunoassay for the specific detection of *Phytophthora cinnamomi* in soils," *Phytopathology* 84:1284–1292.

Carey, J. R. 1996a. "The future of the Mediterranean fruit fly *Ceratitis capitata* invasion of California: A predictive framework," *Biological Conservation* 78:35–50.

———. 1996b. "The incipient Mediterranean fruit fly population in California: Implications for invasion biology?" *Ecology* 77:1690–1697.

Champion, A. G., C. Wong, A. Rooke, D. Dorling, M. Coombes, and C. Brunsdon. 1996. *The Population of Britain in the 1990s: A Social and Economic Atlas.* Clarendon Press, New York.

Christensen, C. M. 1965. *The Molds and Man: An Introduction to the Fungi.* University of Minnesota Press, Minneapolis.

Condit, R., P. S. Ashton, N. Manokaran, J. V. LaFrankie, S. P. Hubbell, and R. B. Foster. 1999. "Dynamics of the forest communities at Pasoh and Barro Colorado: Comparing two 50-ha plots," *Philosophical Transactions of the Royal Society of London Series B, Biological Sciences* 354:1739–1748.

Condit, R., S. P. Hubbell, J. V. Lafrankie, R. Sukumar, N. Manokaran, R. B. Foster, and P. S. Ashton. 1996. "Species–area and species–individual relationships for tropical trees: A comparison of three 50-ha plots," *Journal of Ecology* 84:549–562.

Davis, S. M., and J. C. Ogden, eds. 1994. *Everglades: The Ecosystem and Its Restoration.* St. Lucie, Delray Beach, FL.

Day, J. C. 1996. *Population Projections of the United States, by Age, Sex, Race, and Hispanic Origin: 1995 to 2050.* U.S. Department of Commerce, Economics and Statistics Administration, Bureau of the Census, Washington, DC.

Devenson, E. D., and D. M. Hunter. 2000. "Decision support for Australian locust management using wireless transfer of field survey data and automatic Internet weather data collection," in *Telegeo 2000: Proceedings of the Second International Symposium on Telegeoprocessing,* edited by R. Laurini and T. Tanzi, 103–110. Ecole des Mines de Paris, Sophia Antipolis, France.

Dewey, F. M., C. R. Thornton, and C. A. Gilligan. 1997. "Use of monoclonal antibodies to detect, quantify and visualize fungi in soils," *Advances in Botanical Research* 24:275–308.

Dewey, S. A., K. P. Price, and D. Ramsey. 1991. "Satellite remote-sensing to predict potential distribution of dyers woad (*Isatis tinctoria*)," *Weed Technology* 5:479–484.

Doren, R. F., and D. T. Jones. 1997. "Plant management in Everglades National Park," in *Strangers in Paradise,* edited by D. Simberloff, D. C. Schmitz, and T. C. Brown, 275–286. Island Press, Washington, DC.

Doren, R. F., K. Rutchey, and R. Welch. 1999. "The Everglades: A perspective on the requirements and applications for vegetation map and database products," *Photogrammetric Engineering and Remote Sensing* 65:155–161.

Edmonds, R. L., ed. 1979. *Aerobiology: The Ecological Systems Approach.* Dowden, Hutchinson & Ross, Stroudsburg, PA.

Everitt, J. H., G. L. Anderson, D. E. Escobar, M. R. Davis, N. R. Spencer, and R. J. Andrascik. 1995. "Use of remote sensing for detecting and mapping leafy spurge (*Euphorbia esula*)," *Weed Technology* 9:599–609.

Everitt, J. H., and C. J. Deloach. 1990. "Remote sensing of Chinese tamarisk (*Tamarix chinensis*) and associated vegetation," *Weed Science* 38:273–278.

Forstall, R. L., ed. 1996. *Population of States and Counties of the United States: 1790 to 1990 from the Twenty-One Decennial Censuses.* U.S. Department of Commerce, Bureau of the Census, Population Division, Washington, DC.

Freeman, E. M., and L. W. Melander. 1924. "Simultaneous surveys for stem rust: A method of locating sources of inoculum," *Phytopathology* 14:359–362.

Fry, W. E., and S. B. Goodwin. 1997. "Resurgence of the Irish potato famine fungus," *BioScience* 47:363–371.

Fry, W. E., S. B. Goodwin, A. T. Dyer, J. M. Matuszak, A. Drenth, P. W. Tooley, L. S. Sujkowski, Y. J. Koh, B. A. Cohen, L. J. Spielman, K. L. Deahl, D. A. Inglis, and K. P. Sandlan. 1993. "Historical and recent migration of *Phytophthora infestans:* Chronology, pathways, and implications," *Plant Disease* 77:653–661.

Goding, J. W. 1983. *Monoclonal Antibodies: Principles and Practice.* Academic Press, London.

Hadwen, W. L., A. Small, R. L. Kitching, and R. A. I. Drew. 1998. "Potential suitability of North Queensland rainforest sites as habitat for the Asian papaya fruit fly, *Bactrocera papayae* Drew and Hancock (Diptera: Tephritidae)," *Australian Journal of Entomology* 37:219–227.

Harlan, J. R. 1975. *Crops and Man.* American Society of Agronomy, Crop Science Society of America, Madison, WI.

Harper, J. L. 1977. *The Population Biology of Plants.* Academic Press, New York.

Higgins, S. I., D. M. Richardson, R. M. Cowling, and T. H. Trinder-Smith. 1999. "Predicting the landscape-scale distribution of alien plants and their threat to plant diversity," *Conservation Biology* 13:303–313.

Hill, T. C. J., J. T. Tippett, and B. L. Shearer. 1994. "Invasion of Bassendean dune *Banksia woodland* by *Phytophthora cinnamomi,*" *Australian Journal of Botany* 42:725–738.

Holm, L. G., J. V. Pancho, J. P. Herberger, and D. L. Plucknett. 1979. *A Geographical Atlas of World Weeds.* Wiley, New York.

Humphries, S. E., R. H. Groves, and D. S. Mitchell. 1991. "Plant invasions of Australian ecosystems," Part 1, in *Plant Invasions. The Incidence of Environmental Weeds in Australia.* Australian National Parks and Wildlife, Kowari 2, Canberra.

Hutton, L. D. 1927. "Barberry eradication reducing stem rust losses in wide area," *Yearbook of the U.S. Department of Agriculture,* 114–118, Washington, DC.

Jalas, J., and J. Suominen, eds. 1972. *Atlas Florae Europaeae,* Vol. 1: *Pteridophyta (Psilotaceae to Azollaceae).* Committee for Mapping the Flora of Europe, Helsinki.

Johnson, L. F., and E. A. Curl. 1972. *Methods for the Research on the Ecology of Soil-Borne Plant Pathogens.* Burgess, Minneapolis, MN.

Krumpe, P. F. 1972. *Remote Sensing of Terrestrial Vegetation: A Comprehensive Bibliography.* University of Tennessee, Knoxville.

Kumar, L., S. Sridhara, B. P. Singh, and S. V. Gangal. 1998. "Characterization of Cogon grass (*Imperata cylindrica*) pollen extract and preliminary analysis of grass group 1, 4 and 5 homologues using monoclonal antibodies to *Phleum pratense,*" *International Archives of Allergy and Immunology* 117:174–179.

Lass, L. W., and R. H. Callihan. 1993. "GPS and GIS for weed surveys and management," *Weed Technology* 7:249–254.

———. 1997. "The effect of phenological stage on detectability of yellow hawkweed (*Hieracium pratense*) and oxeye daisy (*Chrysanthemum leucanthemum*) with remote multispectral digital imagery," *Weed Technology* 11:248–256.

Lewis, W. H., P. Vinay, and V. E. Zenger. 1983. *Airborne and Allergenic Pollen of North America.* Johns Hopkins University Press, Baltimore, MD.

Liebhold, A. M., K. W. Gottschalk, E. R. Luzader, D. A. Mason, R. Bush, and D. B. Twardus. 1989. *Gypsy Moth in the United States: An Atlas.* U.S. Department of Agriculture, Forest Service, Radnor, PA.

Lloyd, J. W., E. C. van den Berg, and E. van Wyk. 1999. *The mapping of threats to biodiversity in the Cape Floristic Region with the aid of remote sensing and geographic information systems.* Report GW/A1999/54. Agricultural Research Council, Institute for Soil, Climate and Water, Pretoria, South Africa.

Lonsdale, W. M. 1993. "Rates of spread of an invading species: *Mimosa pigra* in northern Australia," *Journal of Ecology* 81:513–521.

Lovborg, U., P. J. Baker, and E. R. Tovey. 1998. "A species-specific monoclonal antibody to *Cynodon dactylon,*" *International Archives of Allergy and Immunology* 117:220–223.

Mack, R. N. 1997. "Plant invasions: Early and continuing expressions of global change," in *Past and Future Rapid Environmental Changes: The Spatial and Evolutionary Responses of Terrestrial Biota,* edited by B. Huntley, W. Cramer, A. V. Morgan, H. C. Prentice, and J. R. M. Allen, 205–216. NATO ASI series. Series 2: Global Environmental Change, Vol. 47. Springer-Verlag, Berlin.

———. 2000. "Assessing the extent, status and dynamism in plant invasions: Current and emerging approaches," in *The Impact of Global Change on Invasive Species,* edited by H. A. Mooney and R. J. Hobbs, 141–168. Island Press, Washington, DC.

———. 2001. "Motivations and consequences of the human dispersal of plants," in *The Great Reshuffling. Human Dimensions of Invasive Alien Species,* edited by J. A. McNeely, 23–34. IUCN Press, Geneva.

Mack, R. N., and D. A. Pyke. 1983. "The demography of *Bromus tectorum* L.: Variation in time and space," *Journal of Ecology* 71:69–93.

Mack, R. N., D. Simberloff, W. M. Lonsdale, H. Evans, M. Clout, and F. A. Bazzaz. 2000.

"Biotic invasions: Causes, epidemiology, global consequences and control," *Ecological Applications* 10:689–710.

Madden, M., D. Jones, and L. Vilchek. 1999. "Photointerpretation key for the Everglades vegetation classification system," *Photogrammetric Engineering and Remote Sensing* 65:171–177.

Magor, J. I. 1970. *Outbreaks of the Australian plague locust (*Chortacetes terminifera *Walk.) in New South Wales during the period 1937–1962, particularly in relation to rainfall.* Anti-Locust Research Centre, Ministry of Overseas Development, London.

Martyn, R. D., R. L. Noble, P. W. Bettoli, and R. C. Maggio. 1986. "Mapping aquatic weeds with aerial color infrared photography and evaluating their control by grass carp," *Journal of Aquatic Plant Management* 24:46–56.

McCormick, C. 1999. "Mapping exotic vegetation in the Everglades from large-scale aerial photographs," *Photogrammetric Engineering and Remote Sensing* 65:179–184.

McFadyen, R. C., and B. Skarratt. 1996. "Potential distribution of *Chromolaena odorata* (Siam weed) in Australia, Africa and Oceania," *Agriculture Ecosystems & Environment* 59:89–96.

Meier, F. C. 1933. "The stem rust control program," *Journal of Economic Entomology* 26:653–659.

Mooney, H. A. 1999. "A global strategy for dealing with alien invasive species," in *Invasive Species and Biodiversity Management,* edited by O. T. Sandlund, P. J. Schei, and A. Viken, 407–418. Kluwer Academic Publishers, Dordrecht, The Netherlands.

Moore, C. G. 1999. "*Aedes albopictus* in the United States: Current status and prospects for further spread," *Journal of the American Mosquito Control Association* 15:221–227.

Moore, C. G., and C. J. Mitchell. 1997. "*Aedes albopictus* in the United States: Ten-year presence and public health implications," *Emerging Infectious Diseases* 3:329–334.

Ogden, E. C., G. S. Raynor, J. V. Hayes, D. M. Lewis, and J. H. Haines. 1974. *Manual for Sampling Airborne Pollen.* Hafner Press, New York.

Peters, D., and G. Weste. 1997. "The impact of *Phytophthora cinnamomi* on six rare native tree and shrub species in the Brisbane Ranges, Victoria," *Australian Journal of Botany* 45:975–995.

Pimentel, D., L. Lach, R. Zuniga, and D. Morrison. 2000. "Environmental and economic costs of nonindigenous species in the United States," *BioScience* 50:53–65.

Podger, F. D., and D. H. Ashton. 1970. "*Phytophthora cinnamomi* in dying vegetation on the Brisbane Ranges, Victoria," *Australian Forestry Research* 4:33–36.

Price, P. W. 1997. *Insect Ecology,* 3rd ed. Wiley, New York.

Pysek, P., and K. Prach. 1995. "Invasion dynamics of *Impatiens glandulifera:* A century of spreading reconstructed," *Biological Conservation* 74:41–48.

Rafnar, T., M. E. Brummet, D. Bassolino-Klimas, W. J. Metzler, and D. G. Marsh. 1998. "Analysis of the three-dimensional antigenic structure of giant ragweed allergen, Amb t 5," *Molecular Immunology* 35:459–467.

Rauh, W. 1995. *Succulent and Xerophytic Plants of Madagascar.* Strawberry Press, Mill Valley, CA.

Reynolds, D. 1988. "Twenty years of radar entomology," *Antenna* 12:44–49.

Richardson, D. M., B. W. van Wilgen, S. I. Higgins, T. H. Trinder-Smith, R. M. Cowling, and D. H. McKell. 1996. "Current and future threats to plant biodiversity on the Cape Peninsula, South Africa," *Biodiversity & Conservation* 5:607–647.

Ridgway, R. L., R. M. Silverstein, and M. Ń. Inscoe. 1990. *Behavior-Modifying Chemicals for Insect Management: Applications of Pheromones and Other Attractants.* M. Dekker, New York.

Rock, B. N., D. L. Skole, and B. J. Choudhury. 1993. "Monitoring vegetation change using satellite data," in *Vegetation Dynamics and Global Change,* edited by A. M. Solomon and H. H. Shugart, 153–167. Chapman & Hall, New York.

Rutchey, K., and L. Vilchek. 1999. "Air photointerpretation and satellite imagery analysis techniques for mapping cattail coverage in a northern Everglades impoundment," *Photogrammetric Engineering and Remote Sensing* 65:185–191.

Sarukhan, J., and J. L. Harper. 1973. "Studies on plant demography: *Ranunculus repens,* L., *R. bulbosus* L., and *R. acris* L. I. Population flux and survivorship," *Journal of Ecology* 61:675–716.

Schappi, G. F., P. E. Taylor, M. C. F. Pain, P. A. Cameron, A. W. Dent, I. A. Staff, and C. Suphioglu. 1999. "Concentrations of major grass group 5 allergens in pollen grains and atmospheric particles: Implications for hay fever and allergic asthma sufferers sensitized to grass pollen allergens," *Clinical and Experimental Allergy* 29:633–641.

Sculthorpe, C. D. 1967. *The Biology of Aquatic Vascular Plants.* St. Martin's Press, New York.

Shattuck, S. O. 2000. *BioLink: The Biodiversity Information Management System.* CSIRO Publishing, Collingwood, Victoria, Australia.

Sindal, B. M., and P. W. Michael. 1992. "Spread and potential distribution of *Senecio madagascariensis* Poir. (fireweed) in Australia," *Australian Journal of Ecology* 17:21–26.

Smith, A. D., D. R. Reynolds, and J. R. Riley. 2000. "The use of vertical-looking radar to continuously monitor the insect fauna flying at altitude over southern England," *Bulletin of Entomological Research* 90:265–277.

Soberon, J., J. Llorente, and H. Benitez. 1996. "An international view of national biological surveys," *Annals of the Missouri Botanical Garden* 83:562–573.

Southwood, T. R. E. 1978. *Ecological Methods: With Particular Reference to the Study of Insect Populations,* 2nd ed. Chapman & Hall, London.

Spencer, N. R. 1996. "Purge spurge: Leafy spurge database," in *Proceedings of the IX International Symposium on Biological Control of Weeds,* edited by V. C. Moran and J. H. Hoffman, 340. University of Cape Town, Cape Town, South Africa.

Stakman, E. C., and D. G. Fletcher. 1930. *The common barberry and black stem rust.* U.S. Department of Agriculture Farmers' Bulletin 1544.

Stevens, N. E. 1933. "The dark ages of plant pathology in America: 1830–1870," *Journal of the Washington Academy of Sciences* 23:435–446.

Sussman, R. W., G. M. Green, and L. K. Sussman. 1996. "The use of satellite imagery and anthropology to assess the causes of deforestation in Madagascar," in *Tropical Deforestation: The Human Dimension,* edited by L. E. Sponsel, T. N. Headland, and R. C. Bailey, 296–315. Columbia University Press, New York.

Tufte, E. R. 1983. *The Visual Display of Quantitative Information.* Graphics Press, Cheshire, CT.

Ullah, E., R. P. Field, D. A. McLaren, and J. A. Peterson. 1989. "Use of airborne thematic mapper (ATM) to map the distribution of blackberry (*Rubus fruticosus* agg.) (Rosaceae) in the Strzelecki Ranges, south Gippsland, Victoria," *Plant Protection Quarterly* 4(4):149–154.

U.S. Congress. 1840. *Annual Report of the Commissioner of Patents.* U.S. Senate Document 152, Session 26-2, Serial No. 378.

USDA. 1960. *Index of Plant Diseases in the United States.* Agriculture Handbook No. 165. Agricultural Research Service. Crops Research Division, Washington, DC.

Wardle, P. 1991. *Vegetation of New Zealand.* Cambridge University Press, Cambridge, UK.

Welch, R., M. Madden, and R. F. Doren. 1999. "Mapping the Everglades," *Photogrammetric Engineering and Remote Sensing* 65:163–170.

Weste, G. 1997. "The changing status of disease caused by *Phytophthora cinnamomi* in Victorian open forests, woodlands and heathlands," *Australasian Plant Pathology* 26:1–9.

Weste, G., and G. C. Marks. 1987. "The biology of *Phytophthora cinnamoni* in Australasian forests," *Annual Review of Phytopathology* 25:207–229.

Williams, D. G., and Z. Baruch. 2001. "African grass invasion in the Americas: Ecosystem consequences and the role of ecophysiology," *Biological Invasions* 2:123–140.

Williamson, M. 1996. *Biological Invasions.* Chapman & Hall, London.

Wills, R. T. 1993. "The ecological impact of *Phytophthora cinnamomi* in the Sterling Range National Park, Western Australia," *Australian Journal of Ecology* 18:145–159.

Wilson, B. A., J. Aberton, and D. M. Cahill. 2000. "Relationships between site factors and distribution of *Phytophthora cinnamomi* in the Eastern Otway Ranges, Victoria," *Australian Journal of Botany* 48:247–260.

Yli-Panula, E. 1997. "Allergenicity of grass pollen in settled dust in rural and urban homes in Finland," *Grana* 36:306–310.

Zentmyer, G. A. 1980. Phytophthora cinnamomi *and the Diseases It Causes.* American Phytopathological Society Monograph No. 10. American Phytopathological Society, St. Paul, MN.

Zobel, B. J., G. van Wyk, and P. Stahl. 1987. *Growing Exotic Forests.* Wiley-Interscience, New York.

9

Best Practices for the Prevention and Management of Invasive Alien Species

Rüdiger Wittenberg and Matthew J.W. Cock

Invasive alien species (IAS) often alter ecosystem functioning. Competition, predation, and more subtle interactions, such as hybridization, can all decrease biodiversity and cause extinction. Enormous economic losses can arise through direct losses of agricultural and forestry products and through increased production costs associated with control measures (U.S. Congress 1993; Pimentel et al. 2000). To address this problem, populations of existing invasive species must be managed. New introductions must be assessed as to the threat they present and done only on the basis of a pest risk analysis, and new invasions must be minimized.

An ounce of prevention is worth a pound of cure; this maxim of medicine, dictating such measures as quarantine and inoculation, is equally valid for biological invasions. The rapid global increase in trade, travel, transport, and tourism is leading to an increase in introductions of nonindigenous species (Peck et al. 1998). Prevention is the first and most cost-effective line of defense against IAS; once an introduced invasive species has become established, it is extremely difficult, if not impossible, to eradicate. Management of IAS includes eradication and the control of density, range, and spread to acceptable levels.

Prevention

Prevention is the first and most cost-effective defense against IAS. It is essential to prevent alien species that have invasive potential from entering the country, in addition to following a planned risk assessment process for species and pathways. Although significant costs are associated with prevention, failure to prevent a single highly invasive species from colonizing, such as the zebra mussel in North America, might lead to enormous costs, outweighing all prevention efforts. The most obvious cost of prevention is the expense of maintaining the exclusion apparatus (salary and training of interception personnel and facilities such as fumigation chambers, inspection apparatus, and quarantine quarters). A second cost is that affecting people who are not allowed to profit from bringing in alien species (which may or may not be intended for release into the environment); likewise, the public might have benefited from a planned introduction disallowed by the prevention procedures. These costs are offset by the benefits that accrue to society from the prevention of potential invasions. However, prioritizing support for a costly prevention system often is difficult because the impact of alien species and the potential costs of invasions cannot be reliably predicted and calculated against the actual costs of the apparatus. Box 9.1 describes a prevention effort by the Australian Defense Force.

Most prevention measures are focused on certain species known to be pests elsewhere. However, these species are predominantly economically important species for the

Box 9.1. Australian Defense Force Efforts to Keep Alien Species Out

Finding and cleaning every tiny grass seed from an M113 armored personnel carrier sounds like a tough job. Cleaning the same seeds off 1,000 army vehicles—everything from trucks to front-end loaders and water tankers—is even tougher. Then cleaning off every trace of soil, every piece of foliage, insect, and egg off 10,000 pallets of army equipment—everything from generators to tents and refrigerators—is stretching the limits of probability. But it had to be done in Dili, East Timor, before 5,000 Australian peacekeeping soldiers and all their vehicles and equipment could return to Australia.

The likelihood of seeds and plant matter being spread by direct contact with military equipment is high. Weeds and seeds can spread as contaminants in soil stuck to vehicles, machinery, radiators, cuts in tires, equipment, camouflage netting, and personal equipment. Some seeds are light and windborne and are easily trapped in radiator grilles, equipment brackets, and other small areas. Soil generally collects around the wheels and tracks of vehicles and on boots, personal equipment, clothing, tents, packaging boxes, and tent poles.

The job of checking that all the vehicles, equipment, and troops were not carrying pests and diseases into Australia fell to the Australian Quarantine and Inspection Service (AQIS). The job of cleaning all that equipment to AQIS standards fell to the Australian Defense Force.

Captain Kevin Hall was assigned to devise the washing and inspection procedures to comply with the AQIS quarantine requirements. He developed an illustrated 160-page manual that became the bible for the major cleaning operation in Dili that had up to 300 staff operating 20 wash stations 18 hours every day for 3 months.

This cleanup manual covers everything from how to clean soil out of the tires of graders to where insects can lodge in a Unimog. It has photographs of all the army vehicles and equipment with diagrams on how and where to clean them. It lists the equipment needs, from high-pressure water and air hoses to vacuum cleaners, brushes, and even dustpans. All the necessary techniques were developed for the task and documented in the manual, which establishes guidelines that AQIS and the military could use not only for the East Timor operation but also for future operations.

Captain Hall recently received a 2000 National Quarantine Award from AQIS in recognition of his efforts.

Source: AQIS (2000), summarized in Wittenberg and Cock (2001).

agriculture, forestry, or human health sectors. Preventing entry of species on these "black lists" is the conservative goal of quarantine and other measures taken at present. A more recent attempt to incorporate all potentially dangerous organisms, in terms of protecting both economic assets and biodiversity, is a move to using "white lists" (Panetta et al. 1994; U.S. Congress 1993). The approach is often called "guilty until proven innocent" (Ruesink et al. 1995). A proposed intermediate step is the use of "pied lists," which are more realistic to implement when efforts are constrained by a lack of facilities, staff, and funding. The pied list would contain a section of known pest species (equivalent to black lists) with strict regulations and measures to ensure pest-free imports. Another section of the list would describe species cleared for introduction (white lists)—organisms declared as safe. All species not listed on either list would be regarded as potential threats to biodiversity, ecosystems, or economic sectors. A stakeholder proposing an intentional introduction would have to demonstrate beyond reasonable doubt the safety of the proposed introduction in a risk assessment process. Species assessed for their likely invasiveness would be moved to the white or black list depending on the outcome of this investigation.

The principal pests in agriculture and forestry often are nonindigenous species—as are the plants used in agriculture and forestry—and have been dealt with for many decades using specific prevention, eradication, and control methods. Usually these insect pests, pathogens, and weeds are well known in agricultural systems, but until recently little attention has been paid to the species that threaten natural habitats. As a first step, knowledge gained from the experience in agricultural and forestry systems in combating alien species must be used to address the impact of invasive species in natural habitats. For example, quarantine facilities and other related services for agriculture should be used and expanded to address the threat of environmental pests. Facilities may need increases in capacity or changes in design and management to cope with groups of organisms not previously considered a hazard.

There are three principal strategies to reduce further introductions.

- *Interception.* The first step is based on regulations and their enforcement with inspections and fees. Accidental introductions are best addressed before exportation or upon arrival of goods and trade. This approach involves inspection, decontamination, and constraints on specific trade commodities rated as high risk. The illegal importation of prohibited items (smuggling) should also be considered. Wherever regulatory laws are in place, some people will try to evade them. This pathway is a high-risk route for introductions because smuggled items are not inspected, and smugglers are unlikely to take precautions to prevent unwanted introductions as contaminants or "hitchhikers." Staffing and financial constraints limit the control of traffic in illegally introduced commodities. In accord with the precautionary principle, a risk assessment process should be the basis for every proposed intentional introduction unless the species is already on a white list.

- *Treatment.* If goods or their packaging materials are suspected to be contaminated with nonindigenous organisms, or where high security is needed for other reasons, treatment is necessary, including biocide applications (e.g., fumigation, immersion, or spraying), heat and cold treatment, pressure, and irradiation (Sharp and Hallman 1994).
- *Trade prohibition.* Finally, when even strict measures will not prevent unwanted introductions through high-risk pathways, trade prohibition based on international regulations can be set in place. They can be applied with respect to particular products, source regions, or routes. Under the World Trade Organization Sanitary and Phytosanitary Agreement (WTO 1994), member countries have the right to take sanitary and phytosanitary measures to the extent necessary to protect human, animal, and plant life or health if these measures are based on scientific principles and are supported by sufficient scientific evidence.

Public education is an essential part of prevention and management programs. In fact, some scientifically well-devised projects have been delayed or canceled because of public disapproval. Public awareness and support can increase greatly the success of projects to protect and save biodiversity. Travelers often are unaware of laws and regulations to prevent introductions of alien species and the reasons for them. Education should focus on raising awareness of the reasons for the restrictions and regulatory actions and of the environmental and economic risks involved. In addition to printed material (e.g., posters and brochures), video presentations and announcements on airplanes are a promising approach. The public and industries should perceive prevention measures not as an arbitrary nuisance but as a necessary part of travel and trade. Box 9.2 describes a public information campaign designed to stop the spread of *Miconia calvescens* in French Polynesia.

The most common approach for preventing invasive organisms is to target individual species. However, a more comprehensive approach is to identify major pathways that lead to harmful invasions and manage the risks associated with them. Although international trade and travel are believed to be the leading cause of harmful unintentional introductions, very few countries have a good knowledge base on the actual pathways

Box 9.2. Public Awareness and Early Detection of *Miconia calvescens* in French Polynesia

With the recognition by local authorities (the French Polynesian Government and the French High Commission) of the severity of the *Miconia calvescens* invasion on the islands of Tahiti and Moorea (French Polynesia), an *M. calvescens* research and control program was started in 1988.

The Department of Environment published three information and education posters ("Le Cancer Vert" in 1989, "Danger Miconia" in 1991, and "Halte au Miconia" in 1993) and widely distributed them to all 35 high volcanic islands of French Polynesia considered susceptible to invasion by this plant species. Each year, researchers displayed an information board on the *M. calvescens* program during popular events in the town of Papeete, Tahiti ("Environmental Day" in June, "Agricultural Fair" in July, "Science Festival" in October).

Active manual and chemical control operations started in 1991 on the newly invaded island of Raiatea, where the Rural Development Service discovered small infected areas in 1989. To date, six annual campaigns have been organized on Raiatea with the help of hundreds of schoolchildren, nature protection groups, and the French Army. The 5-day campaigns were publicized in local newspapers, on the radio, and on a local TV channel (RFO 1, which is watched in all the inhabited islands of French Polynesia) during the news, in both French and Tahitian languages.

As a direct result, a pig hunter reported a small population of *M. calvescens* found in a remote valley on the island of Tahaa, and local inhabitants noticed *M. calvescens* seedlings on the island of Huahine. In June 1997, during a botanical exploration in the Marquesas Islands conducted by the Research Department and the National Tropical Botanical Garden (Hawaii), a small population was discovered and destroyed on Nuku Hiva. Once again, an article was published in the local newspapers and a talk was broadcast on local radio stations (including the Marquesan radio).

During the 4 days of the first Regional Conference on Miconia Control held in Papeete, Tahiti, in August 1997, local TV, newspapers, and radio were highly involved. As a result, more isolated plants were found and reported in the remote islands of Rurutu and Rapa (Austral archipelago) and Fatu Hiva (Marquesas archipelago) and immediately destroyed by the Department of Agriculture.

Source: J. Y. Meyer, pers. comm. in Wittenberg and Cock (2001).

involved. Exclusion methods based on pathways rather than individual species are a more efficient way to concentrate efforts where pests are most likely to enter national boundaries and avoid wasting resources elsewhere. Entire pathways may also be analyzed for risk, and this may be a more efficient procedure where many possible species and vectors are involved.

Within-country movements as well as between-country movements must be considered. Some species translocated within the same country can disrupt ecological systems, such as fish species introduced into other watersheds or marine species introduced from the Atlantic coast of the United States to the Pacific coast. These species can be just as damaging to biodiversity as completely nonindigenous species. This widens the approach of prevention from country boundaries to ecoregions where natural boundaries normally prevent range expansions.

Human-made structures may enhance subsequent spread of alien species formerly restricted to one area. The completion of the Welland Canal between Lake Ontario and Lake Erie enabled invasive organisms, such as the sea lamprey (*Petromyzon marinus*), to bypass Niagara Falls and subsequently spread to other lakes and river systems (Simberloff 1996). The opening of the Suez Canal initiated an influx of hundreds of Red Sea species into the oligotrophic Mediterranean Sea, which outcompeted and replaced indigenous species (Galil 1999).

In conclusion, prevention is the backbone of alien species management. Its successes are measured in terms of invasions that did not happen, yet resources for prevention are critical. Exclusion methods based on pathways rather than individual species are a more efficient way to concentrate efforts where pests are most likely to cross national boundaries and where several potential invasive species can be linked with a specific pathway. Three major possibilities to minimize further invasions are interception, treatment of suspect imported material, and prohibition of particular commodities under international regulations. Deliberate introductions should all be subject to an import risk assessment.

Early Detection

The longer a species goes undetected in the early, noninvasive stage, the less opportunity there is to intervene, the fewer options remain for its control or eradication, and the more expensive any intervention is (Mack et al. 2000). For example, eradication rapidly ceases to be an option as an alien is left to reproduce and disperse. However, during the lag phase it can be difficult to distinguish doomed populations from future invaders. Because not all alien species necessarily become invasive, species known to be invasive elsewhere under similar conditions are priorities for early detection. The possibility of early eradication or early control of a new colonizer makes investment in early detection worthwhile.

Surveys for early detection should be carefully designed and targeted to answer specific questions as economically as possible. Some invasive species are easily seen, whereas others are cryptic and necessitate special efforts to locate or identify them, particularly when they are in low numbers. Visitors knowledgeable about invasive species in other areas may be the first to draw attention to a new invasive, but waiting for someone to see and report a new invader often means that the invader is well established by the time the authorities become aware of it. Experts should survey certain groups of pests to enable a rapid response before the invasive species becomes well established.

A contingency plan usually is a carefully considered plan of the action that should be taken when a new invasive species is found or an invasion is suspected. The plan may be just a simple document that all staff, selected volunteers, or relevant organizations have written, are aware of, and will act on in a contingency. Alternatively, the plan may be expanded to include comprehensive kits of tools that are stored in a ready-to-use condition at appropriate locations. To prepare the plan, possible contingencies must be considered and possible actions discussed and agreed to by all parties. Contingency funding also must be immediately accessible to deal with species colonization at an early stage. The following are examples of possible contingency plans, using alien plants as models.

- A suspected new invasive is found. It is just one plant. The finder is a botanist and knows that it is a new alien plant with invasive potential. In this case, the plant should be pulled up and put in a secure container on the site to prevent seeds and bits from dropping. Then it should be taken to the quarantine station and burned. The site it came from should be carefully marked and checked every 6 months for the next 2 years.
- A suspected new invasive is found. It is a small patch of plants. The finder is a conservation officer who is not a botanist and is not sure whether it is a new invasive or a rare native plant. In this situation a small piece of the plant should be collected and taken to a botanist. Every endeavor should be made to identify the plant within 3 days of discovery. If it is an invasive, the contingency plan says all flowers and seeds should be removed and put in a secure container on the site to prevent seeds and bits from being dropped. Then they should be taken to the quarantine station and burned. The site it came from should be marked carefully and action to manage the plant considered, following standard assessment procedures.

Management

Even the most effective prevention and early warning systems have leaks, so new introductions must be anticipated and control options investigated. A management project dealing with invasive species must address several important issues, including planning, budgeting, monitoring, analysis, recording, reporting, follow-up, and dissemination of results. Adequate funding must be secured for all steps until the project goal, set before the beginning of the project, is met.

The first step is to determine the management goal for any invasive management project. The target area must be defined. It may be an entire country, all or part of an island, or all or part of a reserve or conservation area.

In some instances regional projects include more than one country, so good coordination between countries is needed. Therefore, it is often advisable to base an eradication or control program on an ecosystem that may cross political boundaries. Sometimes, the political situation might prohibit this approach.

The areas of highest quality for biodiversity and conservation with outstanding natural beauty, species-rich areas, and rare habitats often are protected as national parks, with little human activity besides tourism. The management goal for these kinds of areas is the preservation or restoration of the natural systems, often combined with the development and maintenance of ecotourism.

The management area, as defined in the management goal, must be surveyed for alien as well as native species to assess the potential loss of natural habitat. These surveys include literature search, collection records, and actual surveys in the area. The documentation must include the best available knowledge about the abundance and distribution of alien species, their impact on the habitat, and, when justified (e.g., based on experience in neighboring areas), a prediction of future impact. If earlier data are available, a comparison between past and current species composition and distribution of single alien species can reveal the status and spread of species in that area. Past control actions and their success or failure also should be summarized.

Consider the management options for each target species, using local knowledge, information from databases, and published and unpublished sources. Local circumstances, such as the cultural and socioeconomic features of the area, may affect the suitability of different options. Options for eradication, containment, or control and needs for additional surveys, experimental investigations, and other research all should be evaluated. Eradication, containment, and control options must be evaluated for cost-effectiveness, including possible impacts on nontarget species, other possible detrimental effects, and the likelihood of success, before decisions are made.

Priorities are set in the hope of minimizing the total, long-term workload, and hence the cost of an operation, in terms of money, resources, and opportunities. Therefore, we should act to prevent new infestations and assign highest priority to existing infestations that are expanding most rapidly, are most disruptive, and affect the most highly valued areas of the site. We also should consider the difficulty of achieving satisfactory control, giving higher priority to infestations we think are more likely to be controlled with available technology and resources.

The priority-setting process can be difficult, partly because it is influenced by so many factors. It helps to group these factors into four categories, which can be seen as filters designed to screen out the worst pests:

- Current extent of the species on or near the site
- Current and potential impacts of the species
- Value of the habitats or areas that the species infests or may infest
- Difficulty of control

The categories can be used in any order, but we emphasize the importance of the current extent of the species category and suggest that it be used first. In the long run, it is usually most efficient to devote resources to preventing new problems and immediately addressing incipient infestations. Ignore categories that are unimportant on your site.

All IAS populations must be monitored because many species not yet regarded as invasive may be "sleeping" organisms passing through their lag phase of invasion and will become invasive later on.

There are three main strategies to deal with nonindigenous species that have already established populations in the area of concern: eradication, containment, and control. Eradication is the most desired but often the most difficult approach. Control of alien species after their establishment becomes irreversible can be divided into containment (i.e., keeping species within regional barriers) and control in a stricter sense (i.e., suppressing population levels of alien species to below an acceptable threshold).

Eradication

Eradication is the elimination of the entire population of an alien species, including any resting stages, in the managed area. When prevention has failed, an eradication program can be the preferred method of action to deal with an introduced organism. Eradication, as a rapid response to the early detection of a nonindigenous species, often is the key to a successful and cost-effective solution (Simberloff 1997). However, eradication should be attempted only if it is feasible. Eradication is the type of clear-cut decisive intervention that appeals to politicians and the public, but beware of the temptations of attempting an eradication program that is unlikely to succeed. A careful analysis of the costs (including indirect costs) and likelihood of success must be made rapidly and adequate resources mobilized before eradication is attempted (Sandlund, Schei, and Viken 1999). However, if eradication of the invasive species is achieved, it is more cost-effective than any other measure of long-term control. Efforts to eradicate screwworms from North America and North Africa are described in Box 9.3.

Eradication programs can involve several control methods, used alone or in combination. The methods vary depending on the invasive species, the habitat, and the circumstances. Successful eradication in the past has been based on the following methods:

- Mechanical control (e.g., hand picking of snails or hand pulling of weeds)
- Chemical control (e.g., using toxic baits against vertebrates or spot spraying plants)
- Biopesticides (e.g., *Bacillus thuringiensis* [*Bt*] sprayed against insect pests)

Box 9.3. Eradicating Screwworms from North America and North Africa

Screwworms, the larvae of the screwworm fly, are parasites that cause great damage by entering open wounds and feeding on the flesh of livestock and other warm-blooded animals, including humans. The New World screwworm fly (*Cochliomyia hominivorax*) is native to the tropical and subtropical areas of North, South, and Central America, and similar but less damaging species occur in the Old World.

After mating, the female screwworm fly lays her eggs in open wounds. One female fly can lay up to 400 eggs at a time and as many as 2,800 eggs during its lifespan of about 31 days. The screwworm grows to more than 1 centimeter within a week of entering the wound. The full-grown larva then drops from the wound, tunnels into the soil, and pupates before emerging as an adult screwworm fly. Left untreated, screwworm-infested wounds lead to death. Multiple infestations can kill a grown steer in 5–7 days. Losses to livestock producers in the United States have exceeded $400 million annually.

Screwworms are eradicated through a form of biological control called the sterile insect technique (SIT). Millions of sterile screwworm flies are raised in a production plant located in the southern Mexican state of Chiapas. During the pupal stage of the fly's life cycle, the pupae are subjected to gamma radiation. The level of radiation is designed to leave the fly perfectly normal in all respects but one: it is sexually sterile. Thus, when the artificially raised flies are released into the wild to mate with native fly populations, no offspring result from the matings. These unsuccessful matings lead to the gradual reduction of native fly populations. With fewer fertile mates available in each succeeding generation, the fly breeds itself out of existence.

In the early 1950s, the U.S. Department of Agriculture's Agricultural Research Service developed the SIT for screwworm control. This SIT was used in Florida in 1957, and by 1959 screwworms had been eradicated from the southeastern United States. The technique was next applied in the more extensively infested Southwest starting in 1962. Self-sustaining screwworm populations were eliminated from the United States by 1966. Since then, a cooperative international program has been pushing the screwworm back toward the Isthmus of Panama, with a view to eradicating it from Central America and, in the future, the Caribbean.

Therefore, when an infestation of the New World screwworm appeared in Libya in 1988, the tools for its eradication were already available. Recognizing the enormous threat to humans, livestock, and wildlife, an urgent national and international effort was mounted to prevent its spread to the rest of Africa and the Mediterranean Basin. The SIT campaign was successful in achieving eradication, preventing the enormous losses that would have occurred if the infestation had spread.

Source: Wittenberg and Cock (2001), edited from the USDA-APHIS Web site (http://www.aphis.usda.gov/oa/screwworm.html).

- Habitat management (e.g., grazing and prescribed burning)
- Hunting of invasive vertebrates

Some groups of organisms are more suitable for eradication efforts than others. Some methods used in past efforts are summarized here. Each situation must be evaluated individually to find the best methods for that area under the given circumstances.

- Plants can be eradicated best by a combination of mechanical and chemical treatments (e.g., cutting woody weeds and applying a herbicide to the cut stems).
- Many successful eradication programs have been carried out against land vertebrates on islands. The methods most common were bait stations where toxic substances were offered to the invasive species (e.g., rat control). Bigger animals can be hunted if the ecosystem is of an open kind with less cover. Eradication programs against land vertebrates may elicit public opposition, especially that of animal rights groups.
- Among land invertebrates, only snails and insects have been successfully eradicated on occasion. Snails can be hand picked, whereas the most common options to eradicate insects are based on the use of insecticides or biopesticides, usually by widespread application, or using baits or traps.
- The use of sterile male releases, often in combination with insecticide control, has been effective on several occasions against insects such as fruit flies and the screwworm fly.
- There are two published successful eradications of invasive species in the marine environment to date. An infestation of a sabellid worm in a bay in the United States was eliminated by hand picking of the host (Culver and Kuris 1999), and a mussel species was eradicated in Australia using pesticides (Bax 1999).
- Foreign freshwater fish species have been eradicated in the past by using toxins specific to fish (Courtenay 1997).
- Pathogens of humans and domesticated animals have been eradicated by host vaccination. In general, it seems more feasible to apply eradication methods to the hosts rather than directly to the pathogens.

If an eradication program is feasible, it is the preferred choice for action against a nonindigenous invasive species. The advantage of eradication as opposed to long-term control is the opportunity for complete rehabilitation to the conditions prevailing before the invasion of the alien species. No long-term control costs are involved (although precautionary monitoring for early warning may be appropriate), and the ecological impacts and economic losses diminish to zero immediately after eradication. This method is the only option that totally meets the management goal because the invasive species is completely eliminated.

The major drawback of eradication programs is that they may not succeed, in which case the entire investment will have been largely wasted; at most the spread of the target alien species will have been slowed. Because eradication programs usually are very costly and need full commitment and attention until their successful completion, no eradication program should be started unless an assessment of the available options and methods has shown that eradication is feasible. Thus, eradication should be pursued only when funding and commitment of all stakeholders have been secured. Public awareness of the problems caused by the invasive species should be raised beforehand and public support sought.

A well-designed and realistic eradication approach must be developed to achieve the necessary goal. In most cases, well-established populations and large areas of infestation are unsuitable for eradication programs. Many failed attempts were highly costly and affected nontarget species, as in the case of the attempt to eradicate South American fire ants in the southern United States (Simberloff 1996). The insecticide initially used proved disastrous to wildlife and cattle. The ant bait subsequently developed also had nontarget effects and proved to be more effective against native ant species than the intruder. This increased the populations of the nonindigenous species by decreasing interspecific competition with native ant species. Finally, the eradication efforts had to be abandoned.

The best chances for successful eradication of unwanted species are during the early phase of invasion, while the target populations are small or limited to a small area. Improvements in eradication technology, eradication experience elsewhere, and improved knowledge of the basic ecology of invaders will improve eradication attempts in the future. Eradication efforts have been especially successful in island habitats. These can include ecological islands isolated by physical or ecological barriers, such as forest remnants surrounded by agricultural fields. However, the target species may survive in small populations outside an ecological island and, depending on the degree of isolation, could rapidly reinvade the ecological island after an eradication campaign.

Although eradication methods should be as specific as possible, the rigorous nature of concentrated eradication efforts has incidental effects on nontarget species. In most cases these losses can be seen as inevitable and acceptable costs to achieve the management goal and can be balanced against the long-term economic and biodiversity benefits. Any toxins used should be as specific as possible and their persistence should be of short duration. However, some toxins unacceptable for use in a long-term control program might justifiably be used in an eradication campaign over a short period of time.

Eradication or control of well-established nonindigenous species that have become a major element of the ecosystem will influence the entire ecosystem. Predicting the consequences of the successful elimination of such species is difficult. The relationships of the invasive species to indigenous and nonindigenous species must be considered. A strong carnivore–prey relationship between two invasive species points to the need to investigate the potential for combined methods to eliminate both species at the same time. Control of one species in isolation could have drastic effects on the population dynamics of the second species. Eliminating the normal prey may eliminate the carnivore, or it may cause it to change its behavior and feed on native species. Eliminating an introduced carnivore is likely to allow the introduced prey to increase greatly in numbers and may cause more damage than when both were present (e.g., rabbit and red fox in Australia, both introduced from Europe).

The basic criteria for a successful eradication program are summarized as follows:

- The program must be scientifically based. Unfortunately, most traits rendering species invasive make eradication efforts more difficult (e.g., high reproduction rate and dispersal ability). That means that invasive species are likely to be difficult to eliminate by nature.
- Eradication of all individuals must be achievable. It becomes progressively more difficult and costly to locate and remove the final individuals at the end of the program, when the population is dwindling.
- Support from the public and all stakeholders must be ensured beforehand.
- Sufficient funding must be secured for an intensive program (allowing for contingencies) to make sure that eradication can be pursued until the last individual is removed.
- Small, geographically limited populations of nonindigenous species are easiest to eliminate. Therefore, immediate eradication is the preferred option for most species found in early detection surveys, so any early warning program must have contingency funds available for these actions.
- Immigration of the alien species must be zero; that is, the management area must be completely isolated from other infested areas, as is the case for islands. Potential pathways for the species between infested areas and the management area must be controlled to prevent new invasions.
- All individuals of the population must be susceptible to the eradication technique used. If individuals learn to avoid the technique, they would not be susceptible to the technique and would survive. Perhaps a combination of methods more successful at both high and low densities would be more successful under these circumstances.
- A technique must be designed to ensure detection of the last survivors at very low densities at the end of the program (e.g., pheromone traps installed at high densities in high-risk areas). Organisms that have less obvious stages, such as seed banks of weeds, must be monitored for longer periods.
- A subsequent monitoring phase should be part of the eradication program to ensure that eradication has been achieved.
- Methods for prevention and early detection of the eradicated species should be put in place.

Containment

Containment of nonindigenous invasive species is a special form of control. The aim is to restrict the spread of an alien species and to contain the population in a defined geographic range. The containment methods are the same as those described for prevention, eradication, and control. The invasive species population is suppressed using different methods along the border of the defined area of containment, individuals and colonies spreading beyond this area are eradicated, and introductions into areas outside the defined containment area are prevented.

Species most likely to be successfully contained in a defined area are those that spread slowly over short distances. The nearest suitable habitat for the species should be separated by a natural barrier or an effective artificial barrier. The most suitable cases for containment are habitat islands without suitable connections that would allow the easy spread of invasive species.

If containment of an invasive species in a well-defined area is successful, habitats and native species are safeguarded against the impacts caused by the harmful alien species outside this area. In cases where eradication is not feasible and the range of the invasive species is limited to an isolated area, containment of the species in that area can protect the rest of the country, even if the species is harmful in the containment area. However, a careful analysis of the containment options, their costs, and their likely benefits should always be carried out.

A species confined in a defined area necessitates constant attention, control measures at the border, and prevention measures against spread of the species. Thus, successful containment is difficult to achieve and involves several different costly methods.

Control

A control program against a nonindigenous invasive species should aim for the long-term reduction in density and abundance to below an acceptable threshold. The economic and ecological harm caused by the species under this threshold is considered acceptable. The weakened state of the invasive species allows native species to regain ground, and they may even further diminish the abundance of the alien species. In rare cases this might even lead to local extinction of the nonindigenous species (especially when habitat restoration efforts support native species and put intact natural systems back in place), but this is clearly not the principal goal of control efforts.

If prevention methods have failed and eradication is not feasible, managers must live with the introduced species and can only try to mitigate the negative impacts to biodiversity and ecosystems. All control methods, with the exception of classic biological control (which when successful is self-sustaining), entail long-term funding and commitment. If the funding ceases the population and the corresponding negative impacts will increase.

In the short term, control seems to be a cheaper option than eradication, so it is often the preferred method. The investment in funding and commitment need not be as great as for eradication programs, and those who benefit often bear most of the costs. Also, funding can vary between years depending on the perceived importance of the problem, political pressure, and public awareness. However, the lower recurring costs are deceptive because in the long run effective control is more expensive than a successful eradication campaign.

Mechanical, chemical, and biological control, habitat management, and integrated combinations of methods are all used successfully to control populations of invasive species. In many cases, the best way to manage an invasive species may include a system of integrated management tailored to the species and the location. Therefore, it is important to accumulate the available information, assess all potential methods, and use the best method or combination of methods to achieve the target level of control.

Always bear in mind that managing an invasive species is not the management goal but only a step toward a higher goal, such as habitat restoration, preservation of a "pristine" ecosystem, reestablishment of the natural succession process, or sustainable use of ecosystem services for local people.

Successful control of an invasive species can have indirect effects on native species, the ecosystem, and local biodiversity. The potential effects of reducing or eradicating the invasive species in a habitat should be evaluated beforehand and measures taken to ensure that these effects are largely or solely positive. For example, removal of an aggressive invasive plant from a site might need to be accompanied by planting of indigenous species to fill the gaps and to prevent these gaps from being filled by other unwanted plants.

There are many specific methods to control invasive species. Given the highly complex nature of invasion ecology and the importance of local conditions, general statements about suitable control methods for groups of alien species in specific habitats or world regions should be approached with great caution. Precise predictions of the behavior, spread, and impacts of nonindigenous species introduced into new environments are not available because too many of the relevant parameters are no more than informed guesses. In many cases even the taxonomic status of the invasive species is uncertain. However, descriptions of methods used to control certain species and their effectiveness under specific environmental factors are available. These experience-based reports are essential for invasive species management and must be made readily available (e.g., in databases accessible through the Internet). The goal of anyone involved in invasive species management should be to use the best practices available and to disseminate information to serve the higher goal of preserving the earth's biodiversity and mitigating problems caused by invasive organisms on a worldwide scale.

Mechanical Control

Mechanical control can be carried out by directly removing individuals of the target species by hand or with tools. In many cases introduced pests can be controlled or even eradicated in small-scale infestations by mechanical control, such as hand pulling weeds (Cronk and Fuller 1995) or hand picking animals. An advanced method of mechanical control is the removal of plants by specifically designed machines, such as harvesting vehicles for water hyacinth–infested lakes and rivers. In some cases of very persistent plants in large, open areas such as pastures, bulldozing may be appropriate.

Mechanical control can be used to eradicate IAS or control their densities and abundance. Basically, all organisms can be removed mechanically one way or another. However, control should be carried out or supervised by trained staff who have reviewed the available information to choose and apply the most effective method.

Mechanical control is highly specific to the target, and nontarget effects are largely limited to disturbance by human presence and direct mechanical impact on vegetation and soil. The downside of such methods is that they are highly labor intensive. In countries where human labor is costly, the use of physical methods is limited mainly to volunteer groups. Most manual work is expensive and has to be repeated for several years to remove all individuals. For weeds whose seeds can be dormant in the soil for a long period, monitoring through that potential dormancy period is necessary after local eradication. These methods can be effective when the population of the invader is still small and the population is limited to a small area. Weeds that grow vigorously from cut plant parts or multiply vegetatively are more difficult to control (Cronk and Fuller 1995).

Chemical Control

Chemical pesticides, including herbicides and insecticides, developed to control pest in agricultural production and eliminate vectors of diseases can be used to decrease population levels of invasive organisms below a threshold of ecologically tolerable impact. In the past, extensively used broad-spectrum herbicides such as DDT had far-reaching detrimental impacts on the environment and on human health, but today these herbicides are banned in most countries, and there are more specific products on the market with fewer negative nontarget effects. Some insecticides, such as those based on chemical structures similar to insect hormones, can also be specific to target groups of insects. Nevertheless, major drawbacks are the high costs, the need for repeated applications, and the impacts on other species. Moreover, an additional problem very clearly demonstrated in agriculture and human disease vector control is that repeated use of pesticides can provide selective pressure that enables many target species to evolve increasingly effective resistance to these chemicals. In response, the dosage must be increased or a different group of pesticides used, usually further increasing the control costs. When resistance is anticipated (e.g., based on experience in agricultural systems), the likelihood of its development can be reduced by using chemicals in combination, in sequence, or rotated with other control measures.

Selection of a pesticide to control an invasive species begins with a determination of effectiveness against the target and all appropriate nontarget species that might come in contact with the chemical, either directly or through secondary sources. Additionally, the environmental half-life, method of delivery, means of reducing nontarget species contact, demonstration of efficacy, and collection of data to ensure environmentally safe use must be evaluated. Most countries require pesticides to be registered for specific uses. Once identified, tested, and registered, a pesticide can allow the rapid control of a target species over large areas and thereby reduce the need for personnel and costs for the more traditional methods such as traps and barriers.

Biological Control

Biological control is the intentional use of populations of natural enemies (i.e., species that feed on the target species so as to reduce its fitness) or naturally synthesized substances against pest species. Biological control can be split up in several approaches grouped under two headings: those that are self-sustaining and those that are not. Methods that are not self-sustaining include the following:

- Mass release of sterile males that copulate with the wild females without producing any offspring in the next generation. The very successful screwworm eradication program was carried out using this method. Given the high costs, this approach is perhaps best used in eradication and containment programs.
- Inducing host resistance against the pest. This approach is particularly relevant in agriculture, where plant breeders select (or create) varieties resistant to diseases and insects.
- Biological chemicals, that is, chemicals synthesized by living organisms. This category overlaps with chemical control, and the listing of a particular method in one or the other category is a question of definition (e.g., although applying living *Bacillus thuringiensis* (*Bt*) is undoubtedly a biological control option, the group to which the use of the *Bt* toxins belong is debatable). Other examples of chemicals in this group are rotenone, neem, and pyrethrums, extracted from plants.
- Inundative biological control using pathogens, parasitoids, or predators that will not reproduce and survive effectively in the ecosystem. Large-scale or mass releases are made to react quickly to control the pest population.

Self-sustaining biological control includes the following:

- At its simplest, classic biological control is the introduction of natural enemies from the original range of the target species into new areas where the organism is invasive. IAS often are controlled in their indigenous range by their natural enemies but are usually introduced into new environments without these population-controlling natural enemies. Freed of their parasitoids, parasites, and predators, alien species often grow and reproduce more vigorously in the country of introduction. Natural enemies for introduction are selected on the basis of their host specificity to minimize risks of significant effects on nontarget species. The aim is not to eradicate the invasive alien but to reduce its competitiveness with native species, hence reducing its density and

Box 9.4. Biological Control of an Insect to Save an Endemic Tree on St. Helena

In the 1990s, gumwood, *Commidendrum robustum* (Asteraceae), the endemic national tree of St. Helena, was in danger of extinction because of an alien insect. Orthezia scale, *Orthezia insignis,* is native to South and Central America but is now widespread through the tropics. It was accidentally introduced into St. Helena in the 1970s or 1980s and became a conspicuous problem when it started feeding on gumwood in 1991. Gumwood once formed much of the extensive woodland that used to cover the higher regions of the island but is now limited to two stands of around 2,000 trees. It is a typical example of the remarkable indigenous flora on St. Helena.

Once the gumwoods became infested in 1991, an increasing number of trees were being killed each year, and at least 400 had been lost by 1993. Orthezia scale damages its host primarily through phloem feeding, but colonization of the honeydew that orthezia scale excretes by sooty molds has a secondary effect by reducing photosynthesis. Because orthezia scale is polyphagous, and large populations could be maintained on other hosts such as lantana, it spread easily onto the rare gumwood trees. Gumwoods are susceptible to orthezia scale, and if nothing had been done, gumwood probably would have become extinct in its natural habitat.

The International Institute of Biological Control (now CABI Bioscience) helped the government of St. Helena carry out a biological control program against this pest. There was already an indication that a suitable predator might be available. Between 1908 and 1959, the predatory coccinellid beetle *Hyperaspis pantherina* had been released for the biological control of *O. insignis* in Hawaii, four African countries, and Peru. Substantial control was reported after all the releases.

A collection of *H. pantherina* was obtained from Kenya, where it had been introduced to control orthezia scale on jacaranda, was cultured and studied in UK quarantine. These studies showed that reproduction of the beetle depends on the presence of orthezia scale, that *H. pantherina* normally lays eggs directly onto adult females of *O. insignis,* and that the first two instars of the larvae often are passed inside the ovisac of the female host, after which the host itself is often consumed. An assessment of the St. Helena fauna had also shown that there did not seem to be any related indigenous species (although quite a few exotic pest scales were present), so it was concluded that introduction of this predator would not only be safe in terms of effects on nontarget organisms but also would be likely to control the orthezia scale and save the gumwoods.

In 1993, *H. pantherina* was imported, cultured, and released on St. Helena. It rapidly became established and did indeed control orthezia scale on gumwoods. It was concluded that gumwood had been saved from extinction in its natural habitat. This is probably the first case of biological control being implemented against an insect in order to save a plant species from extinction.

Sources: Booth et al. (1995) and Fowler (1996), summarized in Wittenberg and Cock (2001).

its impact on the environment. Box 9.4 describes the biological control of *Orthezia insignis* on St. Helena.

- Augmentation of natural enemies can be used during critical periods of the season (e.g., early in the season, when natural populations are low) or under pest outbreak conditions to try to achieve immediate control. Once released, these natural enemies can reproduce in the new environment, and an important part of the impact may be produced by the progeny in subsequent generations. The control agent is reared or cultured in large numbers and released.
- Conservation biological control, such as habitat management, can be used to encourage populations of native predators and parasitoids (e.g., by releasing or replanting native alternative hosts and food resources).

Probably the most important of these for management of IAS is classic biological control, which has been used extensively against weeds and insect pests with many examples of successful control (Julien and Griffiths 1998; Greathead and Greathead 1992). Like eradication, successful classic biological control has high up-front costs but low recurring costs and so has an attractive benefit–cost ratio. Conservation managers are coming to realize that this method, if used under modern protocols for careful screening of potential biological control agents, provides a safe and cost-efficient approach that can solve many IAS problems. In comparison with other methods, classic biological control is highly cost-effective and self-sustaining (Tisdell 1990; Fowler 1996).

The main disadvantages are the lack of certainty about the level of control that will be achieved, the delays until the established agents achieve their full impact, and the potential impact on nontarget species, especially indigenous species closely related to the target (Wajnberg, Scott, and Quimby 2001).

There is a risk associated with the introduction of any alien species, whether a new crop, a garden plant, or a biological control agent. These risks must be understood, assessed, and managed. There is growing awareness in the scientific community that introduction of an alien natural enemy can have unanticipated nontarget effects, both by direct and indirect effects on indigenous nontarget species and by spreading to new areas or new countries where they can produce these effects on different indigenous species. It has been demonstrated that some biological control natural enemies can have an impact on nontarget indigenous species (e.g., Louda et al. 1997; Boettner, Elkinton, and Boettner 2000; Pemberton 2000).

Nevertheless, it is generally agreed that if applied carefully using narrowly host-specific natural enemies, classic biological control can be very effective in controlling IAS. It is for this reason that prerelease studies are so important. The risks of direct impact can be assessed to a reasonable degree using methods currently available (lab and field host specificity and impact studies), although indirect effects are much more difficult to evaluate.

A risk assessment exercise should be carried out for any proposed biological control introduction. Lab and field assessments of the actual and potential host range of a proposed biological control agent can be used to assess the risk to nontarget species. Of course, any such risk assessment should also include an assessment of the impact of not doing biological control, that is, the potential impact of the target pest on the economy and environment if no steps are taken to bring it under control. The International Plant Protection Convention's Code of Conduct for the Introduction of Exotic Biological Control Agents (IPPC 1996) provides guidance on procedures for introducing biological control agents in countries that do not already have suitable national protocols in place. In the end, the decision to proceed with a biological control introduction is a national one, taking into consideration the concerns of neighboring countries and based on the values and priorities of society at that time.

Although biological control can be recommended to control an established population of an IAS, especially widespread species not amenable to other control approaches, the theory of natural population regulation underlying biological control does not anticipate eradication with this method (Krieg and Franz 1989). In a successful biological control program, the IAS population is reduced to an acceptable level, but the populations of prey or host and predator or parasitoid will remain in a lower dynamic balance. Biological control can be particularly appropriate for use in nature reserves and other conservation areas because it is more environment-friendly than more intrusive, less selective measures such as large-scale chemical pesticide use.

Habitat Management

Prescribed Burning

In certain environments the practice of prescribed burning can change the vegetation cover in favor of native plant species, thereby decreasing population levels of weeds. Fire has been used frequently to manage IAS in the United States (e.g., to eradicate Australian pine [*Casuarina equisetifolia*] in pine forests and other fire-tolerant communities). However, only trained and experienced people should undertake prescribed burning because of the many health and safety risks involved.

Grazing

Habitat management with grazing mammals can be a suitable option to obtain the desired plant cover. This method works best where the plants that are to be preserved are adapted to grazing (i.e., they are adapted to high populations of large herbivorous mammals or prevalent in human-made habitats such as pastures and heathland) and the invasive species are more palatable than the natives. On the other hand, unmanaged grazing often favors alien plants because grazing can preferentially remove native vegetation, leaving alien plants, especially toxic species, to grow under reduced competition. This twofold enhancement leads to a monotypic stand of an alien plant, such as leafy spurge infestations in the United States.

Hunting

Continuous hunting can be used to control exotic species, such as deer, originally introduced for hunting purposes (Mack et al. 2000). There are two approaches: commercial hunting principally for meat and recreational hunting. Both approaches can generate income for the landowner or the state. Some exotic species are easy to hunt and are favored species for hunters and so should be straightforward to manage by hunting; conversely, more wary species or those less preferred by hunters are less likely to be managed effectively.

Many other invasive species can be eaten or have edible fruits, which can be exploited for human consumption or as fodder for domesticated animals. In many densely populated parts of the world, invasive plants are esteemed for their production of highly valued firewood or other uses. A high percentage of introduced fish and crustacean species make a good meal, so recreational and industrial fishing are helping to control these invasive populations.

However, in the promotion of an alien species as a food resource lurks the danger of providing an incentive for individuals to spread the alien species to as yet uninfested areas or breed them in captivity, from where they may eventually escape. This issue must be evaluated on a case-by-case basis.

Integrated Pest Management (IPM)

A combination of methods often provides the most effective and acceptable control. For example, the benefits of a chemical herbicide application will be much greater if followed by vegetation management and restoration efforts. As a tool to achieve the overall management goal, the regularly monitored, coordinated integration of methods based on ecological research almost always achieves the best results in managing an invasive species' population. This integrated process entails an assessment of the situation and normally complementary experimental studies to establish the best practice to manage the invasive species. The control process can be complicated or can involve only two different methods, one used after the other. Because these control projects depend on so many variables, no general recommendation can be given for any taxonomic group. The strategy has to evolve based on the knowledge available on the invasive organism, the ecosystem invaded, the climatic conditions, and other native and alien species in the same habitat.

Conclusion

Figure 9.1 summarizes the different steps in dealing with alien species. The most useful tools to prevent introductions of invasive organisms are national and international regulations and their enforcement. Intentional introductions must be screened for their

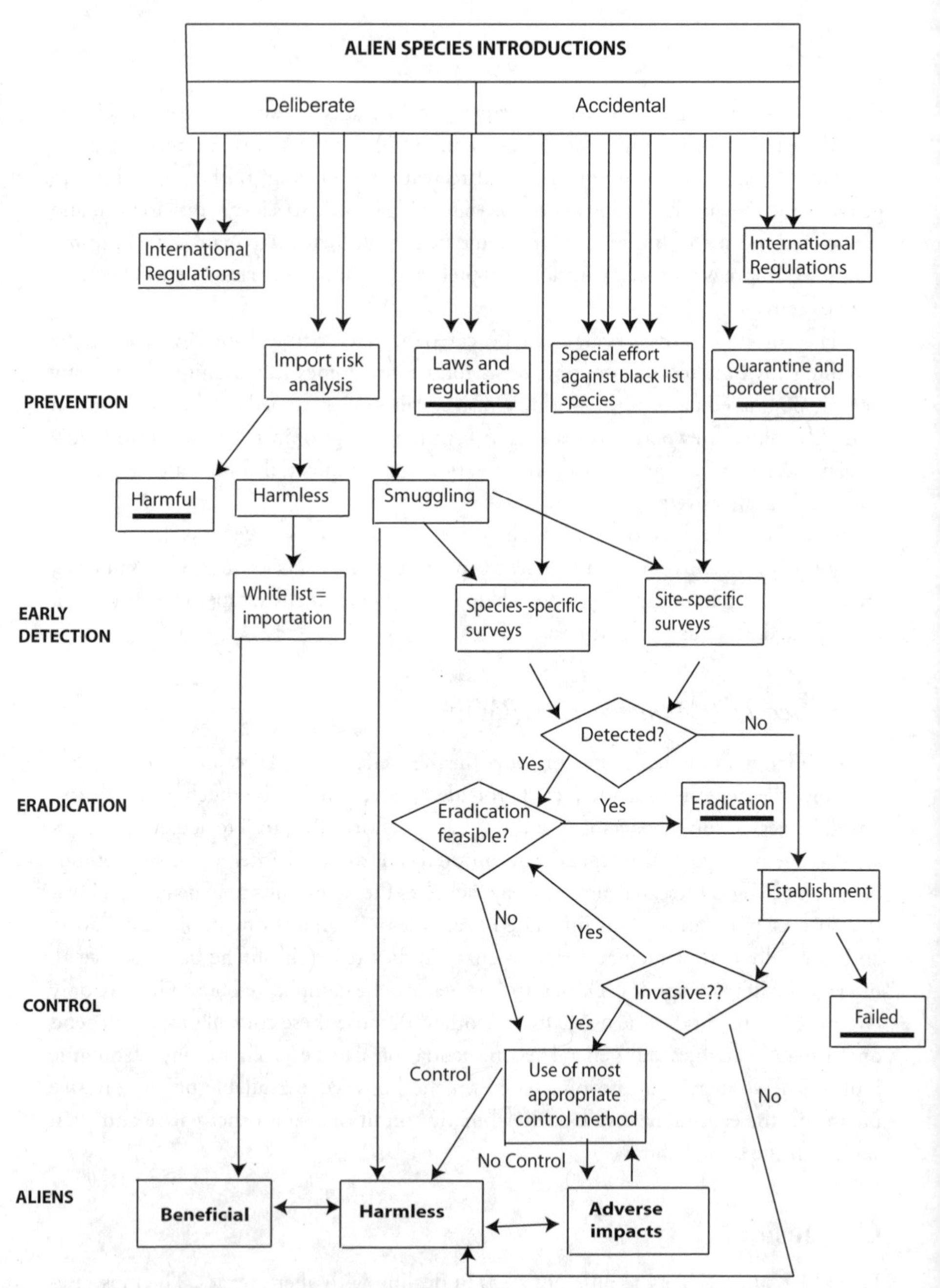

Figure 9.1. Flow chart for addressing alien species introductions. Black bars mark the potential final stages of introduced alien species. Diamonds symbolize important bifurcations and decision points.

invasive potential in a risk assessment process. Accidental introductions must be minimized by effective quarantine inspection services. Raising awareness is a powerful tool for preventing and managing invasive organisms. After an alien species is established, eradication is the best management option, if feasible. If eradication is not possible, or fails, control measures must be used, which often are most effective when used in an integrated way. The only long-term self-sustaining control available today is classic biological control, which, when effective, has an excellent benefit–cost ratio.

References

AQIS. 2000. Media release of the Australian Quarantine and Inspection Service, Department of Agriculture, Fisheries and Forestry, May 23, 2000. Online: http://www.aqis.gov.au/.

Bax, N. J. 1999. "Eradicating a dreissenid from Australia," *Dreissena!* 10(3):1–5.

Boettner, G. H., J. S. Elkinton, and C. J. Boettner. 2000. "Effects of a biological control introduction on three nontarget native species of saturniid moths," *Conservation Biology* 14:1798–1806.

Booth, R. G., A. E. Cross, S. V. Fowler, and R. H. Shaw. 1995. "The biology and taxonomy of *Hyperaspis pantherina* (Coleoptera: Coccinellidae) and the classical biological control of its prey, *Orthezia insignis* (Homoptera: Ortheziidae)," *Bulletin of Entomological Research* 85:307–314.

Courtenay, W. R. 1997. "Nonindigenous fishes," in *Strangers in Paradise,* edited by D. Simberloff, D. C. Schmitz, and T. C. Brown, 109–122. Island Press, Washington, DC.

Cronk, Q. C. B., and L. Fuller. 1995. *Plant Invaders.* Chapman & Hall, London.

Culver, S. C., and A. M. Kuris. 1999. "The sabellid pest of abalone: The first eradication of an established introduced marine bioinvader?" in *Marine Bioinvasions: Proceedings of the First National Conference,* 100–101. MIT Press, Cambridge, MA.

Fowler, S. V. 1996. "Saving the gumwoods in St. Helena," *Aliens* 4:9.

Galil, B. 1999. "The 'silver lining': The economic impact of Red Sea species in the Mediterranean," in *Marine Bioinvasions: Proceedings of the First National Conference,* 265–267. MIT Press, Cambridge, MA.

Greathead, D. J., and A. H. Greathead. 1992. "Biological control in insect pests by insect parasitoids and predators: The BIOCAT database," *Biocontrol, News and Information* 13:61–68.

IPPC. 1996. "Code of conduct for the import and release of exotic biological control agents," *ISPM* 3. FAO, Rome.

Julien, M. H., and M. W. Griffiths, eds. 1998. *Biological Control of Weeds. A World Catalogue of Agents and Their Target Weeds,* 4th ed. CABI Publishing, Wallingford, UK.

Krieg, A., and J. M. Franz. 1989. *Lehrbuch der biologischen Schädlingsbekämpfung.* Paul Parey, Hamburg, Germany.

Louda, S. M., D. Kendall, J. Connor, and D. Simberloff. 1997. "Ecological effects of an insect introduced for the biological control of weeds," *Science* 277:1088–1090.

Mack, R. N., D. Simberloff, W. M. Lonsdale, H. Evans, M. Clout, and F. A. Bazzaz. 2000. "Biotic invasions: Causes, epidemiology, global consequences, and control," *Ecological Applications* 10:689–710.

Panetta, D., P. Pheloung, W. M. Lonsdale, S. Jacobs, M. Mulvaney, and W. Wright. 1994. "Screening plants for weediness: A procedure for assessing species proposed for importation to Australia." A report commissioned by the Australian Weeds Committee, Canberra.

Peck, S. B., J. Heraty, B. Landry, and B. J. Sinclair. 1998. "Introduced insect fauna of an oceanic archipelago: The Galapagos Islands, Ecuador," *American Entomologist* Winter:218–237.

Pemberton, R. W. 2000. "Predictable risk to native plants in weed biological control," *Oecologia* 125:489–494.

Pimentel, D., L. Lach, R. Zuniga, and D. Morrison. 2000. "Environmental and economic costs of nonindigenous species in the United States," *BioScience* 50:53–65.

Ruesink, J. L., I. M. Parker, M. J. Groom, and P. M. Kareiva. 1995. "Reducing the risks of nonindigenous species introductions: Guilty until proven innocent," *BioScience* 45(7):465–477.

Sandlund, O. T., P. J. Schei, and A. Viken, eds. 1999. *Invasive Species and Biodiversity Management.* Kluwer Academic Publishers, Dordrecht, The Netherlands.

Sharp, J. L., and G. J. Hallman. 1994. *Quarantine Treatments for Pests of Food Plants.* Westview Press, Boulder, CO.

Simberloff, D. 1996. "Impacts of introduced species in the United States," *Consequences* The Nature and Implications of Environmental Change, US GCRIO: 2(2). Available at www.gcrio.org/CONSEQUENCES/vol2no2/article2.htm.

———. 1997. "Eradication," in *Strangers in Paradise,* edited by D. Simberloff, D. C. Schmitz, and T. C. Brown. Island Press, Washington, DC.

Tisdell, C. 1990. "Economic impact of biological control of weeds and insects," in *Critical Issues in Biological Control,* edited by M. Mackauer, L. E. Ehler, and J. Roland. Intercept, Andover, UK.

U.S. Congress. 1993. *Harmful Nonindigenous Species in the United States.* Office of Technology Assessment, OTA-F-565. U.S. Congress Government Printing Office, Washington, DC.

Wajnberg, E., J. K. Scott, and P. C. Quimby. 2001. *Evaluating Indirect Ecological Effects of Biological Control.* CABI Publishing, Wallingford, Oxon, UK.

Wittenberg, R., and M. J. W. Cock, eds. 2001. *Invasive Alien Species: A Toolkit of Best Prevention and Management Practices.* CABI Publishing, Wallingford, Oxon, UK.

WTO. 1994. *Agreement on the Application of Sanitary and Phytosanitary Measures.* WTO, Geneva.

10

Legal and Institutional Frameworks for Invasive Alien Species

Clare Shine, Nattley Williams, and
Françoise Burhenne-Guilmin

The Need for Legally Backed Approaches to Invasive Alien Species

The introduction of plant and animal species beyond their natural range has risen sharply as a result of increased opportunities for transport, trade, travel, and tourism between different countries and continents. Newer, faster, and safer methods of transport provide vectors for live plants, animals, and biological material to cross biogeographic barriers that would usually block their way.

A small percentage of introduced alien species go on to become invasive, forming a self-regenerating pollution (de Klemm 1996) that takes no account of political or administrative borders. The causes and impacts of biological invasions thus are by definition international. Globally, invasive alien species (IAS) are now considered the second cause of biodiversity loss after direct habitat destruction.

Tackling the problem is particularly complex because most activities leading to the introduction of potential IAS have legitimate economic and social objectives. These include

- Intentional introductions of species for use in biological production systems (such as agriculture, forestry, fisheries), landscaping, and recreational or ornamental purposes
- Intentional introductions of species for use in containment or captivity (e.g., zoos, aquaculture, mariculture, aquaria, horticulture, the pet trade), from which there is a known risk of escape or release to the wild
- Unintentional introductions of species, organisms, or pathogens through trade and travel pathways (as byproducts, parasites of traded products, or hitchhikers and stowaways in vessels, vehicles, or containers that deliver products)

Introductions may also occur as a result of environmental change at different levels, including ecosystem disruption, changing land use practices, and climate change (Mooney and Hobbs 2000).

Established alien species make a major contribution to the economy in many countries; in some, they also provide indigenous and local communities with alternative forms of subsistence and opportunities to participate in the cash economy.

On the other hand, alien species introductions entail risks that are often poorly understood by policymakers. Environmental, economic, social, and public health impacts that are seen as insignificant in the short term can be extremely serious in the longer term or when cumulative effects across different sectors are taken into account. Costs resulting from biological invasions may include the following:

- Reduction in the value of agricultural land
- Increased operating costs and loss of income
- Collapse of buildings and power failures
- Inefficient irrigation and lowered water tables
- Seed contamination, spread of disease, and incremental pest control costs
- Public health costs associated with spread of disease
- Loss of sport, game, endangered species, and biodiversity
- Ecosystem disturbance and costs associated with protection, monitoring, and recovery
- Loss of opportunity and ecosystem services for future generations
- Loss of equitable access to resources
- Loss of biological components or whole ecosystems on which indigenous and local communities depend
- Loss of traditional knowledge, customs, and practices associated with the native species under threat (adapted from Corn et al. 1999)

At the international, national, and local level, comprehensive policies, tools, and procedures are needed to support assessment of risks associated with alien species introductions and to strike a balance between legitimate socioeconomic activities and appropriate safeguards for the environment, communities, and public health.

The Global Invasive Species Program (GISP) recognizes the need for effective, legally backed approaches as part of a more comprehensive and internationally consistent approach to IAS problems. Priorities identified by the GISP Strategy and Toolkit (Wittenberg and Cock 2001) include building capacity in developing countries for prevention and mitigation of such problems and improved identification, assessment, and management of pathways for introductions.

Policy and Legal Challenges Raised by IAS

Legally backed approaches to IAS have long been fragmented. Measures to exclude unwanted organisms tend to be most developed for economic production sectors (e.g., agriculture) but are often piecemeal and reactive in other sectors. Specific difficulties faced by policymakers and legislators include

- The range of production and trade activities that depend on or involve alien species
- The number of trading partners, sectors, and regulatory bodies concerned
- The number of intercontinental and intracontinental pathways and entry points
- The lack of tools to address state and private accountability for consequences of invasions
- Variations in national capacity and willingness to take steps to prevent or minimize IAS-related problems
- Gaps in data on native species, making it harder to determine what is alien
- Problems in predicting which alien species may become invasive
- The lack of objective criteria or methods for assessing risk
- Difficulty in defining what should be regulated or managed
- Legal difficulties in tackling ongoing invasions, particularly where they result from lawful introductions in the past or affect private land
- The value attributed to alien species by many different groups of stakeholders
- Low political or public awareness of problems posed by IAS

Rationale for International Instruments and Relationship with National Measures

Given the volume of transoceanic movements of potential IAS, unilateral action by a few states can never be enough to prevent or minimize new unwanted introductions. Over the last 50 years, and particularly during the last decade, international support has grown for concerted bilateral, regional, or global measures, as appropriate, to address different IAS-related problems (Glowka and de Klemm 1996).

National initiatives and innovations play a major role in forming new international instruments and coordination processes. A few countries seriously affected by biological invasions, such as Australia and New Zealand, have led the way in developing strong frameworks to tackle existing IAS problems and prevent future invasions. These national experiences and practices help to inform improvements in global and regional frameworks to tackle international and transboundary dimensions of IAS problems. They can also provide a useful point of reference for other countries that are developing or reviewing their domestic arrangements for IAS prevention and control.

The following sections cover

- The scope of existing international requirements and guidelines
- A summary of the legal approaches and tools supported by such instruments
- Strengths, weaknesses, and gaps in the current international regime
- Indicators for designing and strengthening national frameworks and building domestic capacity
- Conclusions and priorities for future action

Existing International Instruments and Programs

More than 50 internationally agreed legal instruments now deal with some aspect of the introduction, control, and eradication of alien species.[1] These set out the international norms or guidelines to which states have agreed, which should form the baseline for national legal frameworks.

Types of International Instruments

Binding instruments (treaties and conventions) are agreements between states with a mandatory character: they must be observed and their obligations performed in good faith. Treaties often are fairly general in character, particularly global treaties on which consensus must be reached between states with very different levels of development and different constitutional, legislative, and administrative systems.

Treaty commitments take the form of performance obligations (*obligations de résultat*), which require parties to take steps to achieve certain goals, or obligations to use prescribed means (*obligations de moyens*) to achieve a prescribed objective. Both types leave the actual mode of implementation to parties, but under performance obligations, parties have greater flexibility in selecting the methods to achieve set goals.

In many cases, treaty provisions are formulated in a way that is not self-executing. These can be made operational in national legal systems only through appropriate legislation and regulations.

Nonbinding instruments, sometimes called soft laws, adopted by intergovernment forums take the form of resolutions, recommendations, action programs, and codes of conduct. Where developed with high participation of the international community, these are evidence of broad consensus on a particular topic and as such are likely to be implemented in good faith by the states and organizations concerned. However, not being mandatory, they cannot be used to implement common norms for legislation and assessment procedures or to require financial or capacity-building contributions.

In addition, nonstate actors, including international nongovernment organizations such as the World Conservation Union (IUCN),[2] may develop guidelines and other advisory material to help states and nonstate actors to formulate policies and actions.

Evolution and Scope of International Instruments

Different multilateral processes have developed rules and guidelines for specific purposes, which affect the way in which they reference alien species, ecosystems, and pathways. This sectoral pattern of development is reflected in current international institutional arrangements.

The earliest agreements were intended to control the introduction and spread of pests and diseases to protect human, animal, and plant health through the establishment of quarantine systems. A series of quarantine agreements now govern sanitary, zoosanitary, and phytosanitary measures to control introductions for such purposes.

From the 1970s, treaties on nature conservation, environmental protection, and sustainable use of natural resources have systematically included requirements to prevent or control alien species introductions.

In the 1990s, instruments began to address the movement and release of living modified organisms (LMOs), including genetically modified organisms (GMOs) resulting from modern biotechnology. These transgenic organisms have certain points of similarity with alien species: they have no normal distribution because they do not occur in the natural environment until released, their release or escape might have severe and irreversible ecological effects, and they might become self-replicating pollutants and interact with ecosystems in unexpected and unpredictable ways.

The next sections outline the scope and requirements of key instruments in the following areas:

- Conservation and sustainable use of biological diversity
- Conservation and sustainable use of aquatic ecosystems and fisheries
- Transboundary movements of living modified organisms
- Sanitary, zoosanitary, and phytosanitary measures
- International transport
- Relationship between these instruments and the multilateral trading system

Instruments for Conservation and Sustainable Use of Biological Diversity

Convention on Biological Diversity (Rio de Janeiro, 1992)

The Convention on Biological Diversity (CBD), which now has 188 parties, is the only global treaty to mandate prevention and mitigation measures for all IAS categories. In the context of in situ conservation, parties are required, as far as possible and as appropriate, "to prevent the introduction of, control or eradicate those alien species which threaten ecosystems, habitats or species" (Art. 8[h]). Several general provisions of the CBD are also relevant to the design of legal measures for this purpose.[3]

Since 1998, CBD institutions have significantly advanced action on IAS in four main ways:

- Designating IAS as a cross-cutting issue to be considered in all thematic work programs under the CBD.[4]
- Urging closer cooperation and collaboration between the CBD and key international institutions.[5] GISP has facilitated this interinstitutional dialogue through a range of workshops and conferences.
- Adopting guidance for implementation of Article 8(h) in the form of Guiding Principles for the prevention, introduction, and mitigation of impacts of alien species that

threaten ecosystems, habitats, or species.[6] Parties, governments, and relevant organizations are urged to promote and implement the Guiding Principles, which support a three-stage hierarchical approach:
  - Priority should be given to preventing the introduction of IAS between and within states.
  - If an IAS has been introduced, early detection and rapid action are crucial to prevent its establishment. The preferred response often is to eradicate the organism as soon as possible.
  - If eradication is not feasible or the necessary resources are not available, containment and long-term control measures should be implemented.
- Commissioning further consideration of practical options for full and effective implementation of Art. 8(h), including the establishment of an ad hoc technical expert group by the Subsidiary Body on Scientific, Technical and Technological Advice (SBSTTA) to further address gaps and inconsistencies in the international regulatory frameworks at global and regional levels. This work builds on review papers prepared by GISP partners at the request of the CBD Secretariat.[7]

In 2002, the CBD Conference of Parties (COP) endorsed the Global Strategy on Plant Conservation, which sets targets for eradication of invasive plants.

Agenda 21 (UN Conference on Environment and Development, Rio de Janeiro, 1992)

The nonbinding Agenda 21 recognizes the threat posed by the introduction of alien species to environmental security and biodiversity and urges states to take action to address impacts of alien species in various activities, including

- Combating deforestation (Chapter 11)
- Managing fragile ecosystems and combating desertification (Chapter 12)
- Conserving biodiversity (Chapter 15)
- Protecting the oceans, seas, and coastal areas (Chapter 17)
- Protecting freshwater resources (Chapter 18)

Convention on the Conservation of Migratory Species of Wild Animals (CMS, Bonn, 1979)

Parties to CMS are required to prevent, reduce, and control the factors endangering migratory species, including "strictly controlling the introduction of, or controlling or eliminating already introduced exotic species" (Art. III[4]).

This provision has been elaborated in the Agreement on the Conservation of African–Eurasian Migratory Waterbirds (The Hague, 1995), concluded pursuant to the CMS. The deliberate introduction of nonnative waterbird species into the environment is prohibited, and all appropriate measures must be taken to prevent the unintentional release of such species if this would affect the conservation status of wild fauna and flora. Where nonnative waterbird species have already been introduced, appropriate measures must be implemented to prevent these species from becoming a potential threat to indigenous species (Art. III[2]). In 2002, parties to this agreement adopted Interim Conservation Guidelines on the Avoidance of Introductions of Non-Native Migratory Waterbird Species (Resolution 2.3).

Regional Biodiversity-Related Instruments

Many regional agreements contain some type of requirement to regulate the introduction of alien species. These vary widely in scope and content: some apply only to intentional introductions, others just to releases in protected areas.

A nonexhaustive list includes the African Convention on the Conservation of Nature and Natural Resources (1968), Bern Convention on the Conservation of European Wildlife and Natural Resources (1979), Protocol for the Implementation of the Alpine Convention in the Field of Nature Protection and Landscape Conservation (1994), ASEAN Agreement on the Conservation of Nature and Natural Resources (1985), Convention on the Conservation of Nature in the South Pacific (1990), and Convention for the Conservation of the Biodiversity and the Protection of Wilderness Areas in Central America (1992).

The Standing Committee to the Bern Convention has played a particularly active role on IAS issues, first through a series of targeted recommendations[8] and more recently through the adoption of the comprehensive European Strategy on Invasive Alien Species in 2003 (Genovesi and Shine 2004). This adapts the CBD Guiding Principles to pan-European needs and priorities.

Because of the region's isolation and vulnerability to invasion, strict rules have been agreed upon under the Antarctic Treaty Regime. The 1991 Madrid Protocol on Environmental Protection provides that no animal or plant species not native to the Antarctic Treaty Area may be introduced onto land or ice shelves, or into water in the Antarctic Treaty Area, except in accordance with a permit.[9]

Some regional economic integration organizations, including the European Community,[10] address potential impacts of alien species on biodiversity. The Southern African Development Community (SADC) has included measures related to alien species in its Protocol on the Conservation, Sustainable Management and Sustainable Development of Forests and Forest Lands in the SADC Region, which was signed by 15 heads of state in October 2002.

Instruments for Conservation and Sustainable Use of Aquatic Ecosystems

In marine environments and inland water systems, alien species can be particularly hard to detect and organisms can disperse rapidly. International instruments dealing with marine and freshwater ecosystems have long focused on the need to prevent such introductions.

Introductions to Marine and Coastal Ecosystems

The UN Convention on the Law of the Sea (UNCLOS; New York, 1982) requires parties to take all measures necessary to prevent, reduce, and control pollution of the marine environment resulting from the intentional or accidental introduction of species, alien or new, to a particular part of the marine environment, which may cause significant and harmful changes thereto (Art. 196).

The CBD COP addresses introductions to marine and coastal ecosystems in its Jakarta Mandate on Marine and Coastal Biological Diversity.[11] The updated Program of Work (Element 5) aims to prevent the introduction of IAS into the marine and coastal environment and to eradicate to the extent possible the IAS that have already been introduced. It establishes three operational objectives:

- To achieve better understanding of the pathways and the causes of the introduction of alien species and the impact of such introductions on biological diversity
- To put in place mechanisms to control all pathways, including shipping, trade, and mariculture, for potential IAS in the marine and coastal environment
- To maintain an incident list on introductions of alien species

At a regional level, many agreements and action plans developed under the United Nations Environmental Programme Regional Seas Program include provisions on alien species. Binding requirements are laid down by the four protected area protocols concluded for the Mediterranean, wider Caribbean area, southeast Pacific, and eastern African region. The Regional Activity Centre for Specially Protected Areas in the Mediterranean developed an Action Plan on Species Introductions and Invasive Species in the Mediterranean Sea in 2002. Under the Convention for the Protection of the Marine Environment of the North-East Atlantic (1992), the new Annex V on the conservation and protection of marine biodiversity and ecosystems identifies introductions of alien or genetically modified species as a potentially harmful activity that should be appropriately managed.

The nonbinding Global Program of Action for the Protection of the Marine Environment from Land-Based Activities (1995) lists the introduction of alien species as a potential threat to the integrity of marine ecosystems (paras. 149–154) but does not provide specific guidance for addressing the problem.

Introductions to Inland Water Systems

The global Convention on the Law of Non-Navigational Uses of International Watercourses (New York, 1997; not in force) requires watercourse states to take all necessary measures to prevent the introduction of species, alien or new, into an international watercourse that may have effects detrimental to the ecosystem of the watercourse, resulting in significant harm to other watercourse states (Art. 22).

Regionally, the Convention on Fishing in the Danube (Bucharest, 1958) was the first instrument to require states to prohibit the introduction of new species into an inland water system. Acclimatization and breeding of new fish species, other animals, and aquatic plants are prohibited except with the consent of the convention's multilateral commission (Art. 10).

In East Africa, two agreements aim to protect Lake Victoria from IAS. The Agreement for the Preparation of a Tripartite Environmental Management Program for Lake Victoria (Dar es Salaam, 1994) supports a 5-year program to strengthen regional environmental management, including the control of alien species such as the water hyacinth (*Eichhornia crassipes*). The Convention for the Establishment of the Lake Victoria Fisheries Organization (Kisumu, 1994) establishes a regional organization with authority to advise on the effects of the direct or indirect introduction of any nonindigenous aquatic animals or plants into the waters or tributaries of Lake Victoria. Parties agree to adopt and enforce laws and regulations prohibiting the introduction of nonindigenous species to Lake Victoria, other than in accordance with the decision of the tripartite Council of Ministers (Art. XII[3]).

Introductions to Wetlands

The Convention on Wetlands of International Importance Especially as Waterfowl Habitat (Ramsar, 1971) does not reference IAS, but its COP has adopted two resolutions on Invasive Species and Wetlands in 1999 and 2002.[12] Parties are urged to address wetland IAS issues in a decisive and holistic manner, using tools and guidance developed by various institutions and under other conventions. They should recognize that terrestrial IAS can affect wetland ecological character (e.g., lowering of water tables, alteration of water flow patterns) and ensure that appropriate measures to prevent or control such invasions are in place.

Technical Guidance for Fisheries and Aquaculture Operations

Aquaculture and mariculture facilities present a known risk of escapes to aquatic ecosystems. Sectoral codes of conduct have been adopted to establish principles and standards and provide best practice guidance for this rapidly growing industry, but there are still significant gaps in the international regulatory framework.

The 1995 UN Food and Agriculture Organization (FAO) Code of Conduct for Responsible Fisheries urges states to adopt measures to prevent or minimize harmful effects of introducing nonnative species or genetically altered stocks used for aquaculture into waters, especially where there is significant potential for such species or stocks to spread into waters under the jurisdiction of other states. Technical guidance has been developed under this code on a precautionary approach to capture fisheries and species introductions and on aquaculture development (FAO Technical Guidelines for Responsible Fisheries 2/1996 and 5/1997).

The 1994 Code of Practice on the Introductions and Transfers of Marine Organisms was issued by the European Inland Fisheries Advisory Commission of the FAO and the

International Council for the Exploration of the Sea (ICES). It establishes procedures and practices to diminish the risk from intentional and unintentional introductions of marine alien species into marine and freshwater ecosystems.

Living Modified Organisms (Including Genetically Modified Organisms)

As noted earlier, LMOs, including GMOs, can disrupt native biodiversity, natural resources, and ecological processes unless appropriately assessed, regulated, and managed (Kinderlerer 2004).

The CBD requires parties to establish or maintain means to regulate, manage, or control the risks associated with the use and release of LMOs resulting from biotechnology that are likely to have adverse environmental impacts that could affect the conservation and sustainable use of biological diversity, taking into account the risks to human health (Art. 8[g]).

The Cartagena Protocol on Biosafety (concluded under the CBD in Montreal, 2000) aims to "contribute towards ensuring an adequate level of protection in the field of the safe transfer, handling and use of living modified organisms resulting from modern biotechnology." It defines an LMO as "any living organism that possesses a novel combination of genetic material obtained through the use of modern biotechnology" (Art. 3[g]). Parties must ensure that the release of LMOs into the natural environment takes place in a way that prevents and reduces risks to biodiversity, taking into account the risks to human health (Art. 2).

Transboundary movements of LMOs for intentional introduction into the environment are subject to advanced informed agreement (AIA) of the importing state. The AIA procedure is the cornerstone of the protocol: it includes notification to the importing party and risk assessment to be carried out before the transboundary movement. Elements of risk assessment are set out in Appendix II. Parties must also take risk management measures that include monitoring of organisms released and the preparation of emergency plans. Transboundary movements of LMOs for food, feed, and processing are subject to a less restrictive procedure. In both cases, specific requirements for identification are laid down, depending on the purpose of the movement.

If a party knows of a release of LMOs that may lead to an unintentional transboundary movement with possible significant adverse effects, it must notify and consult with potentially affected states and relevant international organizations. Parties must cooperate in identifying LMOs that have adverse effects on biodiversity and in taking appropriate measures regarding the treatment of or trade in LMOs having such effects. However, the protocol does not specify the content of such measures.

Some of the aforementioned instruments apply to GMOs as well as alien species. In addition to UNCLOS and the 1997 International Watercourses Convention, these include

- The 1995 FAO Code of Conduct on Responsible Fisheries (recommendations for aquaculture operations cover nonnative species and genetically altered stocks)
- The FAO Code of Conduct on the Introductions and Transfers of Marine Organisms (includes procedures for the release of GMOs into marine and freshwater ecosystems)
- The 1995 Protocol Concerning Specially Protected Areas and Biological Diversity in the Mediterranean. Parties must regulate the introduction of GMOs as well as alien species to specially protected areas. They should also take all appropriate measures to regulate intentional or accidental introductions of GMOs and alien species to the wild and prohibit those that could have harmful impacts on ecosystems, habitats, or species in the protocol area. Similar measures also apply under the Protocol for the Wider Caribbean Region.

Phytosanitary, Zoosanitary, and Sanitary Measures

The main objective of sanitary, zoosanitary, and phytosanitary measures is to protect the health and life of humans, animals, and plants from damage by pests and diseases. Such measures involve use of quarantine controls for these specific objectives, not for environmental protection per se. However, to the extent that such measures limit introductions or movements of certain alien species, they can have benefits for the natural environment and contribute to maintaining wild native biodiversity.

Because quarantine measures may involve trade restrictions, they must be considered not only in their own right but also in the context of the multilateral trading system.

Plant Health and the International Plant Protection Convention

The International Plant Protection Convention (IPPC; Rome, 1951; revised 1997; revised version not yet in force) provides a framework for international cooperation to "secure common and effective action to prevent the spread and introduction of pests of plants and plant products, and to promote appropriate measures for their control" (Art. 1.1). Its objectives include the development and application of international standards in international trade to prevent the introduction and dissemination of plant pests, taking into account internationally approved principles governing the protection of plant, animal, and human health and the environment (Durand and Chiaradia-Bousquet 1999).

The IPPC defines *pest* as "any species, strain or biotype of plant, animal or pathogenic agent injurious to plants or plant products." This definition covers IAS that qualify as pests of plants or plant products but does not extend to invasive species that are nonplant pests (e.g., hitchhiker organisms such as spiders in table grapes, ants in taro). As regards containment of species that have become invasive, IPPC provisions apply primarily to designated "quarantine pests"[13] that are subject to official control.[14]

Parties to the IPPC (currently 127) must establish a national plant protection organization and adopt legislative and technical measures and standards to identify pests that threaten plant health, assess their risks, and prevent their introduction and spread. National phytosanitary systems should include phytosanitary and other import regulations (quarantine and postquarantine controls); compliance systems for all movements that can help the transfer of pests; pest risk analysis to provide technical justification for import restrictions; inspection, reporting, and surveillance systems; eradication and control systems; and export certification systems to ensure that exported products comply with the import requirements of trading partners (Hedley 2004).

The IPPC Secretariat facilitates development of International Standards for Phytosanitary Measures (ISPMs), which are adopted by the IPPC's governing body, the Interim Commission on Phytosanitary Measures. ISPMs are designed to encourage international harmonization of phytosanitary measures to facilitate safe trade and avoid the use of unjustified measures as barriers to trade.

IPPC implementation is supported by nine regional plant protection organizations, established to strengthen the capacity of countries in a region to address phytosanitary issues:

- The Asia and Pacific Plant Protection Commission (established 1956)
- The Caribbean Plant Protection Commission (established 1967)
- The Comité Regional de Sanidad Vegetal para el Cono Sur (established 1980)
- The European and Mediterranean Plant Protection Organization (established 1951)
- The Inter-African Phytosanitary Council (established 1954)
- The Junta del Acuerdo de Cartagena (established 1969)
- The North American Plant Protection Organization (established 1976)
- The Organismo Internacional Regional de Sanidad Agropecuaria (established 1953)
- The Pacific Plant Protection Organization (established 1995)

Regional plant protection organizations vary widely in their operational capabilities. Some have developed ISPMs and submitted them for consideration to the Interim Commission. As revised in 1997, the IPPC provides for regional economic integration organizations to become parties. This should facilitate coordinated regional contribution to the development of new standards.

In 2003, the IPPC in cooperation with other international organizations, including GISP, organized a Workshop on Invasive Alien Species and the IPPC.

Use of Alien Biological Control Agents

The FAO Code of Conduct for the Import and Release of Exotic Biological Control Agents has been adopted as an ISPM under the IPPC. It sets out agreed procedures for the import and release of exotic agents capable of self-replication (parasitoids, predators, parasites, phytophagous arthropods, and pathogens). The code specifies the respective responsibilities of government authorities and exporters and importers of biological control agents.

Animal Health and the Office International des Epizooties

The Office International des Epizooties (OIE) develops standards and guidance on pests and diseases of concern with regard to trade in animals, including aquatic animals, and animal products but not on animals that are potentially invasive in their own right. It may consider risks to wild animals associated with disease transmission to or from livestock (e.g., rinderpest, avian influenza).

The OIE is recognized as the standard-setting body on animal health under the World Trade Organization (WTO) Agreement on the Application of Sanitary and Phytosanitary Measures (SPS Agreement). Standards are set out in the International Animal Health Code for Mammals, Birds, and Bees, including on import risk analysis and import and export procedures, and in the International Aquatic Animal Health Code, which aims to "facilitate trade in aquatic animal products." The latter specifies minimum health guarantees required of trading partners to avoid the risk of spreading aquatic animal diseases. It contains model international certificates for trade in live and dead aquatic animals.

Human Health and the International Health Regulations

Invasive species may serve as hosts or vectors for diseases that affect human health. Therefore, measures are necessary to control the introduction and spread of alien invasive disease organisms. The International Health Regulations (IHRs, Geneva, 1969; amended in 1982) were adopted by the World Health Assembly of the World Health Organization and are designed to ensure maximum security against the international spread of infectious diseases to humans with minimal interference with world traffic. The goals of the IHRs are to

- Detect, reduce, or eliminate sources from which infections spread
- Improve sanitation in and around ports and airports
- Prevent dissemination of vectors

The IHRs are being revised, in response to increased international traffic and communication technology, to address weaknesses identified in current global outbreak alert and response activities and to cover existing, new, and reemerging diseases including emergencies associated with food safety and animal diseases. After worldwide consultations in 2002, an Intergovernmental Working Group on the Revision of the IHRs met again in November 2004.

Technical Guidelines for Transport Operations

Some international organizations have developed sectoral guidance to reduce IAS risks associated with international transport pathways.

Shipping

The International Maritime Organization (IMO) has worked to prevent the spread of marine alien organisms in ballast water and sediments since the mid-1970s. In 1997, the IMO Assembly adopted voluntary guidelines[15] to help governments and appropriate authorities, ships' masters, operators and owners, and port authorities to establish common procedures to minimize the risk of introducing harmful aquatic organisms and pathogens from ships' ballast water and associated sediments while protecting ships' safety.

In February 2004, the IMO Assembly adopted a legally binding International Convention for the Control and Management of Ships' Ballast Water and Sediments. Parties undertake to give full and complete effect to the provisions of the convention and the annex in order to prevent, minimize, and ultimately eliminate the transfer of harmful aquatic organisms and pathogens through the control and management of ships' ballast water and sediments. They have the right to take, individually or jointly with other parties, more stringent measures for this purpose, consistent with international law. Parties should ensure that ballast water management practices do not cause greater harm than they prevent to the environment and to human health, property, or resources. The convention contains specific provisions on sediment reception facilities (article 5), research and monitoring (article 6), survey, certification, and inspection (article 7), and technical assistance and regional cooperation (article 13). The annex sets out management and control requirements for ships.

The IMO has joined forces with the Global Environment Facility and United Nations Development Programme to implement the Global Ballast Water Management Program (Globallast). This technical cooperation program, which uses a demonstration site approach in six countries, is designed to help less industrialized countries to implement the IMO guidelines and prepare for implementation of the new Convention.

Civil Aviation

The International Civil Aviation Organization (ICAO) has adopted two Resolutions on Preventing the Introduction of Invasive Alien Species.[16] ICAO Contracting States are urged to support one another's efforts to reduce the risk of introducing potentially invasive species through civil air transportation to areas outside their natural range. The resolution calls on the ICAO Council to work with appropriate concerned organizations to identify approaches that the ICAO might take to assist in reducing the risk of introducing such species. In 2002, the ICAO carried out a survey of 188 states to gather data for an assessment of whether civil aviation is a significant or high-risk pathway for unintentional introductions. About half of the respondent states that are aware of IAS problems in their respective countries consider air transport to be a contributing factor (the other half lacked the data to respond). The ICAO Assembly will consider the possible need for a formal prevention strategy at its 35th Assembly in 2004.

Relationship between Existing Instruments and the Multilateral Trading System

Global Trade Agreements

International trade in products and services between the 147 members of the WTO is governed by the 1994 Uruguay Round of Agreements. This provides for binding rules, enforced by a compulsory dispute settlement mechanism, to ensure that governments extend nondiscriminatory market access to each other's products and services. These rules are based on principles of nondiscrimination, transparency, and predictability. WTO agreements are not directly concerned with the content of environmental policies and measures, except to the extent that these may have a significant impact on international trade (Downes 1999).

The WTO Agreement on the Application of Sanitary and Phytosanitary Measures (1995) is relevant to alien species characterized as pests or diseases. Under the SPS Agreement, each WTO member may adopt national measures or standards to protect human, animal, or plant health or life from the risks arising from the entry, establishment, or spread of pests, diseases, or disease-causing organisms; and prevent or limit other damage within its territory from these causes.[17] National measures that may affect international trade must be consistent with WTO principles and rules, as expressed through the SPS Agreement. The SPS Agreement requires or promotes the following:

Use of international standards as a basis for national SPS measures

A national measure that conforms to an international standard benefits from a rebuttable presumption of consistency with the agreement. The SPS Agreement currently recognizes international standards set by organizations within the IPPC framework (pests of plants and plant health), the OIE (pests and diseases of animals), and the Codex Alimentarius Commission (food safety and human health).

For matters not covered by these organizations, international standards may include "appropriate standards, guidelines and recommendations promulgated by other relevant international organizations open for membership to all Members, as identified by the Committee." At present, however, no organization has been recognized to set international standards regarding general environmental and biodiversity protection against IAS.

Although the existing standard-setting organizations have mandates broad enough to cover some environmental and societal impacts, most standards adopted to date do not take these aspects adequately into account, and some countries have been reluctant to support such an extension to environmental issues. This attitude partly reflects concern that risk analysis for pests affecting the natural environment is extremely difficult and that operational capacity to conduct pest risk analysis is often limited (Hedley 2004). However, in 2003 the IPPC's Interim Commission on Phytosanitary Measures approved a new standard on the analysis of environmental risks, including coverage of taxa that affect unmanaged systems, directly or indirectly.[18]

If an international standard is not used or is not available, use of scientifically based risk assessment to justify a national measure

A national measure that sets a higher level of protection than an international standard or is adopted in the absence of an international standard should be justified by a scientifically based risk assessment.

The SPS Agreement provides that where relevant scientific information is insufficient, provisional measures may be applied until sufficient scientific evidence is available. Parties concerned have a duty to seek such scientific evidence actively.[19]

Some cases brought before the WTO dispute settlement body[20] relate to import controls on alien species, in particular whether risk assessment before development of such controls was based on sufficient science. These cases reveal differences in perception between importers and exporters and raise difficult issues about the use of precautionary measures in the design and application of trade measures. WTO rulings have held that although precaution is reflected in the SPS Agreement, this does not override the need for risk assessment based on available scientific evidence.

Consistency in the application of appropriate levels of protection

Use of least–trade restrictive alternatives

SPS measures should not arbitrarily or unjustifiably discriminate between members where identical or similar conditions prevail. Therefore, WTO members must be consistent when they deal with risk over a range of measures and products. In addition, the measure must not be more trade restrictive than necessary to achieve the appropriate level of protection. A measure is deemed to be trade restrictive if another SPS measure that is reasonably available, taking into account technical and economic feasibility, would achieve the appropriate level of protection in a less restrictive way than the measure contested.

Acceptance of equivalent measures

Transparency through advance notification of trade measures

WTO members are required to accept the sanitary and phytosanitary measures of other members as equivalent, even if they are different from their own or those of other members. In addition, members must notify other countries in advance, except in emergencies, of any new or changed SPS-related measures that affect trade, establish an office to respond to requests for more information, and solicit comments from trading partners on the proposed measure.

Regional Agreements

At least three regional economic integration organizations have powers to develop regulations or recommendations regarding certain aspects of trade in potentially invasive alien species:

- The North American Free Trade Agreement (1993). The Council of the Commission on Environmental Cooperation has discretion to develop recommendations regarding introduction of exotic species that may be harmful.
- Mercosur, for the Southern Cone countries of South America (1991). Decision 6/96 of the Mercosur Consejo Mercado Común has approved the WTO SPS Agreement.
- The European Community, for the 25 member states of the European Union.

Legal Approaches and Tools Supported by International Instruments

Existing international instruments support the application of certain legal approaches and tools in measures to address IAS issues, outlined briefly in this section. Techniques for their application at a national level are discussed further later in this chapter.

Ecosystem Approach

CBD Guiding Principle 3 provides that measures to deal with IAS should, as appropriate, be based on the ecosystem approach. This approach may be summarized as a strategy for integrated management of land, water, and living resources within a given ecological unit to promote equitable conservation and sustainable use based on the application of appropriate scientific methods (McNeely 1999).

The ecosystem approach has been defined and elaborated in a series of CBD decisions.[21] Those of particular relevance to alien species include

- Decentralized management to the lowest appropriate level
- Consideration of the effects of management activities on adjacent and other ecosystems
- Economic context of ecosystem management
- Conservation of ecosystem structure and functioning
- Recognition of varying temporal scales and lag effects that characterize ecosystem processes
- Involvement of all relevant sectors and scientific disciplines

Implementation of an ecosystem approach presents particular challenges to national and regional frameworks. First, legal systems usually operate within jurisdictional boundaries that seldom correspond to the boundaries of ecological units. Second, sectoral planning is still much more common than integrated (cross-sectoral) planning. For these reasons, there is a critical need to develop more effective interjurisdictional cooperative management agreements and for mechanisms to operate cross-sectoral integration, both within and between states.

International and Transboundary Cooperation

The obligation of states to cooperate with one another derives from the very essence of general international law. Cooperation is particularly important in international environmental law because ecosystems and natural resources often straddle national boundaries and because many threats to ecosystems and natural resources cannot be addressed and regulated by states individually.

For IAS, cooperation is essential to tackle most pathways and activities giving rise to unwanted introductions and to implement control and eradication measures for species that move beyond the boundaries of the state where they were introduced.

Under customary international law, a state that plans to undertake or authorize activities that may have measurable effects on the environment of another state must inform and consult with that state.[22] Several treaties contain equivalent measures. The Convention on Environmental Impact Assessment in a Transboundary Context (Espoo, Finland 1991)[23] sets out substantive and procedural requirements for assessment of transboundary impacts of certain activities. All stakeholders, including the public in the affected area, should be informed and have an opportunity to participate in environmental impact assessment (EIA) procedures and decision making.

Although many environmental agreements have generic requirements for transboundary cooperation,[24] very few specify how neighboring countries should handle risks relating to alien species introductions. The Benelux Convention is the only treaty to mandate consultation with neighboring states before intentional introductions (of alien plants).[25]

At the nonbinding level, CBD Guiding Principle 4.1 calls on states to recognize the risk that activities within their jurisdiction or control may pose to other states as a potential source of IAS and should take appropriate individual and cooperative actions to minimize that risk, including the provision of any available information on invasive behavior or invasive potential of a species. The European Strategy on Invasive Alien Species sets out detailed recommendations for interstate cooperation, notification, and consultation in order to coordinate precautionary and control measures for invasive species.[26]

The IPPC does not appear to require risk analysis of impacts to another country (i.e., where a pest may have transboundary impacts) but does contain measures to prevent or limit the spread of pests between countries.

Interjurisdictional cooperation and consistency are particularly important in regional economic integration organizations that promote free movement of goods within their borders and between their member states to avoid situations in which stringent measures adopted in one country are undermined by weaker measures across a boundary.[27] Consistent approaches are needed to minimize introductions between ecologically distinct parts of the same free trade area (especially geographically or evolutionarily isolated areas such as islands) and identify the least trade restrictive measures necessary to prevent such spread.

Prevention

The objective of almost all international environmental agreements is to prevent environmental deterioration. For IAS, the duty to take preventive measures is enshrined in all relevant instruments (including CBD Guiding Principles 7–9) and should form the cornerstone of all national legal frameworks. There is widespread consensus that prevention is more desirable ecologically and more advantageous economically than measures taken to remedy a damage already caused. In addition, once an introduced species becomes invasive, eradication may be impossible and the ecological damage irreversible.

The principle of prevention thus applies to all situations that may have adverse effects on the environment. Its application involves a prior assessment of the contemplated activity to determine whether the activity exceeds a threshold set to trigger and justify preventive action. Thresholds may differ from one instrument to another; many environmental instruments set a threshold linked to possibility or likelihood that effects will result in significant harm to the environment. Similar formulas may be followed in national laws.

For alien species introductions, the principle of prevention is particularly relevant to situations where the potential impact (risk) associated with a proposed introduction or pathway is sufficiently known and quantifiable for decision makers to decide, on the basis of evidence provided by tools such as risk assessment, whether the introduction contemplated is acceptable, unacceptable, or acceptable under certain conditions. International instruments require, and national legal systems need to provide, legal mechanisms to implement prevention measures.

- For intentional introductions, these legal mechanisms usually take the form of a permit system, enabling total prohibition (refusal of a permit) or restricted entry (issue of a permit with or without specific conditions).
- For unintentional introductions, measures to prevent or minimize unwanted introductions must be targeted at the pathway level. They involve identification of major pathways that lead to harmful invasions and assessment and management of associated risks through appropriate operating standards and best practices.

Precaution

Many situations are characterized by scientific uncertainty about the potential impact on the environment of an activity or process, with the possibility that inaction may lead to potentially irreversible ecological damage. It is in such circumstances that precautionary measures are called for.

Principle 15 of the Rio Declaration proclaims that "in order to protect the environment, the precautionary principle shall be widely applied by States according to their capabilities. Where there are threats of serious or irreversible damage, lack of full scientific certainty shall not be used as a reason for postponing cost-effective measures to prevent environmental degradation."

There has been intense debate about whether the use of precautionary measures in environmental management reflects a recommended policy approach or derives from a legally established principle. What is clear, however, is that an increasing number of international instruments, from the Biosafety Protocol to the FAO Code of Conduct on Responsible Fisheries, mandate or support the use of precautionary measures. Many national legal systems have also established a legal basis for precautionary measures.

CBD Guiding Principle 1 provides that given the unpredictability of the pathways and impacts on biodiversity of IAS, efforts to identify and prevent unintentional introductions and decisions concerning intentional introductions should be based on the precautionary approach, in particular with reference to risk analysis, as set forth in the Rio Declaration. This approach also should be applied when considering eradication, containment, and control measures in relation to alien species that have become established. Lack of scientific certainty about the implications of an invasion should not be used as a reason for postponing or failing to take appropriate eradication, containment, and control measures.

Under the current multilateral trading system, the SPS Agreement also permits application of precautionary measures until sufficient scientific information is available.

There is no difference in kind between precautionary and prevention measures: the former can be considered the most developed form of the latter, and the legal techniques used to implement them are the same (e.g., permits, risk analysis). However, in the interests of legal certainty, it is important to clarify when precautionary measures are justified and how to achieve transparency in the mechanisms or procedures enabling their use, specifically to avoid situations that might create conflicts with the multilateral trading system.

The use of precautionary measures is particularly relevant to alien species issues because of the inherent scientific uncertainty and limitations on predictive capacity. Many unknown variables influence the likelihood that an organism will survive transport to, establish, and spread in a given location. This includes possible time lag before an introduced species shows invasive characteristics.

National legal systems therefore must provide for and enable decision making on introductions of alien species when scientific uncertainty surrounds the potential environmental impacts. Transparency and predictability are important elements of such rule making, and so is the need to ensure harmony with the international conventions to which the regulating state has subscribed.

Public Participation and Access to Information

Public participation in environmental planning and decision making is mandated by many international instruments, notably the 1998 Aarhus Convention on Access to Information, Public Participation in Decision Making and Access to Justice in Environmental Matters (in force since October 30, 2001). This principle is increasingly reflected and organized in national legal systems.

Permits and Prior Informed Consent

Several international instruments require states to establish permit requirements for the introduction of alien species, including UNCLOS, the Antarctic Treaty Regime, and the

ASEAN Agreement. Other instruments, such as the Cartagena Protocol on Biosafety, subject the import of certain goods to the Prior Informed Consent or Advanced Informed Agreement of the importing state.

National permit systems provide a framework within which applications to introduce alien species can be assessed and an informed decision made before an intentional introduction is authorized.

Risk Analysis

Risk analysis procedures are essential tools in addressing alien species introductions. They are mandated in different contexts by several international instruments, notably the SPS Agreement and the IPPC, under which a series of standards on pest risk analysis have been approved (e.g., Supplement on Analysis of Environmental Risks to ISPM No. 11 [*Pest Risk Analysis for Quarantine Pests,* 2001]).[28] The CBD Guiding Principles support the use of risk analysis as part of an evaluation procedure before decisions are made on intentional introductions (GP10) and, where appropriate, in the context of measures to identify and minimize sectoral pathways for unintentional introductions of IAS (GP 11).

Across different sectors, risk analysis is used to inform decision making on proposed introductions, activities, and control strategies. It should comprise three components: risk assessment, risk management, and risk communication. The process seeks to identify the relevant risks associated with a proposed introduction and to assess each of those risks.

Assessing risk involves identifying possible harm and carrying out qualitative analysis and quantitative measurement, including probability of occurrence in comparison with other risks. The analysis should be designed to give decision makers objective information that enables them to make technically justified decisions that are proportionate to the anticipated risk and nondiscriminatory. To promote transparency and accountability, each stage of the risk analysis procedure should be documented and publicly available.

Scientific evidence is of critical importance to risk analysis. However, as noted earlier, uncertainty is an integral part of the scientific evaluation involved in risk analysis related to alien species introductions. Difficult questions can arise with regard to the applicability or feasibility of risk analysis in situations of high scientific uncertainty.

In the context of alien species introductions, risk analysis procedures should aim to

- Identify the alien species concerned.
- Assess the potential ecological, social, and economic consequences associated with the entry, establishment, or spread of the species concerned.
- Assess the likelihood of such entry, establishment, or spread.
- Identify and compare alternative measures, including ecological, social, and economic costs and administrative feasibility.
- Review the choice of management strategies.

- Evaluate the likelihood of introduction, spread, or establishment of the alien species if the proposed control or management measures (including a national SPS measure) are taken. This evaluation should include a review of the scientific literature, use of experts' opinions, and information on risk factors supplied by the applicant).
- Determine how the measures recommended can be implemented, including evaluation, monitoring, and adjustment in light of new information.

CBD Decision VII/13 §8 notes that there is potential for the application of existing risk assessment and risk analysis methods, including those established in the contexts of plant and animal health, to a wider range of issues related to IAS.

Environmental Impact Assessments

EIAs are a familiar component of general environmental law and practice at international and national level. Article 14 of the CBD mandates EIAs not only for specific projects but also for programs and policies that are likely to have significant adverse effects on biodiversity.

In the IAS context, the CBD Guiding Principles call for the use of EIAs as part of risk analysis to support decision making on intentional introductions (GP10) and EIAs of sectoral activities that may provide pathways for unintentional introductions (GP11). Guiding Principle 12 (Mitigation) notes that techniques used for eradication, containment, or control should be safe to humans, the environment, and agriculture as well as ethically acceptable to stakeholders in the areas affected by IAS.

Generally, EIAs should aim to make adequate and timely information available on likely environmental consequences of programs and projects, possible alternatives, and measures to mitigate harm. It should be a prerequisite to decisions to undertake or authorize designated processes or activities, inform decision makers of the environmental implications of decisions, and facilitate the integration of environmental considerations into other spheres of decision making. All stakeholders should be able to participate in EIA procedures and decision making.

In the IAS context, a nonexhaustive list of factors that should be considered in EIAs include

- The cumulative, long-term, long-distance, direct, and transboundary effects of an alien species introduction
- Alternative actions, including prohibition of the proposed introduction
- Measures to avert or minimize the potential impact of the introduction
- Periodic reviews and monitoring to determine whether the introduction complies with the conditions set out in the approval and to evaluate the effectiveness of mitigation measures

Overview of the Existing International Regime

Terminology

Alien species terminology presents major challenges for scientists, policymakers, and lawyers. International legal instruments, including scientific material in this field, use many terms to refer to alien species (*nonindigenous, exotic, foreign, new*) and the subset that cause damage (*pest, weed, harmful, injurious, invasive, environmentally dangerous*).

Zoosanitary and phytosanitary instruments define terms such as *pest* and *weed* without reference to origin, which means that they also cover native pests. They avoid terms such as *alien* and *invasive* because they are considered emotive.

In contrast, environmental agreements usually refer to alien or exotic species in combination with a harm or invasive trigger to identify species that should be subject to control. This approach excludes native species that become invasive (e.g., through changes in land use or other environmental degradation). However, useful clarification for the environmental community has been provided by the CBD Guiding Principles, which defines some key terms. *Alien species* refers to "a species, subspecies or lower taxon, introduced outside its natural past or present distribution; includes any part, gametes, seeds, eggs, or propagules of such species that might survive and subsequently reproduce," and *invasive alien species* means "an alien species whose introduction and/or spread threaten biological diversity".

Another point of inconsistency concerns the term *introduction* (the action or process that should trigger controls). Older instruments take a narrow approach that is often limited to intentional introductions of alien species for release to a protected area. This excludes introductions to containment or captivity and translocations between different parts of the same country. The CBD Guiding Principles define *introduction* as "the movement by human agency, indirect or direct, of an alien species outside of its natural range (past or present). This movement can be either within a country or between countries or areas beyond national jurisdiction." *Intentional introduction* refers to the "deliberate movement and/or release by humans of an alien species outside its natural range," and all other introductions come under the heading "unintentional."[29]

Although the term *accidental* often is used as a synonym for *unintentional,* this term may be inappropriate and even misleading, at least for pathways where risks of unwanted introductions are well known.

Taxonomic Coverage

Biological invasions may be generated by all taxonomic groups and at all taxonomic levels. However, only the CBD covers all aspects of IAS as they relate to all levels of the biodiversity hierarchy. Many other international instruments are limited to higher taxa of alien animals and plants and do not specify coverage below the species level.

Phytosanitary and zoosanitary instruments potentially cover all taxonomic groups and lower taxonomic categories to the extent that they qualify as pests of plants or plant products or animal diseases of agreed concern.

Species-specific instruments, which can facilitate targeted cooperation, are limited to the regional or subregional level and are scarce. Examples include the African Eurasian Migratory Water Bird (AEWA) on nonnative waterfowl, the Lake Victoria agreement on control of water hyacinth, and recommendations on eradication adopted under the Bern Convention.[30] Coverage of alien freshwater species is almost entirely limited to nonbinding instruments.

As noted, some instruments contain provisions applicable to both alien species and LMOs or GMOs. The IPPC holds that if GMOs are found to be plant pests, they may be covered by IPPC rules and standards.[31]

Ecosystem Coverage

Invasion processes affect all ecosystems, but the impact is especially severe on the structure and function of vulnerable and isolated ecosystems. The Cooperative Islands Initiative on Invasive Alien Species has been established by the Government of New Zealand, the IUCN Invasive Species Specialist Group and GISP to focus specifically on island ecosystems. It has established close links with the South Pacific Regional Environment Programme and Pacific Island Countries and Territories and is extending its coverage to other regional programs.

Introductions to marine and coastal ecosystems are covered generally by UNCLOS and some regional sea instruments, notably the Mediterranean Action Plan. Substantive guidance is limited mainly to mariculture operations and ballast water. There are no international prevention measures for hull fouling as an IAS vector. The IMO International Convention on the Control of Harmful Anti-Fouling Systems on Ships (2001) provides for the global phase-out of tributyltin in paints, but this ban is designed to reduce chemical pollution of the marine environment and could even lead to a significant increase in the number of introductions of invasive fouling species such as ascidians.

A serious gap concerns introductions to freshwater ecosystems, which are not covered by binding instruments except for the 1997 International Watercourses Convention (not expected to enter into force in the foreseeable future). Known pathways include the use of nonnative organisms in aquaculture, the restocking of marine and inland water systems for commercial and recreational fisheries, bait and pet releases, and water transfer schemes. CBD Decision VII/13 identifies this area for specific attention while taking account of the contributions of national codes and voluntary international efforts, such as the ICES Codes of Practice on the Introductions and Transfers of Marine Organisms and the FAO Code of Conduct for Responsible Fisheries.

Introductions to terrestrial systems, generated mainly through agriculture, forestry, horticulture, landscaping, and erosion control, are covered mainly by phytosanitary and zoosanitary control frameworks. The CBD COP has identified a broad range of pathways that should be addressed, including international assistance and humanitarian programs, tourism, military actions, scientific research, and cultural and other activities. Attention should also be given to intentional introductions for nonfood purposes, such as some aspects of horticulture and transnational and national ex situ breeding projects using alien species (Decision VII/13).

For dry and subhumid lands, the Joint Work Programme between the CBD and the UN Convention to Combat Desertification (Paris, 1994) includes IAS management in its list of priority actions. However, no UN Convention to Combat Desertification decision has addressed IAS to date, nor has guidance been developed on selecting species and varieties for programs on land degradation (e.g., erosion control, planting of windbreaks and shelterbelts, afforestation) to prevent the introduction of potential IAS. For forest ecosystems, international policy processes have also given little attention to IAS.

Progress on Global and Regional Coordination and Synergies

Most international instruments that address IAS focus on a specific dimension, such as a particular protection objective (e.g., migratory species), kind of activity (e.g., introductions for aquaculture), or potentially damaging organism (e.g., pest). Most have their own institutional mechanisms and decision-making procedures. Cooperation between these organizations has grown significantly in recent years, a process to which GISP actively contributes.

There has been a corresponding increase in tools to operationalize cooperation, such as memoranda of understanding and joint work programs. The IPPC and CBD secretariats are currently developing a joint work plan. The CBD COP has also called for closer linkages with the OIE and closer collaboration with the ICAO, the World Health Organization, and the IMO (Decision VII/13).

Dialogue on trade-related aspects of IAS is seriously underdeveloped. CBD Decision VII/13§5 strongly supports more systematic cooperation between the CBD and the WTO. It requests the executive secretary to collaborate, whenever feasible and appropriate, with the WTO Secretariat in its training, capacity-building, and information activities, with a view to raising awareness of IAS issues and promoting enhanced cooperation on this issue, and to renew his application for observer status in the WTO SPS Committee with a view to enhancing the exchange of information on deliberations and recent developments in bodies of relevance to IAS.

All regions can gain from information sharing, capacity pooling for more efficient risk analysis and mitigation programs, and promoting basic consistency in IAS policies, legislation, and practice. Currently, regions vary widely in the priority they attach to IAS issues: it is lower in Southeast Asia, South Asia, and the Neotropics than in Africa, the Austral-Pacific region, and Europe. In the South Pacific, progress in implementing the

South Pacific Regional Environment Programme (SPREP) Regional Invasive Species Programme and the South Pacific Regional Invasive Species Strategy is being reviewed, in cooperation with the Pacific Pollution Prevention Programme and the Secretariat for the Pacific Community, with a view to possible extension to marine and agricultural IAS issues.

GISP organized a series of regional training and capacity-building workshops between 2001 and 2003 (Nordic-Baltic, Meso-America, South America, Southern Africa, South and Southeast Asia, Austral-Pacific), at each of which participants identified more effective cross-sectoral regional coordination as a priority.

Whereas the IPPC and OIE are well supported by regional organizations or representations and have established links with national quarantine services, the CBD and other environmental agreements have more informal mechanisms for regional cooperation. CBD Decision VII/13§6 calls on parties and national, regional, and international organizations to improve the coordination of regional measures to address transboundary issues through the development and implementation of regional standards, support for risk analysis, cross-sectoral cooperation mechanisms, and information exchange.

Responsibility, Liability, and Redress in the Context of IAS

In international law, states have a general responsibility to ensure that activities within their jurisdiction or control do not cause damage to the environment of other states or of areas beyond the limits of national jurisdiction. At present, however, public international law on the possible liability of states for environmental damage is underdeveloped. The lack of an environmental liability and redress regime means that there is no mechanism to promote compliance with international environmental norms and the implementation of the precautionary approach and the prevention principle, nor to shift the costs of environmental damage from society at large to those responsible for the damage. Prevention and control obligations are not underpinned by a deterrent element.

With regard to IAS, the question is whether states may be liable for transboundary environmental effects of activities involving the export of alien species or the spread of existing IAS or pests. This issue raises complex questions about the way in which different sets of existing international rules (on biodiversity, biosafety, quarantine, and trade) fit together.

Biodiversity-related instruments are basically silent on the question of liability. The negotiators of the CBD were unable to reach consensus on a possible liability regime and postponed its consideration to a later date.[32] The CBD COP has established an ad hoc technical expert group to consider a process for reviewing Article 14.2, which will meet for the first time in January 2006. The group is mandated to take account of the outcome of a CBD workshop on liability and redress (Paris, June 18–20, 2001) and the imminent consideration of liability issues within the framework of the Cartagena Protocol on Biosafety.[33]

With regard to IAS, CBD Guiding Principle 4 (Role of States) does not reference liability but provides that states should recognize the risk that activities within their jurisdiction or control may pose to other states as a potential source of IAS and should take appropriate individual and cooperative actions to minimize that risk, including the provision of any available information on invasive behavior or invasive potential of a species. Examples of such activities include

- The intentional transfer of an IAS to another state (even if harmless in the state of origin)
- The intentional introduction of an alien species into their own state if there is a risk of that species subsequently spreading (with or without a human vector) into another state and becoming invasive
- Activities that may lead to unintentional introductions, even if the introduced species is harmless in the state of origin

To help states minimize the spread and impact of IAS, states should identify, as far as possible, species that could become invasive and make such information available to other states.

The Bern Convention seems to be the only environmental agreement under which possible state liability for invasion impacts has been specifically referenced.[34]

In general terms, it is important to lay the foundations for a system to strengthen responsibility for activities generating biological invasions and, where feasible, to repair the damage caused to the environment of other states by introduced alien species.

The counterpart of state liability for damage caused to the environment in other states is the recognition of the right of victims to seek reparation. Principle 13 of the Rio Declaration calls on states to develop national law regarding liability and compensation for the victims' environmental damage and to cooperate in the development of further international law on the subject.

The Convention on Civil Liability for Damage Resulting from Activities Dangerous to the Environment (Lugano, 1993) establishes a system of strict liability for damage caused to persons, property, and the environment by activities carried out in a professional capacity that are considered as inherently dangerous. These include the production, culturing, handling, storage, use, destruction, disposal, and release or any other operation dealing with genetically modified organisms or microorganisms that present a significant risk for humans, the environment, or property. However, the convention does not apply to introduced species other than GMOs and microorganisms, nor does it cover carriage operations.

Under the convention, a public or private person engaged in inherently dangerous activities as defined is liable for the damage caused by it, even if she or he has committed no fault and is able to prove that she or he has taken all possible precautions to avoid the accident. The few exceptions relate essentially to war or force majeure. Compensation for damage to the environment is limited to the cost of measures of reinstatement actually undertaken or to be undertaken, the cost of preventive measures, and any loss or damage caused by such measures.

The convention recognizes the right of environmental protection organizations to request the prohibition of an unlawful dangerous activity that poses a great threat of damage to the environment or an order to force an operator to take measures to prevent damage or make restitution. It is therefore unfortunate that activities involving all categories of alien species introductions are not covered by the convention.

National and Subnational Legislation, Institutions, and Practice

In the national context, law is used to implement policy objectives and to define principles, standards, and procedures to achieve them. It sets rules for the conduct of human activities and allocates rights and responsibilities among the actors concerned. Legal tools can be designed not only to prohibit or restrict actions but also to promote desired practices by providing incentives to the private and public sectors and to individuals.

An important function of national legislation is to establish institutional mechanisms to develop appropriate implementing regulations and to oversee and, as far as possible, ensure compliance. Institutions are needed to monitor success and failure, promote policies for improved implementation, and generate necessary legislative reforms. Establishing efficient institutions is one of the most important roles of legislation, although this role is often underestimated.

Primary Goals and Components

Well-designed IAS frameworks should aim to prevent or minimize the risk of unwanted introductions, both between and within countries, and to provide a basis for effective eradication and control measures where an introduced species becomes invasive. They should establish objective principles, rules, and criteria to regulate rights, responsibilities, and conduct of all stakeholders, from individuals and communities to commercial bodies and administrative agencies across many sectors.

Consistent with international norms and best practice, comprehensive national frameworks on IAS should equip competent authorities with powers to

- Implement and enforce internationally agreed standards for quarantine and transport
- Screen and exclude, restrict, or authorize introductions of alien species to a country or area, consistent with prevention and precaution and the findings of risk analysis and EIAs

- Prohibit or strictly regulate the use and release of alien species in or near closed or vulnerable ecosystems and protected areas
- Provide for monitoring, early warning, and emergency planning systems to support rapid responses when biological invasions are detected
- Require timely measures for eradication or control of species that become invasive, subject as necessary to prior assessment of techniques to be used
- Secure compliance and greater accountability by public, commercial, and private actors
- Support research, training, public education, and awareness

Common Weaknesses of Existing National Regimes

Case studies during GISP Phase I, backed by a review of national legislation and literature, showed much unevenness in existing frameworks (Shine, Williams, and Gündling 2000).

National law has developed along sectoral lines, usually for historical and administrative reasons. In many countries, relevant provisions are found in quarantine, agroforestry, fishery, nature conservation, and water legislation and in hunting, fishing, and wildlife regulations that cover introduction and release of species for restocking. Different kinds of problems may result.

Fragmented Legal and Institutional Frameworks

- Absence of a strategic approach to the problem, with alien-related issues ignored or underrepresented in national environmental or biodiversity planning processes
- Fragmented and dispersed provisions and inconsistent legislative treatment, confusing to users
- Absence of coordinated planning and consultation between sectors and between different levels or tiers of government

Weaknesses Related to Coverage and Terminology

- Gaps in taxonomic coverage; common omissions relate to alien plants, fish, and microorganisms. The law may not specify whether legislation goes below the species level.
- Gaps in ecosystem coverage, especially for freshwater, marine, and coastal ecosystems.
- Narrow or inconsistent objectives, often focused on protecting economic interests rather than native biodiversity in its own right.
- Nonexistent or inconsistent definitions of key terms.

Problems Related to Compliance, Enforcement, and Remedies

- Exclusive reliance on command-and-control approaches, with little use of economic incentives to deter unwanted introductions and promote use of native species

- Stronger focus on intentional introductions than on pathway management to minimize unintentional introductions
- Cumbersome risk assessment and permit procedures; lack of domestic capacity
- Absence of legally backed requirements for monitoring
- Weak or nonexistent powers and duties for eradication, containment, and control; crisis management approach toward managing invasions
- Enforcement deficit because conventional criminal and civil law procedures are difficult to apply in the aliens context

General Considerations for the Design of National Frameworks

Integrating Alien Species Issues in National Planning and Consultative Processes

National environmental and biodiversity planning processes may handle IAS issues as one component of a comprehensive plan or as a standalone plan linked to a broader framework.

From a legal and institutional point of view, alien species strategies should aim to

- Identify sectors, pathways, and policies associated with alien species introduction or use, using risk analysis and environmental assessment tools as appropriate
- Identify all relevant government agencies and affected stakeholders
- Review policy, legal, and institutional measures to identify gaps, weaknesses, and inconsistencies
- Promote cooperation within and between relevant institutions and sectors
- Identify and provide for progressive elimination of perverse incentives[35] and support incentives for use of native species in preference to alien species, where feasible
- Identify opportunities for cost-effective preventive and mitigation measures, building where possible on the contribution of local communities and other stakeholders

Planning processes should address all conflicts of interest and openly balance positive and negative aspects of alien species introductions. They should give proper consideration to long-standing and legitimate interests of many stakeholder groups (e.g., commercial forestry, horticulture, pet trade) to build awareness and contribute in the long term to improved compliance (Baldacchino and Pizzuto 1996).

Participatory approaches linked to IAS planning and management may include

- Opportunities to participate in consultations related to alien species introductions
- Open and transparent permit-issuing procedures for intentional introductions
- Open and transparent procedures for the development of invasive species mitigation and management strategies
- Opportunities for interested individuals and groups to seek judicial review of administrative decisions related to alien species introductions
- Public awareness and education measures related to introduction and management issues

What Kind of Laws: Unitary or Multiple Approaches?

Various options may be considered to overcome sectoral fragmentation. The first and most ambitious is to review and consolidate existing measures into a unitary legislative framework that covers all categories of species, all sectors, all ecosystems, and the full range of actions to be taken. Legislative reform on this scale is politically and technically complex and likely to generate resistance from powerful administrations with long-established mandates. New Zealand has gone furthest down this line by adopting comprehensive laws on intentional introductions of aliens and GMOs (Hazardous Substances and New Organisms Act, 1996) and unintentional introductions and management and control planning (Biosecurity Act, 1993) (Christensen 2004).

A second option is to enact a core instrument to determine common essential elements for the prevention and control of IAS and to harmonize goals, definitions, criteria, and procedures. This instrument would also appoint or establish a coordinating body (e.g., a lead authority).

A third option, taking a minimalist (but probably realistic) approach, is to harmonize all relevant acts or regulations to remove inconsistencies or conflicting provisions and promote more uniform and consistent practice in the country concerned. A coordinating body would be needed to set indicators for harmonization and provide necessary advice and oversight.

What Kinds of Institutions and Coordination Mechanisms?

In most countries, several sectors share responsibility for IAS control. There is often no coordinating framework to link up the administrations and agencies with relevant mandates or to ensure consistent implementation.

At a minimum, steps should be taken to identify conflicts of interest. For example, a single agency may have a statutory duty to promote agricultural, forestry, or fishery development and to enforce quarantine controls. Sectoral officials may come under pressure from traders to release consignments from postentry quarantine earlier than scientific caution dictates. To the extent possible, regulatory responsibilities should be clearly separated from economic development objectives.

Appropriate institutional arrangements depend on the regulatory structure. Under a unitary framework, the lead authority might be the nature conservation authority, the agriculture department, the public health authority, or a special agency. Quarantine and customs services still retain generic responsibilities for application and enforcement of border controls.

In countries where a mix of sectoral rules remains in place, IAS responsibilities continue to be divided between relevant institutions and agencies. Some kind of coordination mechanism is needed to ensure consistent and efficient practice. For example, the United States established a federal Invasive Species Council, bringing together representatives of eight federal departments, to oversee the development and adoption of the national Invasive Species Management Plan.

A coordination mechanism could take the form of a cross-sectoral commission or committee. Members should be drawn from all relevant government institutions and could also include permanent or ad hoc representatives of nongovernment organization where appropriate. The role of local authorities (municipalities) in conservation, planning, and enforcement should also be recognized.

All countries should consider establishing a scientific authority on alien species issues, which could provide technical assistance for planning and decision-making procedures and application of regulatory controls. Legislation should specify the respective functions of regulatory and scientific authorities and how the two relate.

Few countries have high-level political accountability for alien species issues, although this is now changing. In 1997, New Zealand allocated cabinet-level responsibility to this subject to the minister of food, fibre, biosecurity, and border control. Chief technical officers have also been appointed in all key ministries.

Relationship between National and Subnational Laws and Institutions

Special steps to harmonize IAS policies may need to be taken in countries with a federal or regionalized structure, where law-making and enforcement powers are divided between national and subnational government (Sharp 1999).

In federal states, subjects that are usually the exclusive responsibility of national government include international trade in commodities and species and quarantine and pest control measures that may involve import restrictions. The national government also has the power to negotiate and ratify treaties on all subjects.

Subnational units (provinces, cantons, *Länder*) have varying powers and duties, depending on the country concerned. Devolved powers most relevant to alien species management often relate to domestic trade and transport, infrastructure development, land and water management, and nature conservation.

Consistent rules must be developed and implemented to ensure that prevention and mitigation measures in one subnational unit are not canceled by conflicting approaches across an administrative boundary.

Specific Components of National Frameworks

In moving beyond a piecemeal approach to alien species control, decision makers need to consider carefully the purpose and scope of the policies and laws they adopt.

Objectives

Explicit objectives are necessary to provide a conceptual framework to develop legislation, guide implementation, set priorities, and build awareness. For IAS, these may include

- Protection of animals, plants, plant products, and human health against alien pests, including pathogens

- Protection of species, subspecies, and races against contamination, hybridization, and extinction or extirpation
- Protection of native biodiversity against impacts resulting from IAS (and, if covered by the same legislation, from GMOs)
- Protection against biosecurity threats, defined as matters or activities that, individually or collectively, may constitute a biological risk to the ecological welfare or to the well-being of humans, animals, or plants of a country (IUCN 2000a)

Definitions and Use of Terms

Legal instruments use definitions to provide an agreed meaning for a particular term. These definitions must be precise to provide legal certainty for public and private actors responsible for implementing and complying with the law. As far as possible, they should be consistent with internationally agreed definitions to facilitate communication and common approaches between countries and regions and should be used consistently in all relevant sectoral instruments. Terms used by the scientific community (e.g., *normal distribution*) usually must be defined in legislation or regulations to provide a clear and consistent basis for applying and enforcing the law.

Key terms needing definition include the following:

- *Native* (important if the same text contains conservation and recovery measures for native species and ecosystems). Several laws use a past cutoff date to define *native* for legal purposes; any species not already present in the country that is introduced after that date is deemed to be nonnative. If the date is far back in history (e.g., 1400 in Australia[36]), almost all established or naturalized aliens will be excluded from the definition of *native.*
- *Alien.* This spatial concept is traditionally defined by reference to jurisdictional limits (national or provincial boundaries), but ecological parameters should be used. German legislation defines *alien* as "alien to a region" or "nonlocal,"[37] whereas the United States defines *alien species* with regard to a particular ecosystem as "any species, including its seeds, eggs, spores, or other biological material capable of propagating that species, that is not native to that ecosystem."[38] Where *alien* is defined by reference to *native,* as in the U.S. definition, legislation may need to define *native* to make the definition workable.
- *Threat, harm,* or *pest.* Legal frameworks should be broad enough to apply to all alien species, but it is important to be able to distinguish the subset of alien species assessed as problematic (with negative impact on biodiversity) from those assessed as benign or serving useful purposes. Hungary, New Zealand, and South Africa all make this distinction in their legislation (Shine, Williams, and Gündling 2000).
- *Introduction.* This definition should be broad enough to cover all actions or processes that may lead to introductions, irrespective of intention or negligence, within a country or between countries or areas beyond national jurisdiction.

Geographic Coverage

Because invasion processes can affect all ecosystems, legal frameworks should provide a basis for regulating introductions to any type of ecosystem and for monitoring and managing their use, wherever this takes place.

Because vulnerable and isolated ecosystems need special precautions, states with islands or other vulnerable ecosystems should have a legal basis to strictly control movements to and between islands and between different parts of the same country.

Species Coverage

National legal frameworks should cover all aspects of IAS as they relate to genetic, species, and ecosystem diversity. *Species* should be interpreted broadly to include subspecies and lower taxa, as well as any parts, gametes, or propagules of such species that might survive and subsequently reproduce (IUCN 2000a).

Where national frameworks comprise several laws and regulations, no taxonomic groups should be omitted. Information on the taxonomic coverage of each instrument must be easily accessible.

Whereas most countries regulate GMOs separately from alien species, with different procedures, institutions, and criteria, a few have more streamlined approaches. Options include integration of GMO-related provisions within specific biosecurity legislation (e.g., New Zealand) or alongside alien species under modern framework biodiversity legislation (e.g., Costa Rica, Hungary). Hungary's 1996 Nature Conservation Act covers alien species and also provides for the development of specific regulations on GMOs consistent with the act's objectives.

Measures to Prevent Unwanted Introductions of Alien Species

Preventive frameworks must provide for control measures for intentional introductions (whether for release or for use in containment or captivity) and for management of pathways for unintentional introductions.

Where Should Control Measures Be Applied?

At the point of origin or export

States should recognize the risk they may pose as a source of potentially invasive species and, to the extent possible, take appropriate domestic steps to minimize the risk of transferring alien species to countries or ecosystems in which they may become invasive. Measures of this kind depend closely on information exchange and cooperation in good faith between the trading partners concerned. Suitable approaches, backed by the CBD Guiding Principles, include

• Supply of information on potential invasiveness of the species to the importing or receiving state, particularly where the countries concerned have similar environments

- Development of bilateral or multilateral agreements to regulate trade in certain alien species, with a focus on particularly damaging invasive species
- Support for capacity-building programs for risk assessment of imports in states that lack the necessary expertise and financial or other resources to assess such risks

At the point of import or release

Border control and quarantine measures are necessary to screen and subject intentional introductions to prior authorization and to minimize unintentional introductions and unauthorized introductions (see CBD Guiding Principle 7).

All countries have some form of customs and quarantine legislation. Customs officials generally have powers to prohibit imports, impose restrictions on certain products, execute inspections, detain particular consignments, and treat or destroy living material.

Legal frameworks should be broad enough to provide a basis for restricting imports and internal movements in alien species that may threaten native biodiversity (not just biological production systems). It should be possible to vary the level of restriction, depending on the assessed level of risk. Officials should have adequate powers to intercept potential IAS and to halt unauthorized introductions.

Controls on domestic movements

Alien species on national territory may become invasive for the first time when translocated to a new part of the same country. Legal frameworks should provide a basis for regulating intentional domestic movements of alien species and for assessing projects and programs that may create new pathways for species translocations (e.g., infrastructure development, interbasin water transfers). Domestic controls are also important to contain the further spread of a species that has become invasive elsewhere in the country.

As noted, island states and states with islands should as a priority adopt domestic controls to minimize the risk of interisland or mainland-to-island introduction of alien species.

Special controls for isolated and vulnerable ecosystems

As required under several international instruments, introductions of alien species to protected areas, geographically and evolutionary isolated ecosystems, and other vulnerable ecosystems should be prohibited or subject to extremely strict regulation.

Site-specific controls of this kind are a key component—but not a substitute—for an ecosystem approach to IAS prevention and management. Complementary controls are needed around protected areas to avoid the spread of degraded areas vulnerable to invasion near to refuges of native biodiversity. An integrated approach is particularly important for wetland protected areas because alien aquatic organisms may be easily translocated from beyond the protected area boundaries.

Tools to Screen and Regulate Intentional Introductions

Legal frameworks should provide a basis for controlling the introduction or release of all categories of an alien species, whatever its origin or the purpose of the introduction. In line with CBD Guiding Principle 10, no first-time intentional introduction or subsequent introductions of an alien species already invasive or potentially invasive in a country should take place without prior authorization from a competent authority of the recipient state. An appropriate risk analysis, which may include an EIA, should be carried out as part of the evaluation process before a decision is made on whether to authorize a proposed introduction to the country or to new ecological regions within a country. States should make all efforts to permit only species that are unlikely to threaten biodiversity. The burden of proof that a proposed introduction is unlikely to threaten biodiversity should be with the proposer of the introduction or be assigned as appropriate by the recipient state.

Authorization of an introduction may be accompanied by conditions (e.g., preparation of a mitigation plan, monitoring procedures, payment for assessment and management, or containment requirements).

In the context of intentional introductions, risk analysis should help to distinguish between

- Known or potential problem species, which should be excluded or subjected to rigorous permit conditions and monitoring requirements
- Species that are known not to present significant risks to the state concerned or, in the event of subsequent spread, to another state
- Species for which the risk is uncertain

Species listing, based on prior risk assessment, may be used to support operation of permit systems and build awareness. The main options used in different legal systems are black lists (lists of species that may cause serious biological upheaval and must be prohibited), white lists (lists of species that are deemed not to present a threat and may be introduced), and gray lists (composite lists of species for which the level of risk is less clear). However, listing techniques have major imitations because they are inherently reactive and can never be fully accurate or up-to-date. As a general rule, they should be seen as indicative and not used as a substitute for permit controls and risk analysis, except for black lists supporting the prohibition of known invasives.

Careful consideration should also be given to the legal effects of species lists, specifically the following:

- How does the inclusion or omission of a species from a list affect legal responsibility for the consequences of a subsequent invasion?
- What happens to preapproved introductions if the species is subsequently added to a black or gray list?

Decisions on permit applications should be made in accordance with scientific criteria laid down by technical regulations, which should be periodically reviewed and updated. New Zealand law provides for adoption of decision-making protocols in the form of legislative annexes, available through the ministry Web site; these establish discretionary criteria (to be taken into account when determining a permit application) and mandatory criteria (grounds on which a permit application must be refused).

States should aim to streamline assessment and permit-issuing procedures for greater efficiency. In countries with older legislation, applicants may need to go through separate regulatory procedures under a series of different laws (e.g., technical import assessment, general EIA requirements, and special assessments related to protected species, habitats, or areas). This procedure can be costly and administratively cumbersome. It may even deter applicants from using lawful channels and thus encourage unauthorized introductions.

Where alien species are introduced to containment or captivity, regulations and permit conditions should ensure that the relevant facilities are subject to stringent operating and siting conditions. Strict controls should also apply to internal transport or trade in such species (this is the case in Argentina for aquaculture species (di Paola and Kravetz 2004).

Introductions by private individuals are notoriously difficult to control through regulatory methods. To be effective, all approaches should be developed in close collaboration with the pet breeding and retail industry and supported by public awareness campaigns. One possibility is to introduce a two-tier level of controls. A first list could contain pets that can be adequately controlled through quarantine regulations, without the need to make specific applications and go through EIA and risk assessment procedures. A second list could include pets that should be subject to more stringent assessment, with a total prohibition on importing pets that could establish themselves in the wild if they escaped (e.g., as under Western Australian legislation).

Tools to Minimize the Risk of Unintentional Introductions

Existing national and international frameworks are weak with regard to prevention of unintentional introductions. The GISP Toolkit and Phase II (Wittenberg and Cock 2001) therefore has given priority to better pathway management, based on greater public–private sector cooperation. Because many different sectors are involved, preventing unintentional introductions again raises legal and institutional challenges related to coordination and consistency.

CBD Guiding Principle 11 provides that all states should have in place provisions to address unintentional introductions (or intentional introductions that have become established and invasive). These provisions could include statutory and regulatory measures and establishment or strengthening of institutions and agencies with appropriate responsibilities. Operational resources should be sufficient to allow for rapid and effective action.

The principle calls on states to identify common pathways leading to unintentional introductions and to put in place appropriate provisions to minimize such introductions (e.g., through sectoral activities such as fisheries, agriculture, forestry, horticulture, shipping [including the discharge of ballast waters], ground and air transportation, construction projects, landscaping, aquaculture including ornamental aquaculture, tourism, the pet industry, and game farming). EIAs of such activities should address the risk of unintentional introduction of IAS, and, wherever appropriate, a risk analysis of the unintentional introduction of IAS should be conducted for these pathways.

Ideally, such assessment and analysis should be carried out early in the planning and design stage, before heavy costs have been incurred and it becomes impracticable to develop alternatives.

Consistent with international law and internationally agreed codes of conduct or other guidance on best practice, priority areas for assessment of pathway risks should include

- Development, expansion, and environmental review of international, bilateral, and regional trade arrangements (see CBD Decision VII/13§5)
- Development of new land and water use policies that could enable species to reach new parts of national territory or cross ecosystem or political boundaries (e.g., new transport infrastructure can open up new vectors for introductions and make it possible for alien populations in the new range to be reinforced continuously)
- Other pathways identified in CBD Decision VII/13§7

Companies dealing with transport or movement of living organisms should be required to comply with biosecurity regimes established by governments in exporting and importing countries, and their activities should be subjected to appropriate levels of monitoring and control (IUCN 2000a). It may be appropriate to make the grant or renewal of an operating license subject to proof of compliance with applicable standards.

Border and quarantine controls should be designed to detect stowaway organisms in consignments and substances, with adequate provision made for postquarantine controls. In some countries, it may be necessary to amend legislation to give officials power to control consignments that present a threat to native biodiversity rather than agriculture or forestry.

For construction and infrastructure development programs and projects, risks must be assessed on a case-by-case basis. It may be necessary to amend general EIA legislation and regulations to ensure that all relevant projects are subject to EIAs and that alien species–related criteria are taken into account in any decision-making process.

Competent authorities should have adequate powers and resources to respond rapidly when information becomes available on new pathways and develop appropriate responses in cooperation with relevant international and national organizations.

Monitoring and Early Warning Systems

Legal frameworks should provide a formal basis for monitoring and surveillance of terrestrial and aquatic environments. Early detection and warning systems are essential preconditions for rapid responses to new invasions.

The IUCN Guidelines 2000 recommend that neighboring countries should consider the desirability of cooperative action to prevent alien potentially invasive species from migrating across borders. Cooperation might include agreements to share information and warnings and to consult and develop rapid responses in the event of such border crossings. CBD Guiding Principles 8 and 9 address exchange of information and cooperation.

Monitoring requirements may apply not only to competent agencies but also, with appropriate modifications, to people responsible for intentional introductions and to those owning or occupying relevant land. Objectives of a legally backed monitoring system may include the following:

- To oversee the behavior of intentionally introduced alien species and respond quickly when signs of invasiveness are detected in such species
- To detect the presence on national territory of alien species that have been unintentionally or unlawfully introduced
- To detect the spread of established aliens caused by secondary transfers and spontaneous dispersal processes so that eradication can be initiated while infestations are limited
- To detect the emergence of invasive characteristics of species, particularly woody plants, that were introduced a long time ago

New Zealand's Biosecurity Act establishes a legal basis for gathering, recording, and disseminating information on invasive species present on national territory. This information is used for developing alien pest management strategies at the national and regional level.

Legal Measures for Responding to Invasions

National frameworks tend to be much stronger on preventing introductions than on mitigating their impacts. Very few countries have a comprehensive basis for taking legal measures to tackle accumulated problems linked to past, untreated invasions. Where legislation does provide for eradication and control measures, they are often implemented in a piecemeal way rather than as part of a more strategic ecosystem approach. Available resources are focused more often on IAS that threaten key economic sectors rather than on native biodiversity.

Mitigation measures are rarely complemented by incentives for restoring degraded ecosystems or, where appropriate, for reestablishing native species formerly present on national territory.

Ideally, legally backed mitigation measures should have two strands:

- Short- and long-term measures for eradication, containment, and control of invasives (covered by CBD Guiding Principles 12–15)
- Positive strategies for restoration of native biodiversity

Tools for Eradication and Control

Mitigation measures usually involve killing members of the invasive species or, through chemical or other means, sterilizing them to prevent future reproduction. For these actions to be lawful, the IAS needs to have a legal status compatible with such actions, and the relevant official must have the authority to take the necessary eradication or control measures.

In many countries, this approach is not possible or occurs only in a limited way. Alien species that have become invasive may be legally protected. This outcome occurs where the law confers protection on biodiversity as a whole without making any distinction between alien and native species or where it protects a higher taxon (genus, family, order, or class) that includes an alien species.

To get around this problem, biodiversity and nature conservation legislation must use terminology that excludes alien species from legal protection but retains automatic protection for reintroduced species, future newly described species, and species that occur occasionally in the relevant territory. A possible provision to this effect could provide that legal protection applies only to

> all species that are present, have been present in the past or become present in the future, in a wild state, on the [national] territory, except for species which have been intentionally or unintentionally introduced into that territory as a result of human action after [insert appropriate cut-off date] and species introduced in the same way on the territory of another country and now present on the [national] territory. (de Klemm 1996)

Wording to this effect would exclude introduced species, including those introduced in another state that have extended their range naturally to the territory of the country. It would also exclude GMOs.

Legal frameworks should include specific components to facilitate eradication and control. As noted, rapid response tools are essential. A legal mandate for emergency measures should provide for inspections, confiscation, disinfection of equipment, activity-based restrictions, closure of contaminated areas to traffic or navigation, a ban on anchorage, and provision of alternative buoys or moorings. Such measures should be applicable to any part of national territory.

Longer-term control measures should include powers for competent authorities to

- Regulate possession and domestic translocation of and trade in alien species

- Restrict subsequent releases of alien animals and plants to the wild, whether intentionally or through negligence
- Require land owners and occupiers to notify the relevant authority of the presence of listed alien species on their land and, where required by law, to take specified management measures
- Take eradication and control measures and, where appropriate, develop area-based pest management strategies in consultation with all affected stakeholders

Techniques for eradication or control, including use of alien biological control agents to control target species, should be subject to risk assessment and EIAs, and a permit from the competent authorities should be required. Permit-issuing procedures should be streamlined as far as possible in the interests of administrative efficiency and compliance.

Legal frameworks should support the use of incentives, where appropriate, to promote active participation by indigenous and local communities and landowners in long-term control programs. South Africa's Working for Water Program is an excellent large-scale example of such an approach (Stein 2004).

Legal Tools to Support Restoration of Native Biodiversity

IAS management should form part of a broader suite of policies and measures to conserve biodiversity. Measures to control "negative" biodiversity (IAS) should be combined with positive strategies to restore degraded ecosystems and, if appropriate, reestablish native species formerly present on national territory. Such approaches not only increase native biodiversity but also can increase the resilience of ecosystems against future invasions.

Modern biodiversity or nature conservation laws usually support the use of habitat restoration measures or species recovery plans. In most cases, these can provide a legal basis for controlling invasives in order to achieve the habitat and species conservation objectives.

Legal frameworks should establish stringent conditions for the reintroduction or reestablishment of native species, consistent with applicable international recommendations.[39] These frameworks recognize that reintroductions may be desirable and beneficial but should be controlled in the same way as other types of introduction and that conditions and habitats necessary for the reintroduced species must also be restored (de Klemm 1996).

Consideration should be given to using incentives as part of management and restoration strategies. There are many useful precedents in other areas of environmental management for using tools such as grants, subsidies, tax incentives, contractual management agreements, market-based instruments, and cross-compliance mechanisms to achieve conservation objectives in environmentally sensitive areas under private or communal ownership.

Measures to Enhance Compliance and Promote Accountability

IAS frameworks can be properly effective only if supported by actions to tackle noncompliance and promote a broader culture of accountability. Education and public awareness programs may be necessary to reduce the risks involved in private handling of alien species and to modify certain consumer attitudes and preferences. In the long term, awareness-building strategies among citizens, commercial stakeholders, and administrations may make the biggest contribution to lowering the rate of introductions and effectively controlling invasions.

Criminal Responsibility and Civil Liability

Many invasion-related problems result from actions that are already prohibited or restricted (e.g., unauthorized introductions of alien animals or plants, breach of quarantine regulations). Others result from recklessness or gross negligence and should be subject to appropriate criminal or administrative penalties. Unintentional introductions may be the result of noncompliance with operating regulations (e.g., controls on discharge of ballast water).

Legal frameworks should provide for a full range of controls to reinforce policy on alien species. There should be no gaps in the applicable regulations, and meaningful penalties should be available that reflect the seriousness of such actions. Legislation should include a requirement to restore or otherwise compensate for environmental damage.

An indicative checklist of criminal offenses might include

- Permit-related violations (e.g., failure to obtain, breach of permit conditions)
- Operational violations (e.g., noncompliance with operating rules for breeding and cultivation facilities, safety standards)
- Unlawful international and domestic trade and transport
- Unlawful subsequent releases
- Breach of monitoring and notification requirements
- Failure to take required mitigation measures

Where possible under national legal systems, invasions that result from grossly negligent acts or omissions should be punishable.

As regards civil liability, governments and individuals who have to bear the economic burden of preventing and redressing harm from invasive species may create mechanisms to obtain restitution from those responsible for any damage. In some cases, traditional liability mechanisms may be adequate; these basically hold that the person most responsible for the harm should ensure that the government or private party is compensated.

In both criminal and civil actions, traditional standards of knowledge, intent, and causality can be difficult to apply to damage resulting from biological invasions for the following reasons:

- Intentional introductions are lawful (e.g., a permit is granted because the introduced alien is not identified as potentially invasive).
- Unauthorized introductions are hard to detect, particularly given the number of pathways and private actors involved.
- An unintentional introduction takes place via a pathway that has not been identified as high risk and is not subject to operating regulations or agreed best practices.
- The law does not cover negligent conduct that gives rise to introductions.
- There is often a long time lag before damage occurs or is noticed, by which time it is impossible to determine what caused an introduction or to attribute remediation costs to a particular party.
- The damaged values (native wild species, ecosystems, ecological processes) do not have an "owner" capable of seeking compensation and remediation.
- Financial and technical resources for monitoring (evidence gathering) are scarce.

GISP Phase I confirmed that in many cases it is impossible to identify an "introducer" with the certainty required by law. This conclusion can lead to a de facto legal vacuum surrounding actions that lead, intentionally or unintentionally, to biological invasions. The problem is particularly serious, given the enormous economic costs that may result from a small number of invasive species. GISP recognized the need for new and more innovative approaches to discourage unwanted introductions.

Deterrents and Cost Recovery Mechanisms

There are few deterrents to the export, import, or use of IAS, and those who take the risks are seldom those affected by the consequences of a harmful introduction (Hedley 2004). GISP research found that existing market mechanisms and other economic instruments are not sufficiently developed to provide deterrents for alien species introductions as they relate to biodiversity (Perrings, Williamson, and Dalmazzone 2000).

One way to address the problem of external (uncompensated) damage to the environment is to internalize such costs through various mechanisms based on the polluter-pays principle. This holds that a polluter who creates an environmental harm is responsible for the costs of preventing that harm. The organization or person responsible—who often stands to benefit commercially from the activity or process that generates the pollution—therefore should bear the cost of control measures and should not be subsidized for polluting activities.

Approaches based on this principle are beginning to be applied more widely in environmental management. The underlying concept, that the beneficiary of an activity or process should pay for preventing damage, can be applied *mutatis mutandis* to developers and to consumers of land, water, and other natural resources. It is also applicable to people who introduce alien species and could provide a needed deterrent element.

At present, there are conflicting views about the application of polluter-pays mechanisms to alien species introductions. One view holds that this kind of approach is not applicable to biological pollutants because the invasion ("pollution") is ongoing, not site-limited. An opposing view sees mechanisms based on the polluter-pays principle as a constructive way to introduce and strengthen accountability in an area that is currently very weak (e.g., Jenkins 2000).

Collective Mechanisms to Promote Accountability and Generate Funds for IAS Control

Accountability mechanisms are necessary for all types of introduction but especially for unintentional introductions. Because it is nearly impossible to make a finding of individual responsibility for such introductions, appropriate techniques are needed to support collective responsibility of all actors involved in a particular pathway (e.g., traders of a particular commodity, certain groups of cargo transporters, pet retailers).

Mechanisms of this kind, whether international or national, can be designed to generate sustainable up-front funding not only for remediation but also for prevention (e.g., to fund capacity building for risk assessment and quarantine controls). This up-front loading is critically important because in most countries, and particularly in developing countries, tools are not in place to generate sustainable funding for public investment in IAS prevention and control programs.

Useful precedents exist in other environmental management fields where similar problems of causation or scale of costs arise. Options that might be adapted to alien species control are very briefly outlined in this section.

Mandatory Insurance

Many forms of specialized insurance exist to cover possible damage resulting from risky actions (e.g., car accidents). In the area of nature conservation, Hungarian law requires any legal person, private entrepreneur, or full-time farmer using hazardous substances in protected natural areas or "pursuing activities otherwise dangerous to the character or conditions of the natural value" to provide security or draw up an insurance contract in accordance with special regulations.

Deposit or Performance Bonds

There are now many precedents for the use of environmental insurance or performance bonds for activities known to present a risk. Under construction law, for example, contractors are routinely required to post a bond to ensure that funds are available to pay the costs of completing and cleaning up the project if the contractor fails.

In the Philippines, under legislation applicable to plant imports, permit applicants may be required to deposit a bond equal to the estimated cost of the material to be imported. Under regulations on prospecting for biological and genetic resources, the applicant for a commercial access permit may be required to deposit a performance, compensation, and ecological rehabilitation bond as a condition of the access agreement. In the event of noncompliance, the competent authority may revoke the permit and retain the deposited bond.

One option for alien species control would be to require performance or deposit bonds from commercial permit holders or operators of facilities where alien species are kept in containment or captivity. Deposit bonds could also be adapted to cover private use of alien species; for example, pet purchasers could be required to pay refundable deposits that would be repaid to any pet owner making use of the recovery system.

Charges and User Fees

At a minimum, costs directly linked to permit applications and risk assessments should be met by the applicant. Where possible, the revenue generated (from this and other sources) could be used to fund operating costs of a specialist independent assessment body.

Charges are routinely applied to water or fuel consumption. South Africa's modernized water legislation provides a legal basis for classifying alien species cultivation as a "streamflow reduction activity" for which landowners may be required to pay charges linked to acreage.

In the United States, user fees are widely collected in plant health and to a certain extent in animal health. Much of the U.S. budget for quarantine controls comes from these revenues.

Corrective Taxes and Levies

Several states attach taxes or levies to a particular product, transporter, or passenger to implement domestic or international environmental commitments. Pursuant to the MARPOL treaty, which covers marine pollution, some states levy charges on shipments and passengers (port dues) to cover the cost of port waste treatment facilities and other control and monitoring. The United States levies a tax on key industries to finance the Superfund for cleanup of land contaminated by toxic wastes.

A special tax is levied in Western Australia to fund IAS eradication. In New Zealand, a levy may be imposed to generate financial contributions for pest management strategies.

One option could be for states to impose levies on given pathways, based on risks associated with that pathway. Revenues could be deposited into a fund to cover future mitigation measures, without the entity or person harmed by an IAS having to prove the specific source of the harm or the actor responsible.

Another possibility could be to impose levies or taxes on sales of alien animals and plants made by professional breeders or traders in such species. In practice, such incremental costs are routinely passed on to the ultimate beneficiary (the consumer).

Priorities for Future Action

Significant progress has been made over the last 6 years, since GISP was established, to address the environmental, economic, and social challenges posed by IAS. Although many existing legal and institutional frameworks are still characterized by fragmentation and differences in objectives, scope, and mechanisms, there are now much greater familiarity with the issues at stake and more systematic dialogue and cooperation between key organizations.

Important new instruments, notably the 2004 International Convention for the Control and Management of Ships' Ballast Water and Sediments, must be ratified and implemented as soon as practicable.

At the international level, major gaps in the regulatory framework relate to species that are invasive but do not qualify as plant pests or animal diseases under the rules and standards of the IPPC, the OIE, or other relevant organizations. Priority should be given to the following pathways:

- The use of alien organisms in aquaculture and the restocking of marine and inland water systems for commercial and recreational fisheries, taking account of existing voluntary codes
- Unintentional or opportunistic introductions (e.g., "hitchhiker organisms"), including those through hull fouling, packaging material, import consignments, vehicular transport, and other means
- Unintentional introductions of IAS through international assistance and humanitarian programs, tourism, military actions, scientific research, and cultural and other activities, and through aquaculture escapes, bait and pet releases, and water transfer schemes
- Intentional introductions of alien species for nonfood purposes, including certain aspects of horticulture and trade in pets and aquarium species; as biocontrol agents for control of IAS, pests, or weeds; transnational and national ex situ breeding projects; and international assistance programs, including conservation and development projects

Institutional coordination between relevant organizations needs to be strengthened as follows:

- Fuller consideration of IAS issues in other international forums, including through the Joint Liaison Group of the CBD, UN Framework Convention on Climate Change, the UN Convention to Combat Desertification, and the Collaborative Partnership on Forests.
- Closer collaboration with key bodies such as the World Health Organization, the IMO, the ICAO, the IPPC and OIE, and relevant site- and species-based conventions. (WHO, 1999)
- Closer coordination between national focal points of relevant international instruments, regional institutions, and international conventions and programs.

- New and more systematic links between the CBD and the WTO with regard to trade-related risks from IAS. The CBD should renew its request for observer status in the WTO SPS Committee to enhance the exchange of information on deliberations and developments in the respective bodies relevant to IAS.

Consistent regional approaches are needed to address transboundary issues, particularly on continents with shared land borders where one country's decisions can have immediate risks for its neighbors. Opportunities to develop regional standards and regional support for risk analysis of pathways and proposed introductions should be maximized. Regional institutions with IAS-related functions need to coordinate mandates and work programs to improve prevention and management as cost-effectively as possible.

At national and subnational levels, frameworks for alien species control must be comprehensive and consistent and serve the overall purpose of protecting ecological processes and native wild flora and fauna as well as domesticated animals and cultivated plants against biological invasion by alien species.

Priority should go to strengthening legislation, interagency coordination, and capacity for IAS prevention and management. Customs and quarantine systems play multiple roles in trade facilitation, food security, human health, and environmental protection; additional resources and capacity for border control and quarantine may indirectly contribute to IAS prevention even if this is not the primary objective. Cooperation between biodiversity, agriculture, and land management agencies will be of growing importance in the application of new environmental risk analysis standards. Incentives for use of native species in land management and other programs should be considered, together with more proactive engagement of stakeholder groups.

Notes

1. For a full table of these instruments, see Shine, Williams, and Gündling (2000), Appendix I.
2. IUCN, a partner in GISP, adopted Guidelines for the Prevention of Biodiversity Loss due to Biological Invasion in February 2000 (IUCN 2000a, 2000b).
3. For example, Arts. 6(b), 7(c), 8(l), and 14.
4. Decision of the Conference of the Parties IV/1/C.
5. Most recently, Decision VII/13, which references the UN Food and Agriculture Organization, World Health Organization, International Maritime Organization, Office International des Epizooties, and a range of conventions, including the International Plant Protection Convention, Convention on International Trade in Endangered Species, the Ramsar Convention on Wetlands, the UN Framework Convention on Climate Change, and the UN Convention to Combat Desertification.
6. Annexed to Decision VI/23 (COP6, 2002).

7. See Murphy et al. (2001), Quinlan (2001), Secretariat of the Convention on Biological Diversity (2001), and Shine (2003).
8. Including Standing Committee Recommendations R (84)14, No. 45 of 1995, No. 57 of 1997, and No. 77 of 1999.
9. Annex II, Art. 4 (1).
10. Directive 79/409/EEC on the Conservation of Wild Birds and Directive 92/43/EEC on the Conservation of Natural Habitats and of Wild Fauna and Flora.
11. Decision II/10, 1995; thematic work program adopted in 1998 (Decision IV/5), reviewed and updated in 2004 (Decision VII/5).
12. Resolution VII/14 and VIII/18.
13. Defined under the IPPC as "a pest of potential economic importance to the area endangered thereby and not yet present there, or present but not widely distributed and being officially controlled."
14. The IPPC provides for regulation of nonquarantine pests, such as weeds, pests on propagative material, and diseases that may have indirect effects on plants, but this does not appear to support restrictions on pests with purely environmental impacts.
15. Guidelines for the Control and Management of Ships' Ballast Water to Minimize the Transfer of Harmful Aquatic Organisms and Pathogens (Annex to Resolution A.868[29], Twentieth Assembly, 1997). These guidelines updated the IMO Guidelines for Preventing the Introduction of Unwanted Aquatic Organisms and Pathogens from Ships' Ballast Waters and Sediment Discharges (Assembly Resolution, 1993: Resolution A.774[18]).
16. Assembly Resolution A33-18, which supersedes Assembly Resolution A32-9 (1998).
17. Abridged from Annex A, Definitions.
18. Supplement on Analysis of Environmental Risks to ISPM No. 11 ("Pest Risk Analysis for Quarantine Pests," 2001); Supplement No. 2 on Guidelines on the Understanding of "Potential Economic Importance" and Related Terms Including Reference to Environmental Considerations to ISPM No. 5 ("Glossary of Phytosanitary Terms").
19. In the area of plant health, an IPPC standard provides for states to take emergency measures of temporary application, whose validity must be subject to detailed pest risk analysis as soon as possible (Art. 5.7, ISPM Principles of Plant Quarantine as Related to International Trade).
20. These include EC Measure Concerning Meat and Meat Products (EC-Hormones), WT/DS26/AB/R, WT/DS48/AB/R (January, 19 1996); Australia: Measures Affecting Importance of Salmon (Australia: Salmon), WT/DS18/AB/R (October 20, 1998); and Japan: Measures Affecting Agriculture Products (Japan: Varietals), WT/DS76/AB/R (February 22, 1999).
21. Most recently, Decisions V/6 (2000), VI/12 (2002), and VII/11 (2004).
22. Principle 19 of the Rio Declaration provides that "States shall provide prior and

timely notification and relevant information to potentially affected States on activities that may have a significant transboundary environmental adverse effect and shall consult with those States at an early stage and in good faith."

23. A treaty elaborated in the framework of the UN Economic Commission for Europe by 24 European states, Canada, the United States, and the European Community, in force since 1997 (UNECE 2003).
24. CBD Article 14.1 covers notification, consultation, and emergency planning.
25. Concluded between the Netherlands, Belgium, and Luxembourg (October 17, 1983).
26. Genovesi and Shine (2004); see in particular Part 4.1–3.
27. For example, where a species lawfully imported into one country or province crosses a political boundary and becomes invasive in a neighboring country or province that prohibits its import.
28. The proceedings of the Workshop on Invasive Alien Species and the IPPC (IPPC 2003) provide an excellent information resource on the application of risk analysis to IAS issues.
29. The IPPC uses different terminology for "the entry of a pest resulting in its establishment" and provides clear definitions for each of these established terms.
30. Standing Committee to the Bern Convention, 1999 "Recommendation on the Eradication of Non-Native Terrestrial Vertebrates," No. 77.
31. See Exploratory Working Group on the Phytosanitary Aspects of Genetically Modified Organisms, Biosafety and Invasive Species (FAO, Rome, June 13–16, 2000).
32. This requires the CBD COP to examine, on the basis of studies to be carried out, the issue of liability and redress, including restoration and compensation, for damage to biological diversity, except where such liability is a purely internal matter.
33. Art. 27 of the protocol calls for the COP to set up a procedure to consider liability and redress for damage resulting from transboundary movements on LMOs. The Technical Group of Experts on Liability and Redress under the Cartagena Protocol on Biosafety met in October 2004 to consider the development of rules for this purpose. Results and decisions of this meeting are available at http://www.biodiv.org/biosafety/issues/liability2.aspx.
34. The Standing Committee's nonbinding Recommendation on the Eradication of Non-native Terrestrial Vertebrates (No. 77, 1999) provides that where a species introduced into the territory of a state spreads to neighboring states or entire regions and damages their environment, this should give rise to the liability of the state from which it originated.
35. These are measures instituted for other sectoral objectives that may have the effect of encouraging introduction and spread of unwanted alien species. A 1998 estimate for the United States considered that out of US$1.898 billion spent on agriculture, road transport, fisheries, forestry, and fossil fuels, $1.456 billion was spent on perverse subsidies (Myers and Kent 1998).

36. Environmental Protection and Biodiversity Conservation Act of 1999, s.528.
37. Nature Conservation Act, sec. 20(d)(2).
38. Executive Order 13112, 1999.
39. See in particular Bern Convention Recommendation R 85(15) and IUCN (1995).

References

Baldacchino, A. E., and A. Pizzuto, eds. 1996. *Introduction of Alien Species of Flora and Fauna,* proceedings of a seminar held at Qawra, Malta, on March 5, 1996. Ministry of Environment, Malta.

Christensen, M. 2004. Invasive Species Legislation and Administration: New Zealand, pp. 23–50 in *Harmful Invasive Species: Legal Responses,* edited by M. Miller and R. Fabian. Environmental Law Institute, Washington, DC.

Corn, L. C., E. H. Buck, J. Rawson, and E. Fischer. 1999. *Harmful Non-Native Species: Issues for Congress.* Congressional Research Service Issue Brief RL30123, Resources, Science, and Industry Division, National Council for Science and the Environment, Washington, DC, April 8, 1999.

de Klemm, C. 1996. "Introductions of non-native organisms into the natural environment," *Nature and Environment* No. 73, Council of Europe, Strasbourg.

Di Paola, M., and D. Kravetz. 2004. "Invasive Alien Species: Legal and Institutional Framework in Argentina" pp. 71–88 in *Harmful Invasive Species: Legal Responses,* edited by M. Miller and R. Fabian. Environmental Law Institute, Washington, DC.

Downes, D. 1999. *Integrating Implementation of the Convention on Biological Diversity and the Rules of the World Trade Organization.* IUCN, Gland, Switzerland.

Durand, S., and J. Chiaradia-Bousquet. 1999. *New Principles of Phytosanitary Legislation.* FAO Legislative Study 62. Food and Agriculture Organization, Rome.

Genovesi, P., and C. Shine. 2004. "European strategy on invasive alien species," *Nature and Environment* No. 137, 67 p.

Glowka, L., and C. de Klemm. 1996. "International instruments, processes and non-indigenous species introductions: Is a protocol to the Convention on Biological Diversity necessary?" pp. 211–218 in *Proceedings of the Norway/United Nations Conference on Alien Species,* edited by O. Sandlund, P. Schei, and A. Viken. Directorate for Nature Management and Norwegian Institute for Nature Research, Trondheim.

Hedley, J. 2004. "The International Plant Protection Convention and Invasives," pp. 185–201 in *Harmful Invasive Species: Legal Responses,* edited by M. Miller and R. Fabian. Environmental Law Institute, Washington, DC.

IPPC. 2003. *Proceedings of the Workshop on Invasive Alien Species and the IPPC,* Braunschweig, Germany, September 22–26, 2003. Online: https://www.ippc.int/.

IUCN. 1995. *IUCN Guidelines for Re-introductions:* IUCN, Gland, Switzerland.

———. 2000a. *IUCN Guidelines for the Prevention of Biodiversity Loss due to Biological Invasion* (approved by the IUCN Council, February 2000). IUCN, Gland, Switzerland.

———. 2000b. *Proceedings of the IUCN–ELC Workshop on Legal and Institutional Dimensions of Invasive Alien Species Introduction and Control,* Bonn, Germany, December 10–11, 1999. Many of the papers from this workshop are available at http://www.iucn.org/themes/law/.

Jenkins, P. 2000. *Who Should Pay? Economic Dimensions of Preventing Harmful Invasions through International Trade and Travel.* Paper presented to the GISP Human Dimensions Workshop, Cape Town, South Africa, September 2000.

Kinderlerer, J. and P. Phifer. 2004. "The regulation of genetically modified organisms," pp. 219–236 in *Harmful Invasive Species: Legal Responses*, edited by M. Miller and R. Fabian. Environmental Law Institute, Washington, DC.

McNeely, J. 1999. *The Ecosystem Approach for Sustainable Use of Biological Diversity.* Paper presented at the Norway/UN Conference on the Ecosystem Approach for Sustainable Use of Biological Diversity, Trondheim, September 6–10, 1999.

Miller, M. 2004. "The paradox of U.S. alien species law," pp. 125–184 in *Harmful Invasive Species: Legal Responses*, edited by M. Miller and R. Fabian. Environmental Law Institute, Washington, DC.

Mooney, H. A., and R. J. Hobbs, eds. 2000. *Invasive Species in a Changing World.* Island Press, Washington, DC.

Murphy, S. T., I. S. H. Wilde, M. M. Quinlan, S. Soetikno, and G. Odour. 2001. *Alien Invasive Species: Review of Activities and Programmes on Prevention, Early Detection, Eradication and Control,* Information document commissioned by CBD Secretariat (UNEP/CBD/SBSTTA/6/7), Montreal, Canada.

Myers, N., and J. Kent. 1998. *Perverse Subsidies: Tax $s Undercutting Our Economies and Environments Alike.* IISD, Winnipeg, Toronto, Canada.

Perrings, C., M. Williamson, and S. Dalmazzone. 2000. *The Economics of Biological Invasions.* Edward Elgar, Cheltenham, United Kingdom.

Quinlan, M. 2001. *Report on Procedures, Criteria and Capacities for Assessing Risk on Alien Invasive Species.* Information document commissioned by CBD Secretariat (UNEP/CBD/SBSTTA/6/INF/6), SCBD, Montreal, Canada.

Secretariat of the Convention on Biological Diversity. 2001. Review of the efficiency and efficacy of existing legal instruments applicable to invasive alien species. SCBD, 42 p. *CBD Technical Series* No. 2, Montreal, Canada.

Sharp, R. 1999. "Federal policy and legislation to control invading alien species," *Australian Journal of Environmental Management* 6(3).

Shine, C. 2003. *Invasive Alien Species: Identification of Specific Gaps and Inconsistencies in the International Regulatory Framework.* Information document commissioned by CBD Secretariat (UNEP/CBD/SBSTTA/9/INF/32), SCBD, Montreal, Canada.

Shine, C., N. Williams, and L. Gündling. 2000. *A Guide to Designing Legal Frameworks on Alien Invasive Species.* IUCN Environmental Policy and Law Paper no. 40. IUCN, Gland, Switzerland.

Stein, R. 2004. "Invasive species law and policy in South Africa," pp. 51–70 in Miller, M. and R. Fabian, eds., *Harmful Invasive Species: Legal Responses.* Environmental Law Institute, Washington, DC.

UNECE. 2003. Ministry of the Environment, Finland; Ministry of the Environment, Sweden; Ministry of Housing, Spatial Planning and the Environment, the Netherlands. *Guidance on the Practical Application of the Espoo Convention.* Convention on Environmental Impact Assessment in a Transboundary Context (UNECE). Finnish Environmental Institute (SYKE), Finland, 48 p. Available at http://www.unece.org/env/eia/publications.html/#publications.

Wittenberg, R., and M. J. W. Cock, eds. 2001. *Invasive Alien Species: A Toolkit of Best Prevention and Management Practices.* CABI Publishing, Wallingford, UK.

World Health Organization. 1999. *Public Health and Trade: Comparing the Roles of Three International Organizations*, WHO Weekly Epidemiological Review, pp. 193–200, N. 25. WHO, Geneva.

11

Human Dimensions of Invasive Alien Species

Jeffrey A. McNeely

Human impacts on the ecosystems of our planet continue to grow. Our increasing population and expanding levels of consumption mean that more people are consuming more of nature's goods and services, pushing against the limits of sustainability. Greatly expanding global trade is feeding this consumption, with large containers of goods moving quickly from one part of the world to another by plane, ship, train, and truck.

One critical element in this economic globalization is the movement of organisms from one part of the world to another through trade, transport, travel, and tourism. Many of these movements of organisms into new ecosystems where they are alien (also called nonnative, nonindigenous, or exotic) are generally beneficial to people. But many others have very mixed impacts, benefiting some individuals or interest groups while disadvantaging others. And in a few cases, especially of disease organisms and pests of forests or agricultural crops, the alien species is clearly detrimental to all. This book addresses the latter groups: invasive alien species (IAS), the subset of alien species whose establishment and spread threaten ecosystems, habitats, or species with economic or environmental harm (McNeely et al. 2001).

Farmers have been fighting weeds since the very beginnings of agriculture, and disease organisms have been a major focus of physicians for more than a hundred years. But the general global IAS problem has been brought to the world's attention only recently by ecologists who were concerned that native species and ecosystems were being disrupted (Elton 1958; Drake et al. 1989). Much of the work to date on IAS has focused on their biological and ecological characteristics, the vulnerability of ecosystems to invasions, and the use of various means of control against invasives. However, the problem of IAS is above all a human one, for the following reasons:

- People are largely responsible for moving eggs, seeds, spores, vegetative parts, and whole organisms from one place to another, especially through modern global transport and travel.

- Although some species are capable of invading well-protected, intact ecosystems, IAS more often seem to invade habitats altered by humans, such as agricultural fields, human settlements, and roadways.
- Many alien species are intentionally introduced for economic reasons, a major human endeavor, implying that those earning economic benefits should also be responsible for economic costs should the alien become invasive.
- The dimensions of the IAS problem are defined by people, and the response is also designed and implemented by people, with different impacts on different groups of people.

People introduce organisms into new habitats unintentionally (often invertebrates and pathogens), intentionally (usually plants and vertebrates), or inadvertently when organisms imported for a limited purpose subsequently spread into new habitats (Levin 1989). Many of the deliberate introductions relate to the human interest in nurturing species that are helpful to people for agricultural, forestry, ornamental, or even psychological purposes (Staples 2001). Most human dietary needs in most parts of the world are met by species that have been introduced from elsewhere (Hoyt 1992). It is difficult to imagine an Africa without cattle, goats, maize, and cassava, or a North America without wheat, soybeans, cattle, and pigs, or a Europe without tomatoes, potatoes, and maize—all introduced species. Therefore, species introductions are an essential part of human welfare and local cultures in almost all parts of the world. Furthermore, maintaining the health of these introduced alien species of undoubted net benefit to humans sometimes entails the introduction of additional alien species for use in biological control programs that import natural enemies of, for example, agricultural pests (Waage 1991; Thomas and Willis 1998), but these biological controls may themselves become invasive.

Evidence indicates a rapid recent growth in the number and impact of IAS (Mooney and Hobbs 2000). Trade and, more generally, economic development lead to more IAS; for example, Vilà and Pujadas (2001) found that countries that are more effectively tied into the global trading system tend to have more IAS, being positively linked to the development of terrestrial transport networks, migration rates, numbers of tourists visiting the country, and trade in commodities (Dalmazzone 2000). The general global picture shows tremendous mixing of species, with unpredictable long-term results but a clear trend toward homogenization (Bright 1998; Mooney and Hobbs 2000). The future is certain to bring much more species shuffling as people continue to influence ecosystems in various ways, not least through both purposeful and accidental introduction of species as an inevitable consequence of growing global trade. This shuffling will yield species that become more abundant and many others that decline in numbers (or even become extinct), but the likely overall effect is a global loss of biodiversity at species and genetic levels. But how is the great reshuffling of species being driven by human interests, and how will it affect them? How should people think about the issue? What stakes are involved? Whose interests are being affected? How can scientists, resource managers, and policymakers best address the human dimensions?

These are not trivial questions because the IAS issue has ramifications throughout modern economies. It involves global trade, settlement patterns, agriculture, economics, health, water management, climate change, genetic engineering, and many other fields and concerns. It therefore goes to the very heart of problems policymakers are spending much time debating, usually without reference to IAS. This chapter draws on contributions presented at a Global Invasive Species Program (GISP) workshop held in Cape Town, South Africa, on September 15–17, 2000, to examine some of the ramifications of IAS through many dimensions of human endeavor, including historical, economic, cultural, linguistic, health, psychological, sociological, legal, management, military, philosophical, and political components. It shows that IAS are deeply woven into the fabric of modern life. Although the biological dimensions of IAS are fundamental, more effective responses to the problems they pose must incorporate the kinds of human dimensions that are discussed in this chapter.

Historical Dimensions

Because of a long geological and evolutionary history, our planet has very different species of plants, animals, and microorganisms on the various continents and in the various ecosystems. As a broad illustration, Africa has gorillas, Indonesia has orangutans, South America has monkeys but no apes, and Australia has no nonhuman primates at all. Even within the continents, most species are confined to particular types of habitats; gorillas live in forests, zebras mostly in grasslands, and addaxes in deserts. Oceanic islands and other geographically isolated ecosystems often have their own suites of species, many found nowhere else (called "endemic species"); about 20 percent of the world's flora is made up of insular endemics, found on only 3.6 percent of the land surface area. Geographic barriers have ensured that most species remain within their region, thus resulting in a much greater species richness across the planet than would have been the case if all land masses were part of a single continent. This historical biogeographic framework provides the basis for defining concepts of native and alien species. It is also important to recognize that biogeography is dynamic, as species expand and contract their ranges and the contents of ecosystems change as a result of factors such as climate change (Udvardy 1969).

Humans apparently evolved in Africa, then *Homo sapiens* spread to Europe and Asia more than 100,000 years ago, Australia 40,000–60,000 years ago, the Americas about 15,000–20,000 years ago, and the far reaches of the Pacific less than 1,000 years ago. Our species is a good example of a naturally invasive species, spreading quickly, modifying ecosystems through the use of fire, and driving other species to extinction (Martin and Klein 1984). Wherever people have moved, they have also carried other species with them. The Asians who first peopled the Americas, for example, were accompanied by dogs, and Polynesians sailed with pigs, taro, yams, and at least 30 other plant species (and rats and lizards as stowaways).

Trade is known far back in human prehistory, judging from the discovery of stone tools at a distance from where they were quarried. But as long-distance travel became more regular, trade became more important. Chinese traders have traveled through Southeast Asia for at least several thousand years, and trading routes between India and the Middle East stretch back at least as long. As sailing craft became larger and more reliable, trade increased further and was given a great boost with the voyages of Christopher Columbus, which opened up entirely new sources of species and led to the replacement of the rigid moral strictures of Medieval Europe by a new set of merchant values that stressed consumption (Low 2001).

For at least several thousand years, armies have been an important pathway for moving species from one region to another, with at least some of them becoming invasive. The spread of new diseases by armies is well known. For example, measles was carried into the Americas from Europe by the early conquistadors, and perhaps syphilis went in the opposite direction (McNeill 1976). Rinderpest, a virus that is a close relative of measles and canine distemper, is native to the steppes of Central Asia, but it frequently swept through Europe, being carried by cattle moved to feed armies during military campaigns. Africa remained free of this disease until 1887, when it appeared in Eritrea at the site of the Italian invasion, spreading through Ethiopia in 1888 and conquering the entire continent in less than a decade. In some parts of Africa, rinderpest was followed by wars and cattle raids as the tribal pastoralists sought to maintain their herds (Pearce 2000). Another result was that rinderpest led an ecological revolution against people and cattle and in favor of wildlife species that were resistant to the disease.

The period of European colonialism ushered in a new era of species introductions as the European settlers sought to recreate the familiar conditions of home (Crosby 1986). They took with them species such as wheat, barley, rye, cattle, pigs, horses, sheep, and goats, but in the early years their impacts were limited by the available means of transport. Once steam-powered ships came into common use, the floodgates opened and more than 50 million Europeans emigrated to distant shores between 1820 and 1930, carrying numerous plants and animals that were added to the native flora and fauna (Reichard 2001). More recently, Chinese, Indian, Indo-Chinese, African, and other emigrants have carried familiar species with them to grow in their new homelands in Europe, Australia, and the Americas.

The era of European colonialism also saw the spread of plant exploration, seeking new species of ornamental plants for botanical gardens, nurseries, and private individuals back home (Reichard 2001). The spread of global consumerism was given a significant boost in the early twentieth century through advertising and marketing that was strategically designed to motivate the public to buy more goods (Staples 2001). This ultimately led to an accelerating search to find new species to grow and market, creating consumer demand for products that previously were not present. The invasive characteristics of the newly introduced species often came as a surprise because those responsible for the introduction were unaware of the possible negative ecological ramifications of the species involved.

Many invasive species of plants and animals were carried by the colonial military, especially to Pacific and Indian Ocean islands that had numerous endemic species vulnerable to these invasives. In the seventeenth and eighteenth centuries, navies introduced many plants and animals to remote islands as future food sources, and these often became invasive (Binggeli 2001). The military sometimes brought in exotic plant species to form barriers. For example, the French introduced a cactus (*Opuntia monacantha*) to Fort Dauphin in southeast Madagascar in 1768 to provide an impregnable barrier around the fort. Later, the military also introduced a spineless variety (of *O. ficus-indica*) to feed oxen (Decary 1947). The role of the military in the spread of IAS has continued. World War II was a particularly active time for the introduction of weeds in the Pacific. Some species, such as Bermuda grass (*Cynodon dactylon*), were deliberately introduced to revegetate islands that were devastated by military activity. Many species spread by accident, clinging to military equipment and supplies or sticking to wheels of airplanes. Some grass species were carried from one island to another as seeds adhering to clothing. And because many weeds do best on bare or disturbed ground, war helped to prepare a fertile ground for them. The brown tree snake (*Boiga irregularis*) came to Guam from New Guinea or the Solomon Islands during the war, apparently hitching a ride on either Allied or Japanese ships or planes. Arriving in Guam, the brown tree snake found that the local birds and lizards were not well adapted to such an agile predator armed with poisonous fangs; as a result, the populations of native birds and lizards have plummeted on Guam. Now the brown tree snake is threatening Hawaii, with military transports from Guam being the most likely vector. The U.S. military authorities recognize the danger and are working to minimize the threat.

Thus the faunal and floral assemblages found in any particular location have been profoundly influenced by past human activities, and people are likely to have an even greater impact in the future. This leads to the contemplation of whether the current episode of globalization might lead to increased diversity in at least some places after the dust settles on the current extinction spasm (Parker 2001). As just one example, New Zealand has twice as many plants today as it did when humans first arrived, as well as a whole suite of new mammals; one tragic cost was the loss of an extensive unique fauna of birds. Further development of biotic communities as climates change will depend on organisms invading novel habitats, sometimes hybridizing with the native species, sometimes replacing them, and sometimes adding to the diversity of the ecosystem with new species interactions. By introducing species, humans are creating their own ecosystems (Orr and Smith 1998), often by accident, and disrupting ecosystems that evolved over millions of years.

Human Dimensions of the Causes of Species Invasions

Global trade has enabled modern societies to benefit from the unprecedented movement and establishment of species around the world. Agriculture, forestry, fisheries, the pet trade, the horticultural industry, and many industrial consumers of raw materials today depend on species that are native to distant parts of the world. The lives of people everywhere have been greatly enriched by their access to a greater share of the world's biological diversity, and expanding global trade is providing additional opportunities for further enrichment. Most people warmly welcome this globalization of trade, and growing incomes in many parts of the world are leading to increased demand for imported products. For example, North American nursery catalogues offer nearly 60,000 plant species and varieties to a global market, often through the Internet (Ewel et al. 1999). A generally unrecognized side effect of this globalization is the introduction of alien species, at least some of which may be invasive.

Global mobility has undermined the sense of place that previously provided a psychological anchor to most societies. Modern information technology leaves us "with the disorienting experience that there is no single or universal context with clearly defined borderlines within which we can appeal to reason to settle our differences, but rather a multiplicity of mini-contexts that are not universally shared within which numerous incommensurable interpretations coexist alongside one another" (Hattingh 2001, p. 188). The speed at which many modern people live, including great mobility, seriously compromises their ability to develop a sense of place, a clear vision of the future, or a sense of origin. This makes it much more difficult for people to distinguish between native and alien species or to be concerned about the difference.

Linked to the global marketplace, the world is becoming increasingly urban, with half the world's population living in cities at the turn of the century. Cities tend to be the focal points of the global economy and the entry points for many invasives. Many invasive species are most prolific in urban and urban fringe environments, where long histories of human disturbance have created abundant bare ground and many opportunities for invasion. Many urban dwellers seek ornamentals from a wide range of sources, and these may become invasive. For example, Berlin has 839 native plant species and 593 aliens (Kowarik 1990). Urbanization involves large and mobile populations that can easily escape the environmental consequences of misusing resources. Furthermore, they are seldom aware of the problems of invasive species because they have lost their connections to the natural environment (Staples 2001). Settlement patterns also involve transportation links, and the distributions of many invasives seem to follow transportation corridors. Thus human settlement patterns also are part of the invasive species issue (Marambe et al. 2001).

Many people who seek to introduce a nonnative species into a new habitat do so for an economic reason (McNeely 1999). They may want to increase their profits from agriculture, they may believe that the public will like a newly discovered flower from a distant part of the globe, or they may think that nonnative species will be able to carry out functions that native species cannot carry out as effectively. But few of those introducing alien species have carried out a thorough cost–benefit analysis before initiating the introduction, ignoring (externalizing) the negative impacts that may follow from species introductions because they have not been required to recognize them. They might also be worried that they would be expected to compensate those who are negatively affected.

Similarly, those who have been responsible for inadvertently introducing species into new habitats may not have been willing to make the investment necessary to prevent such accidents from occurring. They may not have realized the dangers, and in any case the dangers would be unlikely to have much economic impact on their own welfare. Rather, the costs of such accidents are borne disproportionately by people other than those who are permitting the accidents to happen. Thus the costs of introducing potentially invasive alien species into new habitats are externalized in considerations of the costs of global trade. The line of responsibility is insufficiently clear to bring about the necessary changes in behavior, so the general public and future generations end up paying most of the costs.

In the early 1990s, Serbian scientists discovered the western corn rootworm (a beetle *Diabrotica vigifera,* whose wormlike larvae feed on the roots of maize plants) near Belgrade airport, apparently inadvertently flown in on military aircraft from the United States. Vigorous international action might have curbed this pest's first known venture outside North America, but the turmoil of war prevented such a collaboration, and now it is too late. By 1995, the pest had spread into Croatia and Hungary, subsequently spreading to Romania, Bosnia–Herzegovina, Bulgaria, and Italy (Enserink 1999). It is likely to spread into every maize-planting country in Europe, and perhaps eventually into Asia, forcing farmers to use chemical pesticides. A problem that would have been easy and cheap to solve if addressed quickly was prevented from being controlled by the human factor of war that blocked the necessary collaboration, and now it has serious economic impacts.

One limitation of human perception of the costs of IAS is that invasions often happen almost invisibly, without any clear responsibility and with very limited initial impacts. Furthermore, monitoring, early detection, and containment of invaders before they cause widespread damage are unlikely to be considered to have a positive cost–benefit ratio because the costs are required now, whereas the main benefits (at least in terms of future costs avoided) remain speculative. On the other hand, where sound cost–benefit studies have been done, they demonstrate the value of control, and prevention is shown to be the best strategy (Jenkins 2001).

All human cultures actively modify their surroundings to achieve an environment that they find pleasing. At least part of the world's cultural diversity results from the local patterns of distribution of plants and animals because the locally available resources and how they are used help to define the character of any particular cultural group. Some

IAS become part of the local culture. For example, Australian Aborigines today are hunting some of the mammals that have invaded Australia (e.g., water buffalo, rabbits, and camels), and they argue that government programs to control these invasive aliens in the name of protecting native fauna and flora are depleting an important food source for them. They contend that they are thus being forced to turn their hunting attention to native species that are already under threat.

The Maori people of New Zealand have similar concerns. For example, some Maori leaders initially opposed eradication of Pacific rats (*Rattus exulans*) from some islands, claiming that they were "a treasure" brought to New Zealand by their ancestors in the course of their migrations (Veitch and Clout 2001). Neither Maoris nor Aborigines have a single perspective on IAS, but they share a concern about the use of poisons to control IAS, potential for pollution of water supplies, and the introduction of yet more alien species for biological control.

Some suggest that people have an innate tendency to focus on life and lifelike processes, a condition Wilson (1984) calls biophilia. This leads many people to value diversity for its own sake, perhaps seeking to enhance the options available for improving their physical or social well-being. One manifestation of this tendency may be a need or desire to have other, nonhuman species living close to us (Mack 2001; Staples 2001). In the United States, Europe, and elsewhere, a thriving pet trade that answers this human need also poses continuous risks of intentional or accidental releases by pet owners (Genovesi and Bertolino 2001). Even people who are professional resource managers, such as the staff at South Africa's Kruger National Park, can be remarkably resistant to the idea of limiting their cultivation of potentially invasive garden plants (Foxcroft 2001). Thus human preference rather than biological traits may have primary importance in determining whether a plant species is introduced.

Human Dimensions of the Consequences of IAS

IAS have many negative impacts on human economic interests. Weeds reduce crop yields, increase control costs, and decrease water supply by degrading catchment areas and freshwater ecosystems. Tourists unwittingly introduce alien plants into national parks, where they degrade protected ecosystems and drive up management costs. Pests and pathogens of crops, livestock, and trees destroy plants outright or reduce yields and increase pest control costs. The discharge of ballast water introduces harmful aquatic organisms, including bacteria and viruses, to both marine and freshwater ecosystems, thereby degrading commercially important fisheries and recreational opportunities. And recently spread pathogens continue to kill or disable millions of people each year, with profound social and economic implications. Although the total economic costs of invasions are uncertain, estimates of the economic costs of particular invasives to particular sectors indicate the seriousness of the problem. Some of these, drawn primarily

Table 11.1. Costs of some alien species invasions

Species	*Economic variable*	*Economic impact (US$)*	*Reference*
Introduced disease organisms	Annual cost to human, plant, and animal health in U.S.	$41 billion/year	Daszak, Cunningham, and Hyatt (2000)
A sample of alien plant and animal species	Economic costs of damage in U.S.	$137 billion/year	Pimentel et al. (2000)
Salt cedar (*Tamarix*)	Value of ecosystem services lost in western U.S.	$7–16 billion over 55 years	Zavaleta (2000)
Knapweed (*Centaurea* spp.) and leafy spurge (*Euphorbia escula*)	Impact on economy in three U.S. states	$40.5 million/year direct costs, $89 million indirect	Bangsund, Leistritz, and Leitch (1999), Hirsch and Leitch (1996)
Zebra mussel (*Driessena polymorpha*)	Damages to U.S. and European industrial plants	Cumulative costs 1989–2000, $750 million to $1 billion	National Aquatic Nuisances Clearinghouse (2000)
Most serious invasive alien plant species	Costs 1983–1992 of herbicide control in Britain	$344 million/year for 12 species	Williamson (1998)
Six weed species	Costs in Australian agroecosystems	$105 million/year	CSIRO (1997), cited in Watkinson, Freckleton, and Dowling (2000)
Pinus, Hakea, Acacia, and lowland acacias	Costs on South African fynbos to restore pristine conditions	$2 billion	Turpie and Heydenrych (2000)
Water hyacinth (*Eichhornia crassipes*)	Costs in 7 African countries	$20–50 million/year	Joffe-Cook (1997), cited in Kasulo (2000)
Rabbits (*Oryctolagus*)	Costs in Australia	$373 million/year (agricultural)	Wilson (1995), cited in White and Newton-Cross (2000)
Varroa mite	Economic cost to beekeeping in New Zealand	$267–602 million	McNeely et al. (2001)
Golden apple snail (*Pomacea canaliculata*)	Impact on rice in the Philippines	$28–45 million/year	Naylor (1996)

from Perrings et al. (2000), are listed in Table 11.1. Many of these estimates remain controversial among economists.

Globalization is bringing with it a series of new medical threats, many of which can be considered a subset of the IAS problem. Viruses are a particular problem because they are so difficult to combat; although vaccines for viruses such as smallpox, polio, and yellow fever have proven effective, cures remain elusive, and large investments to find a cure for AIDS have proven only marginally effective. Even worse, the global changes that are affecting many parts of the world are expected to expand the ranges of many viruses that are potentially dangerous to humans. When people move into formerly unoccupied wilderness areas, they come into contact with a wider range of viruses and bacteria, and air travel carries them around the globe before the symptoms become apparent.

Infectious disease agents often, and perhaps typically, are IAS (Delfino and Simmons 2000). Unfamiliar types of infectious agents, either acquired by humans from domesticated or other animals or imported inadvertently by travelers, can have devastating impacts on human populations. Pathogens can also undermine local food and livestock production, thereby causing hunger and famine:

- The bubonic plague (caused by *Pasturella pestis*) spread from central Asia through North Africa, Europe, and China using a flea vector on an invasive rat species (*Rattus rattus*) that came originally from India.
- The viruses carrying smallpox and measles spread from Europe into the Western Hemisphere shortly after European colonization. The low resistance of the indigenous peoples to these diseases helped bring down the mighty Aztec and Inca empires.
- The Irish potato famine in the 1840s was caused by a fungus (*Phytophtora infestans*) introduced from North America that attacked potatoes, with devastating impacts on the health of local people.
- The influenza A virus originated in birds but multiplies through domestic pigs, which can be infected by multiple strains of avian influenza virus and then act as genetic "mixing vessels" that yield new recombinant DNA viral strains. These strains can then infect pig-tending humans, who then infect other humans, especially through rapid air transport.

At least some invasive plant species may themselves be considered a health hazard in both temperate and tropical regions. Binggeli (2001) reports that large quantities of airborne pollen of *Casuarina equisetifolia* cause respiratory irritations. In proximity to habitations, both *Schinus terebinthifolius* and *Melaleuca quinquenervia* appear to cause respiratory difficulties in many people, and skin contact with leaves and the sap of *S. terebinthifolius* results in red, itching rashes.

The interactions between invasive pathogens, human behavior, and economic development are complex and depend on interactions between the virulence of the disease, infected and susceptible populations, the pattern of human settlements, and their level of development. Large development projects, such as dams, irrigation schemes, land reclamation, road construction, and population resettlement programs, have contributed to the invasion of diseases such as malaria, dengue, schistosomiasis, and trypanosomiasis (WHO 1997). The clearing of forests in tropical regions to extend agricultural land has opened up new possibilities for wider transmission of viruses that carry hemorrhagic fevers that previously circulated benignly in wild animal hosts. Invasive species combined with variations in interannual rainfall, temperature, human population density, population mobility, and pesticide use all contribute to one of the most profound human dimensions of invasive species: the threat to human health.

Components of biological diversity that are threatened or lost as a result of IAS can lead to the loss of traditional knowledge, innovations, and practices. Likewise, customary uses of biological resources in accordance with traditional cultural practices may be inhibited or, in the worst case, discontinued completely. As intimate users of local biological resources, indigenous and local communities potentially are well qualified to monitor the impacts of alien species on local ecosystems and their components (Art. 7 of the Convention on Biological Diversity [CBD]), to identify when those species become invasive, and to be involved in eradication and mitigation programs (Art. 8h of the CBD). But this depends on awareness of the problem. In China, Vietnam, Malaysia, Thailand, Korea, and Cambodia, people "make merit" by releasing captive animals, especially birds, fish, and turtles, but one study in Taiwan found that 6 percent of birds released were exotic, and most of the fish and turtles were captive-bred exotic species that could become invasive (Severinghaus and Chi 1999). Clearly, the cultural process of "making merit" does not intentionally include deleterious impacts on native ecosystems, largely because the people involved often are urban-dwellers who have no concept of IAS.

Human Dimensions of the Response to IAS

There are four main approaches to IAS management:

- Subject all alien species proposed for introduction to expert consideration, following the precautionary principle.
- Improve the scientific basis for predicting which species proposed for deliberate introduction are likely to become invasive and which are likely to be beneficial.
- Improve control of pathways for unplanned introductions (e.g., through ballast water, international trade, and wooden packing material).
- Improve management techniques to eradicate or control IAS once prevention has failed or become impractical.

Human societies seem to have a great capacity for contradiction, with quarantine inspections, for example, being the responsibility of the same governments that promote globalization that undermine government capacity to apply effective quarantine measures (Low 2001). Governments have a responsibility to provide regulations in the public interest, but current economic orthodoxy argues that global trade is fostered by removing regulations that may constrain such trade, such as restrictions that may constrain the introduction of a potentially invasive alien species. These contradictions help underscore both the conflict of interests between global trade and IAS control and the challenges to current management measures and legal frameworks.

The human dimension is the most unpredictable variable in any management program to control IAS. Reaser (2001) and Mack (2001) describe the psychological factors motivating people to import or use alien species that sometimes become invasive, and they show how a more thorough understanding of these psychological factors can slow further invasions and promote the control of the existing ones. They demonstrate that IAS are a byproduct of human values, decisions, and behaviors, suggesting that a focus on human beliefs and resultant behavior might be more effective than focusing primarily on IAS themselves. Resource managers therefore must generate public support and understanding for any control program before a project begins. Thus, social embedding of management actions, as in the Working for Water Programme in South Africa (Noemdoe 2001), can foster effective management intervention.

Economic arguments have much to contribute to programs to address the problems of IAS (Perrings, Williamson, and Dalmazzone 2000). Decision makers often find arguments couched in economic terms to be more convincing than those cast in emotive or ethical terms, and economics-based arguments of costs and benefits can be used to support stronger programs to deal with invasive species.

But although it is important to identify costs and benefits of IAS, such determination does not automatically determine a decision because politically charged value judgments about distribution of benefits are nearly always involved. Furthermore, the costs may be so high as to render an action politically unacceptable, even when the benefits are likely to be even greater; part of the problem is that the benefits may be widely spread throughout the population over a period of many years, whereas the costs of control may need to be paid quickly by taxpayers. It appears that conflicts of interest between various sectors of society regarding the costs and benefits of IAS are an inevitable fact of modern life. Such conflicts might be mediated through a more thorough identification of the full costs of IAS. However, the value of an alien species to any particular interest group may change over time, complicating the determination of costs and benefits.

The distribution of costs and benefits often is more important than their absolute magnitude. A good illustration of the issue is the Nile perch (*Lates niloticus*). Introduced into Lake Victoria for economic reasons, it has led to the extinction of dozens, perhaps hundreds, of cichlid fish species endemic to the lake and has led to deforestation around the lake because firewood is needed to dry the oily perch; forest clearing in turn is leading to siltation and eutrophication, thus adding additional pressure to the continued productivity of the lake (which is also infested with invasive water hyacinth). Although the Nile perch fishery in Lake Victoria generates up to US$400 million per year in export income, few people living around the lake earn these economic benefits. Tons of perch end up on the plates of European diners, while protein malnutrition is a major problem around the lake (WRI 2000). Great economic benefits are flowing to a few people from this IAS, but none of the money is being spent on managing the economic and ecological costs imposed on the poor or on the Lake Victoria ecosystem. The economics of the marketplace have proven more powerful than the ethics of equitable distribution of benefits.

Cultural factors also affect the perceptions different people have of the benefits and costs of IAS. For example, Luken and Thieret (1996) report that within less than a century after the deliberate introduction of Amur honeysuckle (*Lonicera maackii*) into North America to improve habitat for birds, serve ornamental functions in landscape plantings, and stabilize and reclaim soil, the shrub had become established in at least 24 states in the eastern United States. Whereas many resource managers perceive the plant as undesirable, gardeners and horticulturalists consider it useful. Similarly, fishers value the cultural importance of alien trout species (*Oncorhynchus* spp., *Salmo* spp.) in the Sierra Nevada of the American West, despite the powerful negative impact they have on native frog species. And St. John's wort (*Hypericum perforatum*), which is a noxious weed with harmful effects on livestock in North America, is popular in the natural pharmaceutical trade as an antidepressant and is being grown legally as an agricultural crop in the northwestern United States (Reichard 2001). Thus the noxious invasive of one cultural group is the desirable addition of other groups.

The perception that local people have of introduced species may be different from that of conservationists, affecting how they respond. For example, in recent years the people living on Pitcairn Island—descendants of the *Bounty*'s mutineers—have not considered *Lantana camara* as a major weed, as conservationists have done, but believe the shrub to be a soil improver. On the other hand, they view the tree *Syzygium jambos* as a major pest, not because of its impact on the native flora and fauna but rather because of its heavy shading and its spreading, shallow, and dense rooting system, which renders cultivation of gardens an arduous task. Thus, the weed status of a species relates to the way it interferes with day-to-day activities and will change through time as society develops (Binggeli 2001).

The words people use to articulate their concepts and values are often taken for granted, but their linguistic framework contains many assumptions, unarticulated values, implications, and consequences that must be critically scrutinized if people are to be effective in airing their concerns about IAS (Hattingh 2001). The vocabulary used to describe IAS often implies conceptual oppositions, such as native–alien, pure–contaminated, harmless–harmful, original–degraded, and diversity–homogeneity. Ideals such as ecological integrity and authenticity associated with the values implied by such oppositions are undermined by the modern forces of globalization, which use an even more powerful set of oppositions, such as wealth–poverty, freedom–constraint, private–public, and connected–disconnected. Hattingh argues that because globalization tends to promote homogenization, it cannot incorporate many of the basic concepts that are needed to respond to the problem of IAS, such as uniqueness, ecological integrity, and biological diversity.

Some methods of controlling IAS may carry health hazards as well. For example, pesticides can have serious effects on both people and ecosystems. Between 1975 and 1985, forests in Atlantic Canada were sprayed with the insecticide Matacil to control spruce budworm (*Choristoneura fumiferana*). In the late 1990s, fisheries and environmental scientists inferred that the declines in the Atlantic salmon (*Salmo salar*) stocks in the Restigouche River that occurred at that time were related to exposures of the smolt to nonylphenol, used as an inert solvent in the pesticide (Fairchild et al. 1999).

Recognizing the problem of IAS also forces people to face a host of losses, which can be psychologically costly. For people who value a sense of place, perhaps developed during childhood, learning that invasives have become part of their ecosystem can give rise to the following psychic blows:

- What we see is not what we thought it was (i.e., "natural" or wilderness).
- What we love is not deserving (i.e., it is a foreign "weed").
- What we have protected is still at risk (i.e., almost every protected area we have worked so hard to create).
- We must struggle with more complicated trade-offs than we had hoped (i.e., when are pesticides a necessary evil?) (Windle 1995).

Once public enthusiasm to control IAS has been generated, it must be channeled in the right direction. For example, gorse (*Ulex europeus*) has become invasive in montane grasslands of Sri Lanka since its introduction about 150 years ago. Recently, several local nongovernment organizations have launched volunteer programs to remove gorse. However, several endemic reptile and amphibian species have found gorse a congenial habitat, providing food and cover. When the eradication programs removed this habitat, the endemic species were exposed to native opportunistic predators such as crows (Marambe et al. 2001). Therefore, programs to eradicate invasive plant species also must consider restoring the ecological functions of the species that are removed.

More than 40 international conventions, agreements, and guidelines have been enacted for addressing the IAS problem, at least in part, and many more are being prepared (Shine, Williams, and Burhenne-Guilmin 2000). Governments have expressed their concerns about the IAS problem, especially through the CBD, which calls on the parties to "prevent the introduction of, control or eradicate those alien species which threaten ecosystems, habitats, or species" (Art. 8h). But the expanding impact of IAS on both global economies and the environment implies that these international instruments have been insufficient to prevent and combat IAS effectively, suggesting that additional measures, such as a protocol under the CBD, are advisable.

At the national level, those opposed to eradicating IAS on ethical grounds often are prepared to argue their case in court, where litigation can be effective. This challenge calls for a legal framework that clearly recognizes the need to eradicate IAS when they threaten the greater public good and education for judges to ensure that they understand the issues before them. National and even local legislation also must recognize the human dimensions that are identified in this chapter, including ethical concerns, human health, trade, cultural considerations, and even international obligations. Human dimensions are an essential element in trying to determine what existing regulatory, financial, and penal disincentives could be adjusted to deter trade and transport activities that carry high risks and to determine the specific levels of disincentives that will deter invasives (Jenkins 2001). New Zealand has perhaps the most comprehensive modern legislation: its Biosecurity Act of 1993 has led to a strategic approach at the subnational level, involving both invasive species and genetically modified organisms (Warren 2001; Veitch and Clout 2001).

Political support is clearly essential in implementing coherent policies, laws, and regulations to address the IAS problem. This depends in turn on support of the public, which ultimately depends on the quality of information that is provided on the issue and the effectiveness with which such information is disseminated. Advocates need to convince the general public that controlling an invasive species is worthwhile. For example, the program in New Zealand to control the brush-tailed possum (*Trichosurus vulpecula*) was called Operation ForestSave, and the promotional information showed lovely flowering native trees, without a cute, furry possum in sight. In Europe, the issue is presented as Save the Red Squirrels, not Kill the Grey Squirrels. And in Queensland, Australia, the endangered cassowary (*Casuarius casuarius*) is used as the front for controlling feral pigs, fully involving community groups in the program (Low 2001).

The U.S. National Park Service has encountered strong public resistance to efforts to eradicate wild burros (*Equus asinus*) from the Mojave National Reserve, mountain goats (*Oreamnos americanus*) from Olympic National Park, and sheep (*Ovis aries*) from Santa Cruz Island. At least part of this resistance arises because the IAS are large mammals that some interest groups find attractive, but the negative ecological and economic impact of these species might be used to influence the political process. For example, heavy grazing on the native vegetation by feral populations of horses and donkeys allows nonnative annuals to displace native perennials and costs the nation an estimated $5 million per year in forage losses, implying that these species eat forage worth US$100 per animal per year. They also diminish the primary food sources of native bighorn sheep (*Ovis canadensis*) and seed-eating birds, reducing the abundance of these natives (Pimentel et al. 2000). If people are made aware of the ecological and economic impact of the invasive large grazers, public attitudes toward control operations may be changed.

Hattingh (2001) argues that ecological communities are delineated by people, who draw the lines that distinguish between that which is native and that which is alien, These lines are established by the way people perceive such distinctions and communicate these perceptions through narratives or stories that they tell to represent reality.

Philosophers argue that these distinctions exist only insofar as people continue telling these stories. Biologists respond that a species introduced by people on a continent different from where it lives is self-evidently nonnative. The three-tiered policy response developed by GISP to deal with the IAS problem, involving prevention, eradication, and management if the first two do not succeed, depends on these narratives that people have developed about what is native and what is alien, involving value-laden conceptual distinctions. Perhaps the challenge of addressing IAS will raise fundamental questions about the extent to which we define our humanity by our relation to the rest of nature.

Different interest groups may have different ethical positions. For example, some animal rights groups argue that the intrinsic right to exist rests at the level of individual animals, not only of the species as a whole, and therefore strongly resist any measures to control them, much less eradicate entire populations even if they are IAS (although few extend this to pathogenic microorganisms or even insects). Animal rights advocates contend that nature will find its own solution to the new situation and that any human intervention is immoral because we have no right to select one species or individual over another (although of course we do so when introducing an alien species into a new habitat). Public acceptance can be significantly influenced by animal rights groups, whose views must be carefully considered when assessing the feasibility of eradicating an invasive species. Unfortunately, their ethical concerns cannot be answered by scientific evidence of conservation threats or even by economic arguments.

Some widely held ethical values have unrecognized ramifications for IAS. For example, our world has become increasingly interconnected over both time and space, where individuals have come to expect great freedom of individual behavior (Low 2001). But their behavior when introducing alien species has significant, though undefined, influences on many other people, most of whom are unknown to those who are exhibiting that behavior. Ethics of obligations and responsibilities are not always easily understood against the backdrop of the ethic of "consumer freedom" to grow exotic plants or keep exotic pets that might escape captivity to become invasive.

Thus the IAS issue can be seen as ultimately an ethical concern. If people are seeking to maximize their material welfare or even the diversity of species with which they surround themselves, alien species might well be a part of their rational response. But when alien species become invasive, destabilizing ecosystems and reducing diversity, then control is a far more acceptable, even necessary, response. Because invasions invariably involve trade-offs, the determination of costs and benefits of IAS becomes paramount (although this too has its ethical components).

One useful way to build political support is through scenario planning, a way of developing stories of the future that are plausible and meaningful to the intended audience. By describing plausible futures based on options that are available to decision makers, scenarios can provide a means for politicians and other decision makers to deal with the inherent uncertainty of the IAS problem (see Chapman, Le Maitre, and Richardson 2001 for a more thorough discussion of this point). Scenario planning also enables planners to go beyond modeling and biological factors to take into account the extraordinary complexity of human enterprises and human-dominated systems. Scenarios focus on highlighting and understanding the effects of the large-scale forces that push the future in different directions rather than the details of that future.

Thus the IAS concept is not purely dependent on objective ecological criteria but also depends on human concepts used to identify origin, authenticity, and responsibility. Hattingh (2001) advocates an ethics of conceptual responsibility to become more aware of the dominant lines of argument we use in our debates about IAS. Such ethics would include the manner in which IAS function, their history, the mechanisms through which these views have been established as authoritative and through which they have become institutionalized, and their practical policy and political consequences. Hattingh also suggests that we need to be open to alternative narratives that might be more effective in articulating concerns about IAS to policymakers and to the general public.

Conclusion

IAS are able to invade new habitats and constantly extend their distribution, thereby threatening native species, human health, and other economic or social interests. One remarkable human dimension is the fact that a strong consensus can be built that many specific invasions are harmful, including killer bees, water hyacinth, kudzu, spruce budworms, various pathogens, and agricultural weeds. Therefore, the IAS issue can bring together interest groups that might otherwise be in opposition, such as farmers and conservation groups. Bringing in the human dimensions can shift the focus from the IAS itself to the human actions that facilitate its spread or its control and implies that focusing directly on the invasive species is likely to provide only symptomatic relief. A more fundamental solution entails addressing the ultimate human causes of the problem, often the economic motivations that drive or enable species introductions.

This chapter has identified some of the human dimensions involved in IAS. It is apparent that these dimensions are interconnected and are relevant to different degrees in different countries or with different invasive species. But the presence of so many human dimensions implies that approaches to management must involve many sectors of modern society, including trade, tourism, industry, the military, and public health. Addressing the problem will call for more collaboration between ecologists, geographers, land use planners, economists, sociologists, psychologists, and people from other disciplines to investigate the human dimensions of biological invasions.

The complex relationship between globalization and invasion pathways is perhaps the most important human dimension of IAS and should occupy the minds of policymakers in the next few decades (Carlton and Ruiz 2000). Globalization carries with it the rise of transnational corporations, international financing, and multimedia marketing that undermine the political power of most governments, weakening their ability to regulate economic behavior for the public benefit (Hattingh 2001). One important implication is that concern about IAS must be expressed in terms of the threats to the resource base of the global economic system, which translates into monetary figures. Therefore, many of those who are concerned about the problems of IAS have quite properly turned to economics to argue their case.

Humans, with all their quirks, strengths, and weaknesses, are at the heart of the IAS problem and, paradoxically, also at the heart of the solution. Given the ultimate human motivations of survival, reproduction, and perhaps spiritual fulfillment and the more immediate economic motivations, people might be encouraged to contribute to addressing the problem of IAS by the following measures:

- Helping the public to identify and embrace values that have a direct relationship to basic needs and are environmentally sound, thereby also achieving longer-term benefits. This might include promoting the concept of community, including native species, as a value that can balance the powerful economic values of globalized trade.
- Developing conservation practices and ethics that emphasize the importance of natural ecosystems, for example by refining distinctions between natural and anthropogenic conditions, devising ways to use ecosystems without losing biotic diversity, and facilitating shifts in societal values toward more respect for nature.
- Identifying measures that work within existing value systems but encourage people to support conservation measures (e.g., through the use of economic incentives and disincentives).
- Ensuring that the costs of controlling IAS are internalized, paid by those who are benefiting from intentional introduction and those responsible for unintentional introductions.
- Linking the concern about IAS to the drive for development that motivates most people, and almost all governments, today.
- Including human dimensions in the various conventions, agreements, and guidelines on IAS, such as those developed under the CBD.
- Using risk assessment procedures for species introductions that take into account future changes in usage and demonstrate that, to the best of current knowledge, detrimental impacts will be limited.

A fundamental constraint against changing the way people behave in regard to IAS is that few people in any part of the world consciously perceive that they have been affected negatively by IAS, either directly or indirectly. Although the GISP has been reasonably successful in developing technical information for resource managers, the supply of information on IAS to the general public remains poor, so that most people have little idea of which species are invasive, what their impacts are, and what control methods are appropriate. In the absence of such information, inappropriate responses can be expected. On the other hand, human perceptions are filtered by the media, the availability of information, and language, and all of these can be influenced to limit the spread of IAS.

It is remarkable that some agencies that should know better are actually promoting IAS in the name of development. For example, development assistance agencies often seem to prefer to introduce alien species (especially from the country providing the funding for the assistance) rather than promoting native species. Even UN agencies, such as the Food and Agriculture Organization, are widely promoting numerous weedy trees, shrubs, fodder grasses, and legumes that are known to be highly invasive in at least some countries.

Others are guilty of lack of involvement in the issue. For example, if the problems of IAS are to be addressed successfully, more conservation organizations must become more actively involved. Perhaps they have avoided doing so because advocating the eradication of some animals or plants might confuse some of their supporters, especially those who equate conservation with animal rights. This is a serious problem with cute, furry animals such as the invasive American gray squirrel (*Sciurus carolinensis*) in Italy, especially when the Italian press refers to them as "Chip and Dale," after the Walt Disney cartoon characters (ironically using nonnative icons to refer to a problem of nonnatives). In any case, conservation organizations should give this issue much higher priority.

Broader support must be based on a stronger foundation of science. Despite decades of research, scientific knowledge of the biology, ecology, and human dimensions of IAS remains very incomplete. With no more than 20 percent of the world's species scientifically described, scientists simply are unable to predict which species are likely to become invasive or to assess the precise ecological, social, or economic impact they are likely to have. With such incomplete knowledge, we risk unexpected consequences whenever a new species is introduced into an ecosystem. Unpredicted effects, such as ozone depletion, global warming, mad cow disease, pesticide accumulation, and the impacts of hormones in the environment, can result from seemingly beneficial products and procedures. Therefore, it seems sensible to do everything we can to ensure that we err on the side of precaution, perhaps on occasion sacrificing some economic profit for the businesses directly involved while helping to ensure a healthier future for all of society. Thus, we should also strongly support research to assess the risks of IAS and to find effective means of dealing with the risks.

Research priorities for human dimensions of IAS include the following:

- Identifying conflicting interests regarding benefits and risks of introductions, substantiating evaluations of those benefits and risks, and determining the likely distribution of benefits and risks among sectors of society (Ewel et al. 1999).
- Identifying underlying causes for human choice in relation to IAS, including identifying how human beliefs about specific invasive species influence their actions to promote or limit the spread of that species.
- Ascertaining what is known scientifically about the ubiquitous human affinity for other species. Is biophilia an innate behavior? Is it a conditioned response? Does it lead to more alien species being imported? Can the human behavior that stems from this attraction to other species be modified and redirected? If so, how?

- Evaluating potentially useful indigenous organisms rather than nonindigenous ones, thereby reducing incentives for introductions.
- Elucidating the interactions between the media, the public, scientists, and conservationists.
- Identifying the views of indigenous peoples and other interest groups about IAS.
- Carrying out a predictive modeling exercise to project what might be the outcome if we are unable to slow or stop the spread of IAS.

This chapter has sought to elucidate basic economic, social, psychological, ethical, and political aspects of IAS, but each case must be considered on its own merits. That said, here are some human dimensions elements to consider in addressing any IAS problem:

- Ensure that those who are most directly affected by the IAS are involved in decisions about how to manage the problem.
- Build sufficient public information programs into each effort, investing more in this regard where the problem is likely to involve controversial techniques (e.g., use of poisons).
- Conduct a detailed analysis of human dimensions as they affect the interested parties, including the general public and decision makers.
- Build links between the management of IAS and development, involving economic sectors such as health, energy, agriculture (food security), forestry, and fisheries.
- Establish general principles for guiding policies that explicitly promote the identities and values that motivate and direct people to minimize the spread of IAS.

Gould (1998) argues that the preference for native species provides "the only sure protection against our profound ignorance of consequences when we import exotics" because we can never be certain about the behavior of an alien species imported into a new environment. Thus we should do everything possible to prevent unwanted invasions, carry out careful assessments before intentionally introducing an alien species into a new environment, build a stronger awareness among the general public about the problems of IAS, mobilize conservation organizations to address the problems, and build an ethic of responsibility among those most directly involved in the problem. The global trading system brings many benefits, but it needs to be managed in a way that minimizes any deleterious impacts of IAS on ecosystems, human health, and economic interests. Acknowledging the human dimensions of this exchange are central in doing so.

Acknowledgments

The overall management of the GISP process was in the capable hands of Professor Harold Mooney of Stanford University, with able support from Laurie Neville. Veronique Plocq-Fichelet provided steadfast support from ICSU-SCOPE in Paris. The workshop that led to this chapter would not have been possible without the support of Jamie Reaser of the U.S. Department of State. The U.S. Department of State provided the main financial support for the workshop, and the Swiss Development Cooperation provided additional financial support. At the Cape Town workshop, Guy Preston of the Working for Water Programme was responsible for making the facilities available and providing logistical support. Karoline Hanks, Dumi Magadlela, and Simone Noemdoe of the Working for Water Programme played essential roles in making the Cape Town workshop a success. Phyllis Windle kindly acted as rapporteur and synthesizer at the Cape Town workshop. The participants are all very grateful to Brian Huntley for his outstanding hospitality at Kirstenbosch National Botanical Garden in Cape Town. The production of this chapter depended very much on Sue Rallo, who managed the production process and handled all correspondence. I received useful comments from Channa Bambaradeniya, Pierre Bingghi, Maj de Poorter, Llewellyn Foxcroft, Piero Genovesi, Johan Hattingh, Peter Jenkins, David Le Maitre, Tim Low, Dick Mack, Jamie Reaser, Dave Richardson, George Staples, Mark Williamson, and Phyllis Windle. To all of these people, and of course to the participants at the workshop, I owe a personal debt of thanks. I hope that the ideas put forward in this chapter will help ensure that IAS issues begin to get the appropriate level of attention in the ongoing debate about the social, economic, ecological, and political impacts of globalization.

References

Bangsund, D. A., F. L. Leistritz, and J. A. Leitch. 1999. "Assessing economic impacts of biological control of weeds: The case of leafy spurge in the northern Great Plains of the United States," *Journal of Environmental Management* 56:35–43.

Binggeli, P. 2001. "The human dimensions of invasive woody plants," pp.145–160 in *The Great Reshuffling: Human Dimensions of Alien Invasive Species,* edited by J. A. McNeely. IUCN, Gland, Switzerland.

Bright, C. 1998. *Life Out of Bounds: Bioinvasion in a Borderless World.* W.W. Norton, New York.

Carlton, J., and G. Ruiz. 2000. "The vectors of invasions by alien species," in *Best Management Practices for Preventing and Controlling Invasive Alien Species,* edited by G. Preston, G. Brown, and E. van Wyk, 82–89. Symposium Proceedings. The Working for Water Programme, Cape Town, South Africa.

Chapman, R. A., D. C. Le Maitre, and D. M. Richardson. 2001. "Scenario planning: Understanding and managing biological invasions in South Africa," pp. 195–208 in *The Great Reshuffling: Human Dimensions of Alien Invasive Species,* edited by J. A. McNeely. IUCN, Gland, Switzerland.

Crosby, A. W. 1986. *Ecological Imperialism: The Biological Expansion of Europe, 900–1900.* Cambridge University Press, New York.

Dalmazzone, S. 2000. "Economic factors affecting vulnerability to biological invasions," in

The Economics of Biological Invasions, edited by C. Perrings, M. Williamson, and S. Dalmazzone, 17–30. Edward Elgar, Cheltenham, UK.

Daszak, P., A. Cunningham, and A. D. Hyatt. 2000. "Emerging infectious diseases of wildlife: Threats to biodiversity and human health," *Science* 287:443–449.

Decary, R. 1947. "Epoque d'introduction des Opuntias monocantha dans le Sud de Madagascar," *Revue Internationale de Botanique Appliquée et d'Agriculture Tropicale* 27:455–457.

Delfino, D., and P. Simmons. 2000. "Infectious diseases as invasives in human populations," in *The Economics of Biological Invasions,* edited by C. Perrings, M. Williamson, and S. Dalmazzone, 31–55. Edward Elgar, Cheltenham, UK.

Drake, J. A., H. A. Mooney, F. D. Castri, R. H. Groves, F. J. Kruger, M. Rejmánek, and M. Williamson, eds. 1989. *Biological Invasions: A Global Perspective.* Wiley, Chichester, UK.

Elton, C. S. 1958. *The Ecology of Invasions by Plants and Animals.* Wiley, New York.

Enserink, M. 1999. "Biological invaders sweep in," *Science* 285:1834–1836.

Ewel, J. J., and 20 others. 1999. "Deliberate introductions of species: Research needs," *BioScience* 49(8):619–630.

Fairchild, W. L., E. O. Swansburg, J. T. Arsenault, and S. B. Brown. 1999. "Does an association between pesticide use and subsequent declines in catch of Atlantic salmon represent a case of endocrine disruption?" *Environmental Health Perspectives* 107:349–358.

Foxcroft, L. C. 2001. "A case study of human dimensions in invasion and control of alien plants in the personnel villages of Kruger National Park," pp. 127–134 in *The Great Reshuffling: Human Dimensions of Alien Invasive Species,* edited by J. A. McNeely. IUCN, Gland, Switzerland.

Genovesi, P., and S. Bertolino. 2001. "Human dimension aspects in invasive alien species issues: The case of the failure of the grey squirrel eradication project in Italy," pp. 113–120 in *The Great Reshuffling: Human Dimensions of Alien Invasive Species,* edited by J. A. McNeely. IUCN, Gland, Switzerland.

Gould, S. J. 1998. "An evolutionary perspective on strengths, fallacies, and confusions in the concept of native plants," *Arnoldia* 58(1):3–10.

Hattingh, J. 2001. "Human dimensions of invasive alien species in philosophical perspective: Towards an ethics of conceptual responsibility," pp. 783–94 in *The Great Reshuffling: Human Dimensions of Alien Invasive Species,* edited by J. A. McNeely. IUCN, Gland, Switzerland.

Hirsch, S. A., and J. A. Leitch. 1996. *The Impact of Knapweed on Montana's Economy.* Agricultural Economics Report 355, Department of Agricultural Economics, North Dakota State University, Fargo.

Hoyt, E. 1992. *Conserving the Wild Relatives of Crops,* 2nd ed. IBPGR, IUCN, and WWF, Rome.

Jenkins, P. T. 2001. "Who should pay? Economic dimensions of preventing harmful invasions through international trade and travel," pp. 79–88 in *The Great Reshuffling: Human Dimensions of Alien Invasive Species,* edited by J. A. McNeely. IUCN, Gland, Switzerland.

Kowarik, I. 1990. "Some responses of flora and vegetation to urbanization in Central Europe," in *Urban Ecology: Plants and Plant Communities in Urban Environments,* edited by H. Sukopp, S. Mejny, and I. Kowarik, 45–74. SPB Academic Publishing, The Hague.

Levin, S. A. 1989. "Analysis of risk for invasions and control programs," in *Biological Invasions: A Global Perspective,* edited by J. A. Drake, H. A. Mooney, F. di Castri, R. H. Groves, F. J. Kruger, M. Rejmánek, and M. Williamson, 425–432. Scope 37. Wiley, New York.

Low, T. 2001. "From ecology to politics: The human side of alien invasions," pp. 35–42 in *The Great Reshuffling: Human Dimensions of Alien Invasive Species,* edited by J. A. McNeely. IUCN, Gland, Switzerland.

Luken, J. O., and J. W. Thieret. 1996. "Amur honeysuckle, its fall from grace," *BioScience* 46(1):18–24.

Mack, R. N. 2001. "Motivations and consequences of the human dispersal of plants," pp. 23–34 in *The Great Reshuffling: Human Dimensions of Alien Invasive Species,* edited by J. A. McNeely. IUCN, Gland, Switzerland.

Marambe, B., C. Bambaradeniya, D. K. Pushpa Kumara, and N. Pallewatta. 2001. "Human dimensions of invasive alien species in Sri Lanka," pp. 135–144 in *The Great Reshuffling: Human Dimensions of Alien Invasive Species,* edited by J. A. McNeely. IUCN, Gland, Switzerland.

Martin, P. S., and R. G. Klein, eds. 1984. *Quaternary Extinctions: A Prehistoric Revolution.* University of Arizona Press, Tucson.

McNeely, J. A. 1999. "The great reshuffling: How alien species help feed the global economy," pp. 11–32 in *Invasive Species and Biodiversity Management,* edited by O. T. Sandlund, P. J. Schei, and A. Viken. Kluwer Academic Publishers, Dordrecht, The Netherlands.

McNeely, J. A., H. A. Mooney, L. E. Neville, P. Schei, and J. K. Waage, eds. 2001. *Global Strategy on Invasive Alien Species.* IUCN on behalf of the Global Invasive Species Programme, Gland, Switzerland, and Cambridge, UK.

McNeill, W. H. 1976. *Plagues and Peoples.* Anchor Press, Garden City, New York.

Mooney, H. A., and R. J. Hobbs. 2000. *Invasive Species in a Changing World.* Island Press, Washington, DC.

Naylor, R. L. 1996. "Invasions in agriculture: Assessing the cost of the golden apple snail in Asia," *Ambio* 25:443–448.

Noemdoe, S. 2001. "Putting people first in an invasive alien clearing programme: Working for water programme," pp. 121–126 in *The Great Reshuffling: Human Dimensions of Alien Invasive Species,* edited by J. A. McNeely. IUCN, Gland, Switzerland.

Orr, M. R., and T. B. Smith. 1998. "Ecology and speciation," *Trends in Evolution and Ecology* 13(12):503–506.

Parker, V. 2001. "Listening to the earth: A call for protection and restoration of habitats," pp. 43–54 in *The Great Reshuffling: Human Dimensions of Alien Invasive Species,* edited by J. A. McNeely. IUCN, Gland, Switzerland.

Pearce, F. 2000. "Inventing Africa: Which is more authentic: A game-rich wilderness or cattle pasture?" *New Scientist* 2251:30–33.

Perrings, C., M. Williamson, and S. Dalmazzone. 2000. *The Economics of Biological Invasions.* Edward Elgar, Cheltenham, UK.

Pimentel, D., L. Lach, R. Zuniga, and D. Morrison. 2000. "Environmental and economic costs of non-indigenous species in the United States," *BioScience* 50:53–65.

Reaser, J. 2001. "Invasive alien species prevention and control: The art and science of managing people," pp. 89–104 in *The Great Reshuffling: Human Dimensions of Alien Invasive Species,* edited by J. A. McNeely. IUCN, Gland, Switzerland.

Reichard, S. H. 2001. "Horticultural introductions of invasive plant species: A North American perspective," pp. 161–170 in *The Great Reshuffling: Human Dimensions of Alien Invasive Species,* edited by J. A. McNeely. IUCN, Gland, Switzerland.

Severinghaus, L. L., and L. Chi. 1999. "Prayer animal release in Taiwan," *Biological Conservation* 89(3):301–304.

Shine, C., N. Williams, and F. Burhenne-Guilmin. 2000. *Legal and Institutional Frameworks on Alien Invasive Species: A Contribution to the Global Invasive Species Programme Global Strategy Document.* IUCN Environmental Law Programme, Bonn, Germany.

Staples, G. W. 2001. "The understorey of human dimensions in biological invasions," pp. 171–182 in *The Great Reshuffling: Human Dimensions of Alien Invasive Species,* edited by J. A. McNeely. IUCN, Gland, Switzerland.

Thomas, M. B., and A. J. Willis. 1998. "Biocontrol: Risky but necessary?" *Trends in Ecology and Evolution (TREE)* 13(8):325–329.

Turpie, J., and B. Heydenrych. 2000. "Economic consequences of alien infestation of the Cape Floral Kingdom's fynbos vegetation," pp. 152–182 in *The Economics of Biological Invasions,* edited by C. Perrings, M. Williamson, and S. Dalmazzone. Edward Elgar, Cheltenham, UK.

Udvardy, M. 1969. *Dynamic Zoogeography.* Van Nostrand Reinhold, New York.

Veitch, R., and M. Clout. 2001. "Human dimensions in the management of invasive species in New Zealand," pp. 63–74 in *The Great Reshuffling: Human Dimensions of Alien Invasive Species,* edited by J. A. McNeely. IUCN, Gland, Switzerland.

Vilà, M., and J. Pujadas. 2001. "Socio-economic parameters influencing plant invasions in Europe and North Africa," pp. 75–78 in *The Great Reshuffling: Human Dimensions of Alien Invasive Species,* edited by J. A. McNeely. IUCN, Gland, Switzerland.

Waage, J. K. 1991. "Biodiversity as a resource for biological control," in *The Biodiversity of Micro-organisms and Invertebrates: Its Role in Sustainable Agriculture,* edited by D. L. Hawksworth, 149–163. CABI Publishing, Oxford, UK.

Warren, P. 2001. "Dealing with the human dimensions of invasive alien species within New Zealand's biosecurity system," pp. 105–112 in *The Great Reshuffling: Human Dimensions of Alien Invasive Species,* edited by J. A. McNeely. IUCN, Gland, Switzerland.

Watkinson, A. R., R. P. Freckleton, and P. M. Dowling. 2000. "Weed invasion of Australian farming systems: From ecology to economics," pp. 94–116 in *The Economics of Biological Invasions,* edited by C. Perrings, M. Williamson, and S. Dalmazzone. Edward Elgar, Cheltenham, UK.

White, P., and G. Newton-Cross. 2000. "An introduced disease in an invasive host: The ecology and economics of rabbit calcivirus disease (RCD) in rabbits in Australia," pp. 117–137 in *The Economics of Biological Invasions,* edited by C. Perrings, M. Williamson, and S. Dalmazzone. Edward Elgar, Cheltenham, UK.

WHO. 1997. *Health and Environment in Sustainable Development.* World Health Organization, Geneva.

Williamson, M. 1998. "Measuring the impact of plant invaders in Britain," in *Plant Invasions. Ecological Mechanisms and Human Responses,* edited by S. Starfinger, K. Edwards, I. Kowarik, and M. Williamson, 57–70. Backhuys, Leiden, The Netherlands.

Wilson, E. O. 1984. *Biophilia.* Harvard University Press, Cambridge, MA.

Windle, P. 1995. "The ecology of grief," in *Ecopsychology: Restoring the Earth, Healing the Mind,* edited by T. Roszak, M. E. Gomes, and A. D. Kanner, 136–145. Sierra Club Books, San Francisco.

WRI. 2000. *World Resources 2000–2001.* World Resources Institute, Washington, DC.

Zavaleta, E. 2000. "Valuing ecosystem services lost to *Tamarix* invasion in the United States," pp. 261–300 in *Invasive Species in a Changing World,* edited by H. A. Mooney and R. J. Hobbs. Island Press, Washington, DC.

12

Invasive Species in a Changing World: The Interactions between Global Change and Invasives

Richard J. Hobbs and Harold A. Mooney

In this chapter we examine the relationships between invasive species and other types of global change. Humans are currently changing the earth in unprecedented ways, particularly by modifying the composition of the earth's atmosphere, transforming the earth's ecosystems, and using and altering the composition of the world's freshwater supplies, marine fisheries, and other natural resources (Turner et al. 1991; Vitousek et al. 1997a). We examine the proposition that the invasive species problem will be worsened by global change. Increasingly, species are being moved around the globe both deliberately and inadvertently, and this, coupled with increased opportunities for species to invade modified ecosystems, makes it likely that local species mixes will change dramatically. Invasive species have the potential for significant impact on both natural and managed systems. Indeed, increasing evidence indicates that invasive species have become increasingly problematic at local, regional, and continental scales (see Simberloff et al. 1997; Cox 1999; Meinesz 1999). Therefore, we support the view of Vitousek et al. (1996, 1997b) that invasives themselves are a global change element and that their extent and impact should be considered in global change scenarios.

Our focus is on what we see today in terms of invasive species, globally, and what we may expect to see in the next hundred years, not only with changes in climate but also with changes in atmospheric composition, including CO_2 concentration and nitrogen deposition, climate, commerce, land use patterns, and fire regimes. We examine not only how drivers of global change can have an impact on invasives but also how invasives can affect our health, welfare, and economy.

Atmospheric Composition

CO_2 Concentrations

Much scientific effort over the past two decades has gone into understanding the impacts of increasing atmospheric CO_2 concentrations, both on individual species and on intact communities and ecosystems (Bazzaz 1990; Koch and Mooney 1996; Körner and Bazzaz 1996). The rise in CO_2 concentrations directly affects photosynthesis and can result in a wide range of physiological and morphological responses in plants, depending on differences in photosynthetic pathways and intrinsic growth rates (Dukes 2000).

How these responses translate into changes in the success of individual species and in species abundance and composition in ecosystems is difficult to predict because of the myriad likely interactions between species and ecosystem components. In addition, it is clear that species responses to CO_2 are mediated by other factors, such as nutrient and water availability and temperature regimes, all of which are likely to change with changing climatic conditions. Therefore, the likely impact on invasive species of increasing CO_2 concentrations is unclear. For instance, the European annual *Chenopodium album,* a common weed throughout most of North America, responds positively to elevated CO_2 when grown individually but does not respond positively when grown in a Canadian pasture community, even in disturbed (and thus low-density) sites (Hunt et al. 1991; Taylor and Potvin 1997).

Despite the uncertainties involved, Dukes (2000) suggests that increasing CO_2 concentrations may allow increases in plant water use efficiency, which will allow some species, particularly annual grasses, to extend their ranges further into more arid regions. He also suggests that leguminous shrubs may become more invasive because increased CO_2 stimulates nitrogen fixation, whereas C4 grasses may perform less well with increasing CO_2 concentrations, rendering C4 grasslands particularly open to invasion. However, he also points out that the success of particular invasive species is likely to depend on a range of other factors, as discussed in subsequent sections.

Nitrogen Deposition

Another aspect of atmospheric composition that may have an important influence on invasive species, particularly plants, is the greatly elevated level of anthropogenic atmospheric nitrogen that occurs in highly urbanized and industrialized areas of the world and as a result of increased fertilizer use in agriculture. This leads to elevated levels of atmospheric nitrogen deposition. For instance, in Germany deposition rates can be 30–60 kg N/ha/yr, two orders of magnitude greater than the natural background rate (Scherer-Lorenzen et al. 2000). In addition to the direct impacts of this increased nitrogen load, especially in naturally nutrient-poor soils, the potential for invasion is increased by fast-growing grasses and other species. Several studies have indicated the potential for nitrogen fertilization to increase the invasion of natural systems by nonnative grass species (Hobbs and Atkins 1988; Hobbs et al. 1988; Huenneke et al. 1990), and elevated levels of atmospheric nitrogen deposition have been implicated in the invasion of ecosystems in Central Europe (Pysek et al. 1995; Scherer-Lorenzen et al. 2000) and California (Weiss 1999).

Climate Change

Human-induced changes in atmospheric concentrations of various compounds are expected to lead to changes in global and regional climates. Although there is still debate as to the magnitude of these changes and it is still difficult to predict changes in regional climate with any degree of certainty, there is increasing acceptance that accelerated changes in global climate are inevitable and may already be in progress. Of particular importance are changes in temperature and rainfall regimes and in the incidence of extreme events, such as major storms. Biological impacts from these changes may include alterations in species distributions and changes in abundance within existing distributions, resulting from direct physiological impacts on individual species, changed opportunities for reproduction and recruitment, and altered interactions between species (Karieva et al. 1993). Invasive species thus may respond both to the direct changes in climate, which may produce more conducive conditions for establishment or spread of the species itself, and to the indirect effects of changing suitability of local climates for native species and changing biotic interactions between native biotic communities.

Sutherst (2000) provides a conceptual framework for assessing likely impacts of climate change on invasive species by considering three elements of the invasive process: sources, pathways, and destinations. Every invasive species has a source location and ecosystem and is transported via one or numerous pathways to a destination. Each of these is likely to be affected in some way by global climatic changes. Sources of invasive species are of particular concern in considering pest species of agriculture and forestry, with produce often being acceptable on the international market only if it comes from a known pest-free area. With changing climates, the distribution of these pest-free areas may change, and certainty about the status of different areas may decline.

Pathways are particularly important in considering the major routes by which invasive species are dispersed. Although many of these pathways are determined more by trade patterns than climate, changes in major weather systems have the potential to alter the opportunities for medium- to long-distance dispersal of propagules of potentially invasive species, particularly in relation to storm tracks and ocean currents. In addition, Sutherst (2000) points out that increased incidence of drought in some areas of the world may lead to increased political instability and mass movement of people, leading to new pathways of species transport over long distances.

However, it is the destination that is most likely to be affected by potential changes

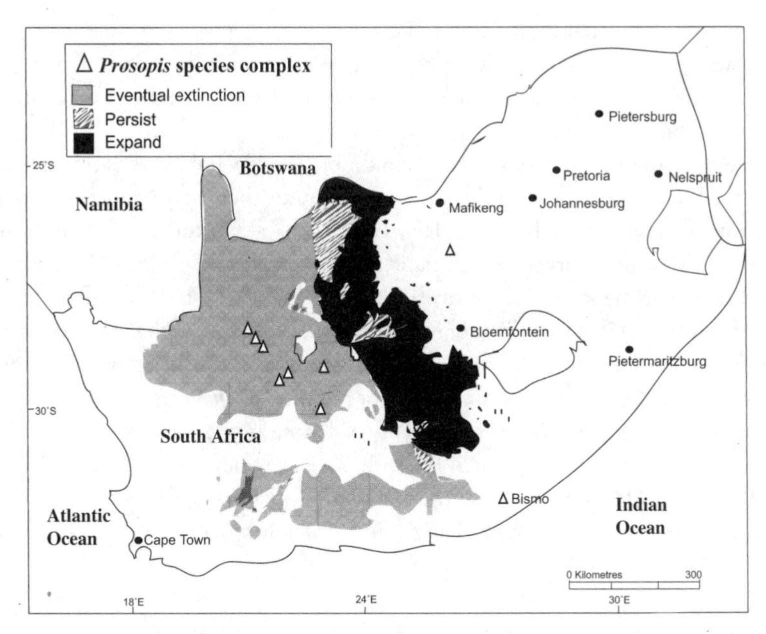

Figure 12.1. Projected shift in the distribution of *Prosopis* species in South Africa with climate change using biogeoclimatic modeling (based on limited data). Indicated are areas into which the species is expected to expand because of newly suitable environmental conditions ("Expand"), areas in which the species should persist because the environment remains within the species' tolerance range ("Persist"), and areas in which the environmental conditions may become unsuitable for the species and lead to its eventual local extinction ("Eventual extinction"). Triangles represent recorded localities for the species. Figure from Richardson et al. (2000).

in climate. As discussed earlier, changes in climatic patterns may alter the distributions of climatic envelopes for species, rendering areas either more or less suitable for colonization. The success of alien species depends on their interaction with their new environment, particularly climate. The impact of these species at their destinations depends on their initial success in establishment, their direction and rate of spread, and their population dynamics and geographic distributions, all of which can be affected by changes in climate.

Richardson et al (2000) provide several examples of modeled predictions of how ranges of existing invasive species in South Africa might change under predicted climate change scenarios. An example of one such prediction, for *Prosopis,* is given in Figure 12.1.

Carlton (2000 p. 39) discusses the potential impact of climate change on marine invasions, particularly in relation to changes in water temperature. However, he points out, "It is worth noting that at the close of the 20th century there is no close finger kept on the pulse of distributional shifts in most marine organisms. Range shifts in marine vertebrates are somewhat more likely to be noted, but . . . portions of the intertidal or shallow-water biota could be undergoing major geographic shifts and yet remain undetected." Mack (2000) also emphasizes the need for effective means of detecting and assessing establishment and spread of invasive species.

If climatic conditions have rendered conditions less favorable for native species or altered the biotic interactions in the native community, there may be increased opportunity for invasion. Similarly, if the pattern of large-scale disturbances such as major storms increases, or if large-scale climatic patterns such as El Niño increase in frequency, there may be increased opportunities for invasion (Hobbs and Mooney 1995, 1996; Horovitz 1997). These disturbances disrupt the structure of the native ecosystems and provide openings for invasion. Episodic flooding is another potential cause of establishment or spread of invasive species. For instance, *Mimosa pigra* escaped from the Darwin botanical gardens, after 80 years of residence, during a major flood that took seed into the catchment of the north Adelaide River, which traverses Kakadu National Park (Lonsdale 1993). It may be that the increased incidence of such extreme events will be the major influence of climate change on the establishment and spread of invasive species.

Land Use Change

Vitousek et al. (1997a) suggest that land transformation, one of the predominant human impacts on ecosystems, probably is the most difficult form of global change to quantify sensibly at a global scale. Although changes can be measured at any given site, the aggregation of these changes regionally and globally presents real challenges. Many different land uses can be identified and categorized by the extent to which they modify the ecosystem. Discussion of land transformation therefore has to capture the richness of the potential changes from one land use to another. Certainly, broad categories of change are apparent globally, such as increasing urbanization, deforestation, and ecosystem fragmentation and agricultural intensification in some areas and abandonment of agricultural land in others. Hobbs (2000) has examined such changes in relation to the effect they have on ecosystems, as illustrated in Figure 12.2.

Possible changes include both ecosystem decline and recovery. A particular human activity may result in a sudden or more gradual change in ecosystem properties. For instance, conversion by urban development or clearance for agriculture results in a sudden state change (A), whereas more chronic impacts from livestock grazing or pollution result in more gradual degradation (B). Another set of transformations result from the impact of individual disturbances such as fire or shifting agriculture. In this case, a sudden state change is followed either by a recovery to a similar state, as was present before disturbance (C), or by a recovery to a different state (D). A final set of dynamics involves the recovery of ecosystems from a damaged or altered state after cessation of a disturbing factor or stressor. This may occur either as the result of normal succession or

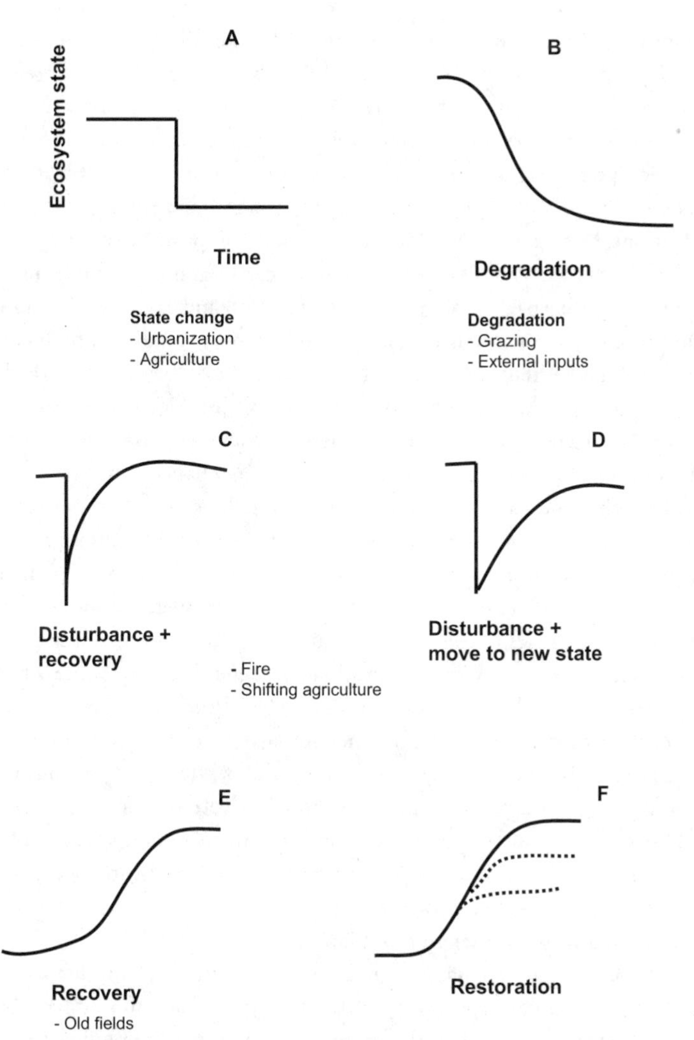

Figure 12.2. Different categories of land transformation, with examples of each; see text for details. Figure from Hobbs (2000).

ecosystem development, as in the case of old field dynamics (E), or through deliberate actions to encourage ecosystem restoration or repair (F). Numerous outcomes are possible, depending on the goals of the restoration efforts and the resources available to achieve them (Hobbs and Norton 1996; Hobbs 1999).

Vitousek et al. (1997a) suggest a two-way interaction between ecosystem transformation and invasions. Human-induced changes to ecosystems are a special subset of the range of potential ecosystem dynamics. Ecologists are increasingly recognizing that ecosystems are dynamic entities in which change is the rule rather than the exception. The species making up ecosystems respond to such change in different ways, so through time the composition of any given ecosystem will not remain the same. Species come and go, depending on the current abiotic conditions and interactions with other species. If ecosystems are dynamic and species compositions are variable, the stage is set for species to appear and disappear in any given system, depending on the current abiotic conditions, levels and types of disturbance, and composition of the regional species pool. Such appearances and disappearances have been ongoing over millennia. However, two major changes have taken place recently. One is the increasing level of human transformation of ecosystems discussed earlier. The second is the dramatic increase in the deliberate and inadvertent transport of biota across the globe. When both changes are put together, they produce the potential for a radical alteration of ecosystem dynamics.

In general, any change in ecosystem properties provides opportunities for species colonization or population expansion. The types of system change illustrated in Figure 12.2 all provide such opportunities. Although disturbance is a natural part of ecosystem dynamics in many systems, human alteration of disturbance regimes and the introduction of novel disturbances change system settings and increase opportunities for invasion (Hobbs and Huenneke 1992). In any ecosystem there are species that can take advantage of disturbances to colonize or expand their populations. These are either disturbance specialists (e.g., species that germinate preferentially after a fire) or generalists, which can tolerate a wide range of conditions. The disturbance provides a window of opportunity for such species, which generally lasts a short time. However, humans have transported species across the globe, both deliberately (for agriculture, forestry, recreation, horticulture, or other pursuits) and inadvertently (e.g., as seed contaminants). Disturbances and land transformations afford these new species opportunities to colonize and spread, and because many are preadapted to disturbance, they are often able to do so as well as or better than the species native to the area. Kalin Arroyo et al. (2000) illustrate this well in relation to plant invasions in Chile.

In addition, land use changes often are brought about by the use of introduced species, such as new forage species and plantation trees. Such species have been transported around the globe with little attention paid to their potential to spread and become problems elsewhere. A classic case is the extensive use of Northern Hemisphere pine species in the Southern Hemisphere, where they are undoubtedly producing valuable timber income but are also causing major problems when they invade adjacent systems and diminish their value in terms of conservation or ecosystem services (Richardson et al. 1994; van Wilgen et al. 1996).

Land transformation thus acts to encourage biotic change by causing system changes that provide the opportunity for biological invasion and by bringing new species from different biogeographic regions into contact with these altered systems.

As well as being a result of land transformation, invasion by nonnative species can drive land transformation. This occurs when the invading species results in the sorts of ecosystem change indicated in Figure 12.2. Such changes are possible when an invading plant species becomes dominant and changes the type of vegetation present. For instance, invading trees can transform a grassland or shrubland into a forest, and invading grasses can change a woody perennial system into an open grassland via land clearance or an altered fire regime (Figure 12.3; D'Antonio and Vitousek 1992; Richardson et al. 1994). Often, the causal link between land transformation and invasions is convoluted. A change in land use, the continuation of an inappropriate land use, or the continuation of inappropriate levels of use can provide the conditions necessary for an invading species to become established. Thereafter, the invading species initiates further system change that precipitates the need for a change in land use or increased management to maintain the existing land use. Thus, the change in system state initiated by the disturbance or management regime is enhanced or accelerated by the invasion of nonnative species. The links between land transformation and invasions are illustrated in Figure 12.4.

Transformation of land from one type of system to another and changes in land use are pervasive across the globe. The areas involved in transformations such as deforestation are enormous, and they are happening rapidly. Although it is possible to obtain some estimates of the more obvious changes caused by deforestation and land clearing (see Skole and Tucker 1993), it is more difficult to assess the totality of land use changes. Also, if one starts considering the likely interactions between land use change

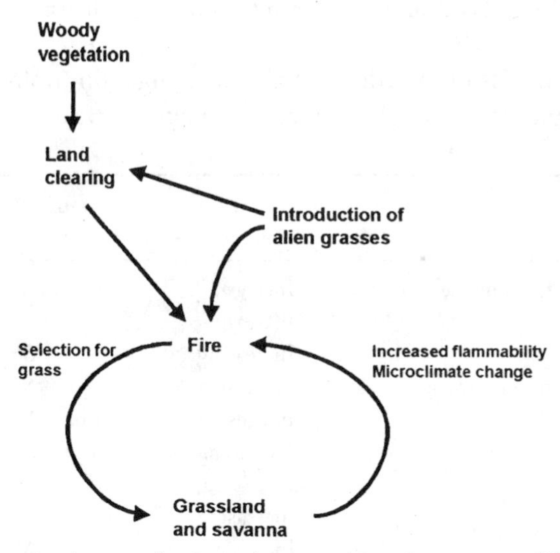

Figure 12.3. Interaction between land use change and invasion, as exemplified by the conversion of woody vegetation (woodland, shrubland) to grassland, which may be initiated by the purposeful clearing of land or after invasion of the woody vegetation by introduced grasses. In either case, a grass–fire feedback system is initiated that prevents the regeneration of woody species. Figure modified from Vitousek et al. (1996).

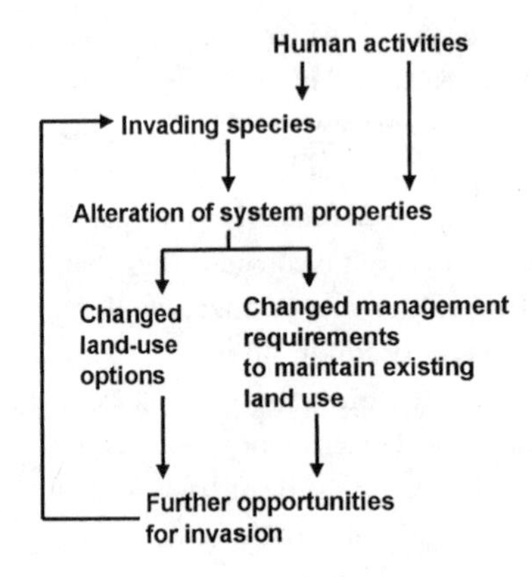

Figure 12.4. Relationships between human activities (management of ecosystems) and invasions. Figure from Hobbs (2000).

and climate change, predictions of future trends become difficult (Leemans and Zuidema 1995; Parry et al. 1996). In addition many trends in land use change are driven primarily by economic, political, and social, rather than biophysical, factors.

In summary, the relationship between land use change and invasions is a two-way street, with land transformation providing opportunities for invasion and, conversely, invasions enhancing and driving land transformations. Kolar and Lodge (2000) discuss a similar set of issues for freshwater systems, indicating that invasive species are likely to interact closely with other global change elements in freshwater systems worldwide (Table 12.1).

Table 12.1. Trends in the effects of global changes on freshwater ecosystems and how strongly each global change increases the number or impact of invasive species

	Trend in global change	*Positive interactions with invading species*
Globalization of commerce	Increasing	Strong
Shipping	Increasing	Strong
Bait trade	Increasing	Strong
Aquarium and pond trade	Increasing	Strong
Aquaculture	Increasing	Strong
Waterway engineering	Increasing	Strong
Canals	Increasing	Strong
Dams	Increasing	Strong
Land use changes	Increasing	Medium
Siltation, eutrophication	Increasing	Medium
Water withdrawal	Increasing	Weak
Climatic and atmospheric changes	Increasing	Weak
Intentional stocking	Increasing	Strong

Source: Kolar and Lodge (2000).

Changed Fire Regimes

D'Antonio (2000) summarizes the current knowledge on the likely interactions between fire, plant invasions, and global change. She concludes that most studies show that fire increases invasion by introduced species. Even where fire was used to control an invasive species, either the species was not effectively reduced or other nontarget invasives increased after the fire. She also concluded that naturally fire-prone ecosystems, such as the fynbos of South Africa or coastal chaparral in California, can be heavily invaded if fire-responsive propagules of invasive species are available or if the natural fire regime is severely altered.

In addition to responding to fire or changed fire regimes, invasive species can also directly alter the fire regime. For instance, invasive species can create conditions that favor the spread of fire, particularly where they are different in their life form or phenology from natives and therefore increase both the continuity and biomass of fire fuels. The most obvious occurrence of this is grass invasion of desert shrubland, dry forest, or woodland ecosystems (Whisenant 1990; D'Antonio and Vitousek 1992) (see Figure 12.3). Invasive species can increase fire intensity in fire-prone systems or can introduce fire into systems where it was previously uncommon. This second scenario is likely to be much more detrimental to the native biota.

Other elements of global change have the potential to affect fire regimes and their interaction with invasive species. For instance, some areas may experience changing frequencies of weather conditions that facilitate the ignition of vegetation or fire spread, and this in turn may be modified depending on the potential pool of invasive species. Alternatively, increasing atmospheric nitrogen deposition may act as a fertilizer and contribute to the spread of fast-growing invasive grasses, and this in turn may influence fire frequency and severity.

However, there is unlikely to be a further feedback from invasive-induced fires to changing composition of the atmosphere. Biomass burning does contribute significantly to the global rise in greenhouse gases, but fires involving invasive species are only a tiny fraction of this increase (Crutzen and Andreae 1990; D'Antonio and Vitousek 1992).

Global Commerce

Recent popular accounts have drawn attention to the trend toward an increasingly global society and economy that is breaking down the traditional boundaries both to trade and to movement of species around the globe (Bright 1998; French 2000). This move toward increased mobility of people and goods brings with it the increased likelihood of movement of species around the planet, either deliberately in the form of commodities such as livestock, pets, nursery stock, and produce from agriculture and forestry or inadvertently as species are transported as unwitting passengers in packaging, in ballast water, and on the aforementioned commodities.

Global trade has greatly increased in recent decades (McNeely 2000). The value of total imports increased from US$192 billion in 1965 to $3 trillion in 1990, a 15-fold increase in 25 years (WRI 1994). Imports of agricultural products and industrial raw materials, those that have the greatest potential to contribute to the problem of invasive species, amounted to $482 billion in 1990, up from $55 billion in 1965. The trade-based global economy stimulates the spread of economically important species, such as rubber, oil palm, pineapples, and coffee, and fields of soybeans, cassava, maize, sugarcane, and wheat. It also stimulates the accidental spread of species through a variety of pathways (Office of Technology Assessment [OTA] 1993; Jenkins 1996; McNeely 2000).

Carlton (1989) and Carlton and Geller (1993) discuss one of these pathways in detail: the transport of marine organisms in ships' ballast water. Ships that take on ballast water in one port and dump it in another part of the world can transport large numbers of species around the globe, and many of these have proved to be important invasive pest species in their destinations. Although many of the problems with ballast water could be overcome with practices such as changing ballast in midocean, ballast water remains an important vector of invasive species.

McMichael and Bouma (2000) also point out that long-distance shipping apparently introduced the cholera bacterium into Peruvian coastal waters in 1991, causing the first outbreak of cholera in South America for nearly 100 years. Similarly, in the 1980s intercontinental trade in used car tires introduced the East Asian mosquito vector for dengue fever, *Aedes albopictus,* into South America, the southern United States, and Africa.

McNeely (2000) notes that agreements under the World Trade Organization could offer some help in dealing with exotic species but points out that bans and restrictions must be founded on science-based risk assessment. However, such risk assessment often is expensive and is still problematic; for example, a risk assessment for proposed importation of raw Siberian larch cost the U.S. government about $500,000 (Jenkins 1996).

Yu (1996) further suggests that the General Agreement on Tariffs and Trade contains important provisions to protect the environment and human health, and these might be expanded to deal with exotic species. These provisions include the Agreement on Sanitary and Phytosanitary Measures, the Agreement on Technical Barriers to Trade, and Article 20: General Exceptions, which protects the right of members to take any measures "necessary to protect human, animal, or plant life or health." McNeely (2000) challenges those concerned about invasive species to bring the invasives issue to the notice of the World Trade Organization and others responsible for setting world trade policy so that the issue can be dealt with more effectively.

Interactions

Drivers of change rarely act independently, and often it is the interaction of numerous factors that influences directions and rates of change. Sala et al. (2000) point out that these interactions represent one of the largest uncertainties in projections of future biodiversity change. Thus, for instance, increased transport of organisms via trade corridors couples with modification of local environments to facilitate invasion by suites of nonnative species. These nonnative species further modify the invaded system and may facilitate further invasion. The potential for invasional meltdown is discussed by Simberloff and Von Holle (1999) and supported by data from the North American Great Lakes (Ricciardi and MacIsaac 2000).

An example of how different drivers interact can be found in the Bay Area of California, where increasing nitrogen emissions from vehicles are deposited on native ecosystems downwind of the major metropolitan areas and act as a slow-release fertilizer (Weiss 1999). On the naturally low-nutrient serpentine soils found in the area, this fertilization allows the invasion and increased growth of nonnative grasses, which eliminate native low-growing forbs. However, cattle grazing maintains a more diverse native community by keeping the nonnative grasses in check. The serpentine grasslands are the only habitat for the federally listed bay checkerspot butterfly (*Euphydryas editha bayensis*), and continued cattle grazing is considered an essential component of habitat management for this species. When cattle grazing was removed from one of the prime habitat areas, increased nonnative grass growth led to a near extinction of the butterfly in that area. Thus, atmospheric pollution, land management, invasive species, and endangered species conservation are all intimately interlinked in this system. Although Weiss's studies are preliminary, it seems likely that many other such interactions are likely to occur in an increasingly urbanized and altered world.

Carlton (2000) summarizes the various factors that are likely to influence invasions in a marine setting; these are listed in Box 12.1. Similarly, Richardson et al. (2000) pro-

Box 12.1. Potential Responses of Biological Invasions to the Drivers of Global Change in the Oceans

Fisheries: Overfishing and Its Collateral Impacts

- *Enhance or depress invasions:* Reduction of keystone species reduces pressure on prey populations; resulting cascading effects (including shifts in abundance and diversity of multiple species) make the community more or less susceptible to invasion, depending on resulting matrix of competitors, predators, resource availability, and other factors.

Chemicals: Chemical Pollution and Eutrophication (Land and Ocean Derived)

- *Enhance invasions:* Local or regional extinction of species and depressed abundances of other species create opportunities for successful establishment of new invasions; eutrophic conditions specifically permit establishment of new invaders such as dinoflagellates and other phytoplankters.
- *Depress invasions:* Native weedy species enhanced by eutrophic conditions prevent establishment of new invasions.

(*continues*)

Box 12.1. Continued

Physical Destruction: Habitat Destruction and Fragmentation

- *Enhance invasions:* Local or regional extinction of species and depressed abundances of other species create opportunities for successful establishment of new invasions.
- *Depress invasions:* Native weedy species enhanced by disturbed conditions prevent establishment of new invasions.

Biological Invasions

- *Enhance or depress invasions:* New invasions either aid the establishment of other invasions (invasion facilitation) or lead to the establishment of a new competitor or predator that depresses the ability of additional new congeners, confamilials, or ecoequivalents to invade (invasion inhibition).

Global Climate Change

- *Enhance invasions under warmer conditions (A):* Warmer-water nonindigenous species become more abundant where established and expand ranges to now-warmer higher latitudes. Invasions newly entering higher latitudes may interact with cold-adapted neogenotypes of nonindigenous species, leading to their extinction (genetic swamping) or continued existence only in higher-latitude refugia.
- *Enhance invasions under warmer conditions (B):* Conversely, lower-latitude nonindigenous populations may become extinct as waters become too warm, permitting new invasions of other warmer-water or eurythermal taxa.
- *Enhance or depress invasions under changing patterns of primary production, altered salinity regimes from changing precipitation patterns, and other changes:* New primary trophodynamic regimes, new patterns and processes of estuarine oceanography (e.g., relative to altered salinity dynamics, particularly the scale of horizontal intrusion of salt wedge), and other physicochemical conversions either enhance or depress new invasions.

Source: Carlton (2000).

vide a conceptual model of the likely interactions between different elements of change and the range and density of a given invasive species, as summarized in Figure 12.5. Barrett (2000) adds the further potential complication that genetic changes may occur more rapidly under changing climatic conditions, and suggests the production of a "hybrid soup" from which could arise genetic combinations with novel phenotypes. These may add further opportunities for the arrival or development of novel invasive species.

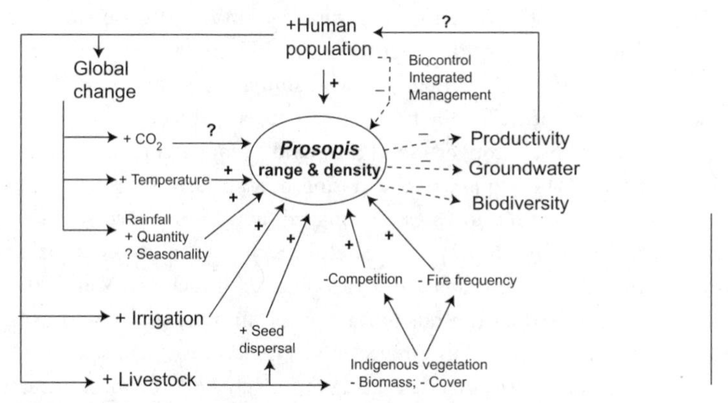

Figure 12.5. Factors leading to increases or decreases in the range and abundance of *Prosopis* spp. in South Africa. Figure from Richardson et al. (2000).

Invasives and Human Health

McMichael and Bouma (2000) discuss the three main pathways by which invasive species can affect human health:

- Changes in exposure to infectious disease agents, either by the emergence or spread of infectious agents or, in the case of vectorborne infections, by the spread of vector species or alternative host species.
- Damage by invasive species to natural or managed food-producing ecosystems.
- Exposure to exo-biotoxins produced by invasive species.

McMichael (1997), McMichael et al. (1999), and McMichael and Bouma (2000) show how human activities are significantly changing the risks associated with human disease organisms. They indicate that climate change will increase some infectious diseases, including malaria, and that land use change and greater free trade are also important drivers increasing infectious diseases. Various seemingly new infectious diseases have appeared in the past few decades in response to land clearance, extensions of agriculture and irrigation, intensive food production, biomedical technology, and human mobility. Meanwhile, the geographic range of many established infectious diseases, such as cholera, dengue, malaria, and various types of encephalitis, has been extended by land use, regional climate change, long-distance trade and travel, demographic shifts, and social (especially urban) behaviors. Invasive species also have the potential to decrease food yields and food storage or to produce foodborne biotoxins, which also pose risks to human health.

McMichael and Bouma (2000) provide numerous examples in which recent land use changes have facilitated the invasion and spread of novel disease organisms. For instance, forest clearance in South America has mobilized various viral hemorrhagic fevers that previously circulated benignly in wild animal hosts. For example, the Junin virus, which causes Argentine hemorrhagic fever, naturally infects wild mice (*Callomys callosus*). However, extensive conversion of grassland to maize cultivation stimulated a population increase of the mouse, thus exposing human farm workers to this "new" virus. In the past 35 years, the land area carrying this new human disease has expanded sevenfold, and several hundred people are infected each year, up to one-third of whom die. More recently, the virulent Guaranito virus entered humans in Venezuela, probably via chronically infected cotton rats that increased in numbers as land use changes provided an extended habitat and improved food supplies for the rats.

The human role in inadvertently decreasing the effectiveness of antibiotics, through overuse in clinical medicine and animal husbandry, is another growing threat to the vulnerability of humans to infectious diseases. McMichael and Bouma (2000 p. 208) conclude that "continued human encroachment on and disruption of the natural environment, and extended physical mobility and dissemination of people, goods, insects, and microbes around the world, are creating a spectrum of new and resurgent infectious disease hazards for human populations." Daszak, Cunningham, and Hyatt (2000) also point out that similar processes provide increasing threats of "pathogen pollution" not only for human populations but also for wildlife, with potentially important impacts on regional biodiversity.

Economics and Ecosystem Services

Accounting for the relative costs and benefits of invading species is a complex process that is still largely undeveloped. Much work has been done in the United States, at least, on the cost–benefit ratios for various forms of invasive specie management (although the distributions of these species continue to be ignored). The OTA (1993) summarizes these ratios and indicates a very wide range of cost–benefit ratios, although in nearly all cases the benefits of control far outweigh the costs involved. Pimentel et al. (2000) estimate the environmental damage and loss caused by invasive species in the United States at $137 billion per annum. This strongly suggests that significantly increased investment in managing invasive species is economically justified. However, those paying the costs, often the taxpaying public, may not always be the primary beneficiaries, and those who earned the benefits from the invasives in the first place end up paying almost none of the costs. Naylor (2000) further argues that there are significant economic benefits of both increased prevention of further invasions and early control of existing invasions.

Naylor (2000) and Zavaleta (2000) discuss novel approaches to considering the economics of invasive species. Naylor (2000) proposes some innovative approaches and outlines the additional complexities of dealing with invasives under global change scenarios. She discusses the complex issue of how we attach value to things, and points out that a wide array of possible valuations exist, including direct use values, which arise from the use of an asset (including consumptive and nonconsumptive uses); indirect use values, which include ecosystem services that are lost as a result of invasive species; and nonuse values, which include such things as existence value (e.g., the value placed on an ecosystem free of invasive species). The analyses of these values and determination of benefits and costs are difficult but not hopeless, and in order to make progress we need greater collaboration between resource managers, ecologists, and economists. These analyses will be of increasing importance as we decide where to put resources into dealing with a problem that will continue to grow.

Zavaleta (2000) provides a detailed case study of the economic benefits and costs of a woody tree, the tamarisk, that has invaded the southwestern United States. She makes a compelling case that even today there would be considerable benefits in controlling this pernicious invader. With global change scenarios of increased demands for water, decreasing precipitation, and increased warming, the benefits of a vigorous control effort become even more persuasive. Such analyses will become increasingly important for deploying limited resources in control efforts. Zavaleta (2000) notes that economic analyses are not the only tool that we will use in decision making relating to control, but they are an important one, and they must address all aspects of the problem.

Richardson et al. (2000) discuss the social and economic dimensions of invasive species issues in the South African context, particularly addressing the issues of who gets the benefit and who bears the cost. Currently, the agency or individual involved in introducing species stands to gain significantly in terms of benefits but is not required to take responsibility for any follow-on costs, such as those of subsequent control if the species becomes invasive or of compensation for lost production by other land users.

McNeeley (2000) points out that although it is important to identify the costs and benefits, such determination does not automatically constitute a decision because value judgments and distributional questions are nearly always involved. Furthermore, the magnitude of the costs may be so high as to render an action politically unacceptable, even when the benefits are likely to be even greater. Part of the problem is that the benefits are likely to be widely distributed throughout the population over a period of many years, whereas the costs of control may need to be paid up front over a short period of time.

Clout and Lowe (2000) examined invasions in New Zealand, one of the world's most severely affected regions. Here, the impacts are so obvious and detrimental that the public has agreed to draconian measures to eliminate or control invaders where possible and to monitor and limit further invasions. In this case, economic, aesthetic, and conservation issues are involved. It is often stated that people are motivated primarily by financial considerations, but the New Zealand experience indicates clearly that decisions are not always driven by economics alone. Considerations of local identity and preservation of endemic biota, for instance, can be of primary importance.

Clearly, the economic and social dimensions of invasive species play an important role in how they are viewed and dealt with by society. The examples provided here indicate that it is possible to demonstrate clear economic benefits from controlling invasive species and hence for ensuring that further potential invaders are not transported into new areas. Social and political recognition of invasive species as a problem worthy of concern is a key factor in determining the extent to which they are dealt with effectively.

Conclusion

In this chapter we have briefly outlined the ways in which invasive species are likely to interact with other components of global change. Clearly, the picture we have drawn is one that includes much uncertainty and reflects our inability to predict the outcomes of novel changes to the world's ecosystems and of numerous complex interacting processes. Nevertheless, we have shown that invasive species will continue to be transported around the globe and invade both natural and managed systems and often are likely to be aided in this by climatic, land use, and other changes. Our capacity to deal with these problems depends not only on the development of sufficient understanding and tools to detect, prevent, and control invasions but also on the degree of political and social will that develops at all levels from local to global. These are all large tasks, but their importance cannot be overstated.

So what can we do about invasive species in a context of global change? We believe several concrete actions are necessary to ensure that the mistakes of the past are not repeated continually in the future:

- A more rigorous process of risk assessment is needed in relation to any deliberate introduction of species to new areas (not just between countries but between states or regions). This, coupled with benefit–cost analyses, allows more informed decision-making in relation to species introductions.
- Invasive species must be factored into the decision-making processes surrounding land use planning and development. Given that land use changes provide prime opportunities for invasion, the potential risks, costs, and benefits must be more explicitly considered.
- Invasive species must be brought increasingly to the attention of the World Trade Organization, governments, and multinational organizations and companies so that they too can make rational decisions on trade and transportation in relation to the risks posed by invasive species.
- Considerations of potential impacts of invasive species must include projections based not just on current environmental conditions but also on likely changes in climate and land use.
- Scientists need to become smarter at considering potential scenarios based on multiple levels of uncertainty. Even qualitative analyses of likely outcomes can provide useful input to decision-making processes.

Acknowledgments

This chapter is based on a synthesis of material presented in Mooney and Hobbs (2000), and all figures used here were originally published in that volume. We thank all the contributors to that volume for their input and inspiration.

References

Barrett, S. R. 2000. "Microevolutionary influences of global change on plant invasions," in *Invasive Species in a Changing World,* edited by H. A. Mooney and R. J. Hobbs, 115–140. Island Press, Washington, DC.

Bazzaz, F. A. 1990. "The response of natural ecosystems to the rising global CO_2 levels," *Annual Review of Ecology and Systematics* 21:167–196.

Bright, C. 1998. *Life Out of Bounds.* W.W. Norton, New York.

Carlton, J. T. 1989. "Man's role in changing the face of the ocean: Biological invasions and implications for conservation of near-shore environments," *Conservation Biology* 3:265–273.

———. 2000. "Global change and biological invasions in the oceans," in *Invasive Species in a Changing World,* edited by H. A. Mooney and R. J. Hobbs, 31–54. Island Press, Washington, DC.

Carlton, J. T., and J. B. Geller. 1993. "Ecological roulette: The global transport of nonindigenous marine organisms," *Science* 261:78–82.

Clout, M. N., and S. J. Lowe. 2000. "Invasive species and environmental changes in New Zealand," in *Invasive Species in a Changing World,* edited by H. A. Mooney and R. J. Hobbs, 369–384. Island Press, Washington, DC.

Cox, G. W. 1999. *Alien Species in North America and Hawaii: Impacts on Natural Ecosystems.* Island Press, Washington, DC.

Crutzen, P. J., and M. O. Andreae. 1990. "Biomass burning in the tropics: Impact on atmospheric chemistry and biogeochemical cycles," *Science* 250:1669–1678.

D'Antonio, C. M. 2000. "Fire, plant invasions and global changes," in *Invasive Species in a Changing World,* edited by H. A. Mooney and R. J. Hobbs, 65–94. Island Press, Washington, DC.

D'Antonio, C. M., and P. M. Vitousek. 1992. "Biological invasions by exotic grasses, the grass/fire cycle, and global change," *Annual Review of Ecology and Systematics* 23:63–87.

Daszak, P., A. Cunningham, and A. Hyatt. 2000. "Emerging infectious diseases of wildlife: Threats to biodiversity and human health," *Science* 287:443–449.

Dukes, J. S. 2000. "Will the increasing atmospheric CO_2 concentration affect the success of invasive species?" in *Invasive Species in a Changing World,* edited by H. A. Mooney and R. J. Hobbs, 95–114. Island Press, Washington, DC.

French, H. 2000. *Vanishing Borders: Protecting the Planet in the Age of Globalization.* W.W. Norton, New York.

Hobbs, R. J. 1999. "Restoration of disturbed ecosystems," in *Ecosystems of the World* 16: *Disturbed Ecosystems,* edited by L. Walker, 673–687. Elsevier, Amsterdam.

———. 2000. "Land use changes and invasions," in *Invasive Species in a Changing World,* edited by H. A. Mooney and R. J. Hobbs, 55–64. Island Press, Washington, DC.

Hobbs, R. J., and L. Atkins. 1988. "Effect of disturbance and nutrient addition on native and introduced annuals in plant communities in the Western Australian wheatbelt," *Australian Journal of Ecology* 13:171–179.

Hobbs, R. J., S. L. Gulmon, V. J. Hobbs, and H. A. Mooney. 1988. "Effects of fertiliser addition and subsequent gopher disturbance on a serpentine annual grassland community," *Oecologia* (Berlin) 75:291–295.

Hobbs, R. J., and L. F. Huenneke. 1992. "Disturbance, diversity and invasion: Implications for conservation," *Conservation Biology* 6:324–337.

Hobbs, R. J., and H. A. Mooney. 1995. "Spatial and temporal variability in California annual grassland: Results from a long-term study," *Journal of Vegetation Science* 6:43–57.

———. 1996. "Effects of episodic rainfall events on Mediterranean-climate ecosystems," in *Timescales in Biological Responses to Water Constraints,* edited by J. Roy, J. Aronsen, and F. DiCastri, 71–85. SPB Academic Publishing, Amsterdam.

Hobbs, R. J., and D. A. Norton. 1996. "Towards a conceptual framework for restoration ecology," *Restoration Ecology* 4:93–110.

Horovitz, C. C. 1997. "The impact of natural disturbances," in *Strangers in Paradise: Impact and Management of Nonindigenous Species in Florida,* edited by D. Simberloff, D. Schmitz, and T. Brown, 63–74. Island Press, Washington, DC.

Huenneke, L. F., S. P. Hamburg, R. Koide, H. A. Mooney, and P. M. Vitousek. 1990. "Effects of soil resources on plant invasion and community structure in Californian serpentine grassland," *Ecology* 71:478–491.

Hunt, R., W. Hand, M. A. Hannah, and A. M. Neal. 1991. "Response to CO_2 enrichment in 27 herbaceous species," *Functional Ecology* 5:410–421.

Jenkins, P. T. 1996. "Free trade and exotic species introductions," *Conservation Biology* 10:300–302.

Kalin Arroyo, M. T., C. Marticorena, O. Matthei, and L. Cavieres. 2000. "Plant invasions in Chile: Present patterns and future prediction," in *Invasive Species in a Changing World,* edited by H. A. Mooney and R. J. Hobbs, 385–424. Island Press, Washington, DC.

Karieva, P. M., J.G. Kongsolver, and R. B. Huey, eds. 1993. *Biotic interactions and global change.* Sinauer Associates, Sunderland, MA.

Koch, G. W., and H. A. Mooney, eds. 1996. *Ecosystem Responses to Elevated* CO_2*.* Academic Press, London.

Kolar, C. S., and D. M. Lodge. 2000. "Freshwater nonindigenous species: Interactions with other global changes," in *Invasive Species in a Changing World,* edited by H. A. Mooney and R. J. Hobbs, 3–30. Island Press, Washington, DC.

Körner, C., and F. A. Bazzaz, eds. 1996. *Carbon Dioxide, Populations and Communities.* Academic Press, San Diego, CA.

Leemans, R., and G. Zuidema. 1995. "Evaluating changes in land cover and their importance for global change," *Trends in Ecology and Evolution* 10:76–81.

Lonsdale, W. M. 1993. "Rates of spread of an invading species: *Mimosa pigra* in northern Australia," *Journal of Ecology* 81:513–521.

Mack, R. N. 2000. "Assessing the extent, status, and dynamism of plant invasions: Current and emerging approaches," in *Invasive Species in a Changing World,* edited by H. A. Mooney and R. J. Hobbs, 141–170. Island Press, Washington, DC.

McMichael, A. J. 1997. "Global environmental change and human health: Impact assessment, population vulnerability, and research priorities," *Ecosystem Health* 3:200–210.

McMichael, A. J., B. Bolin, R. Costanza, G. C. Daily. 1999. "Globalization and the sustainability of human health," *BioScience* 49:205–210.

McMichael, A. J., and M. J. Bouma. 2000. "Global change, invasive species and human health," in *Invasive Species in a Changing World,* edited by H. A. Mooney and R. J. Hobbs, 191–210. Island Press, Washington, DC.

McNeely, J. A. 2000. "The future of alien invasive species: Changing social views," in *Invasive Species in a Changing World,* edited by H. A. Mooney and R. J. Hobbs, 171–190. Island Press, Washington, DC.

Meinesz, A. 1999. *Killer Algae: The True Tale of a Biological Invasion* (translated by D. Simberloff). University of Chicago Press, Chicago.

Mooney, H. A., and R. J. Hobbs, eds. 2000. *Invasive Species in a Changing World.* Island Press, Washington, DC.

Naylor, R. L. 2000. "The economics of alien species invasions," in *Invasive Species in a Changing World,* edited by H. A. Mooney and R. J. Hobbs, 241–260. Island Press, Washington, DC.

OTA. 1993. *Harmful Non-Indigenous Species in the United States.* U.S. Government Printing Office, Washington, DC.

Parry, M. L., et al. 1996. "Global and regional land-use responses to climate change," in *Global Change and Terrestrial Ecosystems,* edited by B. Walker and W. Steffen, 466–483. Cambridge University Press, Cambridge, UK.

Pimentel, D., L. Lach, R. Zuniga, and D. Morrison. 2000. "Environmental and economic costs of nonindigenous species in the United States," *BioScience* 50:53–65.

Pysek, P., K. Prach, and P. Simlauer. 1995. "Relating invasion success to plant traits: An analysis of the Czech alien flora," in *Plant Invasions: General Aspects and Special Problems,* edited by P. Pysek, M. Rejmánek, and M. Wade, 39–60. Academic Publishing, Amsterdam.

Ricciardi, A., and H. J. MacIsaac. 2000. "Recent mass invasion of the North American Great Lakes by Ponto-Caspian species," *Trends in Ecology and Evolution* 15:62–65.

Richardson, D. M., et al. 2000. "Invasive alien species and global change: A South African perspective," in *Invasive Species in a Changing World,* edited by H. A. Mooney and R. J. Hobbs, 303–350. Island Press, Washington, DC.

Richardson, D. M., P. A. Williams, and R. J. Hobbs. 1994. "Pine invasions in the Southern Hemisphere: Determinants of spread and invadability," *Journal of Biogeography* 21:511–527.

Sala, O. E., et al. 2000. "Global biodiversity scenarios for the year 2100," *Science* 287:1770–1774.

Scherer-Lorenzen, M., A. Elend, S. Nollert, and E. D. Schulze. 2000. "Plant invasions in Germany: General aspects and impact of nitrogen deposition," in *Invasive Species in a Changing World,* edited by H. A. Mooney and R. J. Hobbs, 351–368. Island Press, Washington, DC.

Simberloff, D., D. Schmitz, and T. Brown, eds. 1997. *Strangers in Paradise: Impact and Management of Nonindigenous Species in Florida.* Island Press, Washington, DC.

Simberloff, D., and B. Von Holle. 1999. "Positive interactions of nonindigenous species: Invasional meltdown?" *Biological Invasions* 1:21–32.

Skole, D., and C. Tucker. 1993. "Tropical deforestation and habitat fragmentation in the Amazon: Satellite data from 1978 to 1988," *Science* 260:1905–1910.

Sutherst, R. W. 2000. "Climate change and invasive species: A conceptual framework," in *Invasive Species in a Changing World,* edited by H. A. Mooney and R. J. Hobbs, 211–240. Island Press, Washington, DC.

Taylor, K., and C. Potvin. 1997. "Understanding the long-term effect of CO_2 enrichment on a pasture: The importance of disturbance," *Canadian Journal of Botany* 75:1621–1627.

Turner, B. L. I., W. C. Clark, R. W. Kates, J. F. Richards, J. T. Matthews, and W. B. Meyer, eds. 1991. *The Earth as Transformed by Human Action.* Cambridge University Press, New York.

van Wilgen, B. W., R. M. Cowling, and C. J. Burgers. 1996. "Valuation of ecosystem services. A case study from South African fynbos ecosystems," *BioScience* 46:184–189.

Vitousek, P. M., C. M. D'Antonio, L. L. Loope, and R. Westbrooks. 1996. "Biological invasions as global environmental change," *American Scientist* 84:468–478.

Vitousek, P. M., H. A. Mooney, J. Lubchenco, and J. M. Melillo. 1997a. "Human domination of Earth's ecosystems," *Science* 277:494–499.

Vitousek, P. M., C. M. D'Antonio, L. L. Loope, M. Rejmánek, and R. Westbrooks. 1997b. "Introduced species: A significant component of human-caused global change," *New Zealand Journal of Ecology* 21:1–16.

Weiss, S. B. 1999. "Cars, cows, and checkerspot butterflies: Nitrogen deposition and management of nutrient-poor grasslands for a threatened species," *Conservation Biology* 13:1476–1486.

Whisenant, S. G. 1990. "Changing fire frequencies on Idaho's Snake River plains: Ecological and management implications," in *Proceedings of a Symposium on Cheatgrass Invasion, Shrub Die-off, and Other Aspects of Shrub Biology and Management,* edited by E. D.

McArthur, E. M. Romney, S. D. Smith, and P. T. Tueller, 4–10. U.S. Forest Service Gen. Tech. Rep. INT-276. Intermountain Forest and Range Experiment Station, Ogden, UT. United States Forest Service, Washington, DC.

WRI (World Resources Institute). 1994. *World Resources: 1994–95.* Oxford University Press, New York.

Yu, D. W. 1996. "New factor in free trade: Reply to Jenkins," *Conservation Biology* 10:303–304.

Zavaleta, E. 2000. "Valuing ecosystem services lost to *Tamarix* invasion in the United States," in *Invasive Species in a Changing World,* edited by H. A. Mooney and R. J. Hobbs, 261–302. Island Press, Washington, DC.

13

A Global Strategy on Invasive Alien Species: Synthesis and Ten Strategic Elements

Jeffrey A. McNeely, Harold. A. Mooney, Laurie E. Neville, Peter Johan Schei, and Jeffrey K. Waage

The spread of invasive alien species (IAS) is creating complex and far-reaching challenges that threaten both the natural biological riches of the earth and the well-being of its citizens. Although the problem is global, the nature and severity of the impacts on society, economic life, health, and natural heritage are distributed unevenly across nations and regions. Thus, some aspects of the problem necessitate solutions tailored to the specific values, needs, and priorities of nations, whereas others call for consolidated action by the larger world community. Preventing the international movement of IAS and coordinating a timely and effective response to invasions will entail cooperation and collaboration between governments, economic sectors, nongovernment organizations (NGOs), and international treaty organizations. The Global Strategy on Invasive Alien Species is based on contributions from the team leaders of the main components addressed under Phase I of the Global Invasive Species Program (GISP) and the contributions of a wide constituency of experts. This contribution summarizes key findings that address the threats that must be considered for dealing with the complex problems caused by IAS.

IAS Threaten Biodiversity, Food Security, Health, and Economic Development

The spread of IAS is recognized as one of the greatest threats to the ecological and economic well-being of the planet. These species are causing enormous damage to biodiversity and the valuable natural agricultural systems on which we depend. Direct and indirect health effects are increasingly serious, and the damage to nature is often irreversible. The effects are exacerbated by global change and chemical and physical disturbance to species and ecosystems.

Continuing globalization, with increasing trade, travel, and transport of goods across borders, has brought tremendous benefits to many people. However, it has also facilitated the spread of IAS, with increasing negative impacts. The problem is global in scope, and international cooperation is needed to supplement the actions of governments, economic sectors, and individuals at national and local levels. IAS are found in nearly all major taxonomic groups of organisms. Although only a small percentage of species that are moved across borders become invasive, they may have extensive impacts. These effects can be devastating; studies in the United States and India show that the economic costs of IAS in these countries amount to approximately US$130 billion per year.

Consolidated Action for Preventing the Spread of IAS Is Urgent

Preventing introduction of potentially invasive alien species is by far the preferred strategy. To prevent spread, every alien species should be treated as potentially invasive unless convincing evidence indicates that this is not so. For deliberate introductions it is recommended that standardized risk analysis (RA) and risk management procedures (RMPs) be developed, perhaps based on the RISK ANALYSIS and RMPs developed under the Cartagena Protocol of the Convention on Biodiversity (CBD). Preventive measures must be taken at both the source and the destination.

For inadvertent introductions, new and innovative strategies and actions must be developed in cooperation with the trade, travel, tourism, and transport sectors. Awareness raising, legislation, information, education, and training are essential areas to address. Harmonized standards for preventive measures in practical operations in each economic sector should be developed at the international and national level. Cross-sectoral coordination and cooperation are imperative. Mechanisms, procedures, and regulatory measures for achieving synergies and efficiency are key strategic tools for achieving the goal of national biosecurity. The authorities responsible for biodiversity management should cooperate with the sectors of health and primary production to seek synergy in preventive actions.

Eradication Is Difficult and Expensive but Possible; Rapid Response Is Crucial

Because immediate response is more cost-effective and more likely to succeed than action after a species has become established, we recommend an early warning system for IAS. Containment action often is needed for a successful eradication program. Such a program must be science based and have a reasonable chance of success. The involvement of all relevant stakeholders is essential, and public support and acceptance of eradication methods are also important. Monitoring and control after initial efforts often are necessary, and restoration of affected systems is an important consideration.

Containment, Suppression, and Control Are Second Options, but Benefits Often Exceed Costs

Given the high complexity of the ecological characteristics of both IAS and the habitats and species they affect, control measures must be developed and applied on the basis of the best current scientific understanding. Specific cost–benefit analyses should be developed and applied for eradication and control programs for IAS.

Selection of control methods must also be based on thorough scientific knowledge. For chemical control the possible problem of negative effects on nontarget species and the potential development of resistant types and strains must be addressed carefully. For biological control the possibility of the control agent itself becoming invasive must be avoided. An integrated management approach to IAS involving a combination of mechanical, chemical, and biological control measures often is most appropriate. Careful monitoring and coordination are needed. Because the cost and benefit factor influences decisions that are often very difficult politically, the criteria for making such decisions should be developed carefully.

Comprehensive International and National Action Is Needed

Numerous international and regional agreements, regulations, decisions, and recommendations are already addressing the IAS problem. However, coordination of implementation and practical cooperation between those responsible for these instruments are insufficient. Practical prevention, eradication, and control measures are also inadequate. Therefore, we recommend a consolidated action plan. The CBD and the International Plant Protection Convention (IPPC) could take the lead, but trade, transport, travel, and other economic sectors must be closely involved. Other institutions, including the United Nations Environment Programme (UNEP), World Trade Organization (WTO), Food and Agriculture Organization (FAO), and International Maritime Organization (IMO) are key components at the international level. These institutions are supported by other international NGOs, such as the World Conservation Union (IUCN), the World Wildlife Fund (WWF), Wetlands International, Conservation International, and The Nature Conservancy.

Likewise, at the national level, consolidated and coordinated action is needed. This could be part of a national biodiversity strategy and action plan, with close involvement of the economic sectors and identifying people responsible for operative actions involving potential IAS as a key prerequisite. Clear responsibilities for each relevant sector should be identified.

Insurance mechanisms and clear liability regulations for the spread of IAS are almost nonexistent, presenting a major deficiency for controlling the problem. Governments should cooperate with the insurance sector to find solutions, beginning with feasibility studies.

Capacity and expertise to deal with IAS are insufficient in many countries. Capacity building and further research on the biology and control of IAS and biosecurity issues therefore should be given attention and priority. This also relates to financial institutions and other organizations responsible for environment and development cooperation at national and international levels.

An international information system regarding the biology and control of IAS is urgently needed. Tools, mechanisms, best management practices, control mechanism techniques, and resources should be provided and exchanged. The information system must be linked to the Clearing House Mechanism of the CBD.

Awareness raising and education regarding IAS should be given high priority in action programs, and economic tools and incentives for prevention are urgently needed.

GISP has contributed extensively to the knowledge and awareness of IAS and has developed a management toolkit to address the problem. However, many challenges remain to be addressed, and a continuation of the program is recommended. We propose activities to implement five global initiatives:

- Global access to information on IAS threats and their prevention and management
- Directed action at key pathways of IAS introduction, through public and private sector cooperation
- Acceleration of critical research and its dissemination
- Awareness raising and support to policy development
- Building cooperation between institutions towards a global biosecurity platform to mitigate the threat of IAS

We also propose activities to develop national capacity and regional cooperation in the prevention and management of IAS. This program proposes to help governments and development agencies identify and initiate national and regional projects to mitigate threats posed by IAS. It will support existing projects and initiatives and will develop national and international capacity and international networking. This will be achieved through the use and adaptation of the GISP Management Toolkit, Databases, and Early Warning System developed in Phase I and will use outputs generated by international GISP activities. Major components of regional and national initiatives will include national strategy development; surveys, inventory, and taxonomic support; pilot projects on invasive alien prevention and management including habitat restoration; and public outreach and capacity-building programs.

Ten Strategic Responses to Address the IAS Problem[1]

Under the CBD, the Conference of Parties agreed to a set of Guiding Principles, to which GISP has contributed. With these principles in mind, and incorporating the efforts of numerous experts who contributed to the final reports of the 11 GISP components during the GISP Synthesis conference held in Cape Town in September 2000, we have developed 10 strategic responses. These elements are intended to guide policymakers and managers in responding to the growing challenge of invasive alien species.

Element 1: Build Management Capacity

Successfully addressing the problem of IAS entails both national will and capacity to act. Because the problem is a global one, involving almost all nations, an adequately funded and vigorous international response to building management and operational capacity in all countries clearly is a high priority. At a national level, capacity building initiatives can draw on the experience of other countries, the GISP Toolkit of Best Prevention and Management Practices, and other resources. Such initiatives should include the following elements:

- Designing and establishing a rapid response mechanism to detect and respond immediately to the presence of potentially invasive alien species as soon as they appear. At a national level, this entails establishment of easily accessible funds for emergency actions, regulatory support, and interdepartmental coordination on IAS that can quickly identify and give authority to a lead agency or agencies.
- Designing educational programs to build capacity, including training courses aimed at agency field staff, managers, specialists, and policymakers.
- Building the capacity to formulate and implement educational programs aimed at community empowerment (e.g., in early detection and control) and at developing school and university curricula and creating academic chairs and student fellowships in IAS biology.
- Developing national institutions that bring together biodiversity specialists and agricultural quarantine specialists to cooperate in addressing the provisions of the CBD and other relevant agreements (e.g., building environmental elements into pest risk assessment). Existing staff may need retraining in IAS prevention and management skills.
- Establishing IAS specialist positions in national resource management agencies.
- Building border control and quarantine capacity, ensuring that all those involved in agricultural quarantine, customs, or food inspection are made aware of the provisions of the CBD and its Cartagena Protocol on Biosafety and the implications of these provisions for their work.

Element 2: Build Research Capacity

The current knowledge on IAS must be further developed with a cross-sectoral and multidisciplinary approach in order to provide the tools needed to address this pervasive issue. Considerations for such an approach may include the following foci.

Institutional Framework and Collaboration

- Strengthening infrastructure for research on IAS (e.g., systematics, taxonomy, and ecology) at national and regional levels. An international committee to correlate and manage updated taxonomic nomenclature for all IAS would be a useful resource.
- Directing existing relevant research resources and products toward a focus on IAS by informing and engaging academic and national research institutions and exchange programs.
- Building academic groups, or "centers of excellence," on IAS biology and encouraging exchange and collaboration in the formulation of research approaches.
- Developing a mechanism for IAS mitigation and monitoring through cross-sectoral and interagency integrative research and management on an international scale.

Assessment and Prediction

- Building the capacity to identify, record, and monitor invasions and provide current lists of potential and established IAS.
- Determining the relative contribution of anthropogenic factors, natural factors, and their interaction to the spread of IAS.
- Improving the understanding of how and why species become established, investigating species that have the potential to become invasive and ecosystems that may be particularly vulnerable to invasion, and building an understanding of the mechanisms controlling lag times in the development and establishment of IAS.

Management: Early Detection, Assessment, Prevention, and Control

- Building research networks that incorporate the risk assessment, risk management, and research approach.
- Developing and improving techniques to eradicate and control IAS. Considerations include developing species-specific toxins and diseases, improving the basis on which biological control strategies are evaluated, and considering the limiting factors that affect the spread and geographic distributions of taxa.
- Developing better methods for excluding or removing alien species from traded goods, packaging material, ballast water, luggage, aircraft, ships, and other methods of transport.
- Developing methods for ecosystem restoration and sustainability after control meas-

ures, considering the use of native taxa in erosion control and restoration efforts in planning and collaboration with appropriate sectors (e.g., agroforestry, horticulture) and agencies.

Element 3: Promote Information Sharing

Much information about IAS is available. GISP has identified nearly 120 major sources of information on IAS that are accessible electronically. The information that could alert management agencies to the potential dangers of new introductions is not well known or is not widely shared or available in an appropriate format to enable governments to take prompt action (assuming they have the resources, necessary infrastructure, commitment, and trained staff to do so). Therefore, information sharing is essential. The following actions will facilitate information sharing:

- Building a distributed information system of linked regional and national databases on IAS, building on various sources of information (e.g., the Inter-American Biodiversity Information Network and the IUCN/Invasive Species Specialist Group Aliens Listserv). The Global Invasive Alien Species Information System (GIASIS) should serve a distributed network, set data standards, and facilitate the input and sharing of data. It should work in multiple languages and promote wide distribution of information to all interested parties using all available technology.
- Developing the GISP early warning system, including notification of new and political occurrences of IAS.
- As part of GIASIS, establishing a database of failure and success of different IAS eradication and control methods to ensure that all can learn from the experience and linking this database to the GISP toolkit.

Element 4: Develop Economic Policies and Tools

Species invasions are a consequence of economic decisions and have economic impacts. However, the costs of invasions are seldom reflected in market prices. Although prevention, eradication, control, mitigation, and adaptation all yield economic benefits, they are public goods. If left only to the market, the control of IAS, like the control of communicable human diseases, will be inadequately provided for. Because biological invasions often indicate market failure, an important part of any strategy to manage IAS is to make markets work for conservation wherever possible and to provide alternative solutions if markets do not exist and cannot be created. Therefore, GISP encourages countries to incorporate economic principles into their national strategies for addressing IAS, building on the following principles:

- *User pays:* Make those responsible for the introduction of economically harmful IAS liable for the costs they impose.

- *Full social cost pricing:* Ensure that prices of goods and services whose production or consumption worsens the damage of IAS reflect their true cost to society.
- *Precautionary principle:* Because of the potentially irreversible and high costs of IAS, it is important to base management and policy on the precautionary principle, which states that when an activity raises threats of harm to human health or the environment, precautionary measures should be taken even if some cause-and-effect relationships are not fully established scientifically.
- *Protection of the public interest:* Because the control of IAS yields benefits that are a public good, the public must invest in prevention, eradication, control, mitigation, and adaptation.
- *Subsidiarity:* Operate policies and management at the lowest level of government that can effectively deal with the problem.

Particular policies that governments may want to develop to reflect these principles include the following:

- *Developing appropriate property rights:* Ensure that use rights to natural or environmental resources include an obligation to prevent the spread of potential IAS.
- *Estimating social costs:* Assess the economic costs of actual or potential IAS.
- *Assigning liability:* Require importers and users of potential IAS to have liability insurance to cover the unanticipated costs of introductions or of activities that risk introductions.
- *Promoting empowerment:* Enable people injured by the spread of IAS to seek redress.
- *Applying price-based instruments:* To ensure that importers and users of known IAS take account of the full social cost of their activities, apply economic instruments such as commodity taxes, differential land use taxes, user charges, or access fees.
- *Applying precautionary instruments:* Where the risk of damage depends on the behavior of importers and users of IAS, apply precautionary instruments such as deposit refund systems or environmental assurance bonds.

Element 5: Strengthen National, Regional, and International Legal and Institutional Frameworks

Until recently, national legal measures have evolved in a reactive and piecemeal manner, responding to new problems and pathways relating to IAS. However, isolated unilateral action by individual states can never be sufficient to manage the full range of activities and processes that generate invasions. Coordination and cooperation between the relevant institutions are necessary to address possible gaps, weaknesses, and inconsistencies and promote greater harmonization between the many international instruments that address IAS. Strategies should aim to develop or strengthen legal and institutional frameworks at two major levels: national and regional or global.

Developing and strengthening national legal and institutional frameworks should include the following:

- A review of relevant policies, legislation, and institutions to identify conflicts, gaps, and inconsistencies and strengthen or develop effective national measures for IAS prevention, eradication, and control
- The establishment of a coordinating mechanism and process between different levels and departments of government
- Participation and access to relevant information by all stakeholders, including local communities, in the development and implementation of laws and policies
- Ensure control measures are in place to regulate and minimize the introduction of IAS at the point of origin (export), destination (import), or both
- Strict regulation of the movement and release of alien species domestically, especially in or near vulnerable ecosystems, between islands, and to protected areas
- Surveillance, monitoring, and early warning systems to detect the introduction of IAS and take emergency action, as necessary and appropriate
- Establishing an appropriate set of rights and responsibilities to address the impact of IAS along with supporting institutions, compensation mechanisms, and incentives and disincentives

Promoting coordination and cooperation at the international and regional level should involve the following:

- Encouraging a detailed review of possible differences, inconsistencies, or gaps between the mandates of major international and regional instruments relevant to IAS, with a view to encouraging their resolution
- Continuing to integrate and promote biodiversity in international standards and processes, including risk analysis
- Continuing to develop international guidance on standards and methods applicable to IAS
- Encouraging full discussion of a more comprehensive international approach
- Supporting the work of the IMO to develop a legal instrument on marine IAS and encouraging similar developments in other sectors

Element 6: Institute a System of Environmental Risk Analysis

Risk analysis and environmental impact assessment (EIA) procedures have already been adopted in many countries and mandated by certain international instruments. The challenge now is to apply them to address IAS prevention, eradication, and control. Risk analysis measures should be used to identify and evaluate the relevant risks of proposed activity regarding alien species and determine the appropriate measures that should be adopted. EIA plays an important role in the decisions to undertake specific processes or activities. Decision makers should ensure the use of strategic or project-specific EIAs in assessing the long-term and short-term impact of alien species introductions. To ensure the effective use of risk analysis and EIA, decision makers should consider the following:

- Examining methods of the WTO, IPPC, IMO, and other partners to promote the extension of risk analysis criteria and methods to all invasive taxa.
- Building on work undertaken by the plant and animal protection community to develop a rigorous process of risk analysis in relation to any deliberate introduction of species (not just between countries but within a country or region as well), including detailed analysis of the balance between benefits and costs. This assessment would allow more informed decision making in relation to IAS introduction, control, and management.
- Developing criteria to measure and classify impacts of IASs on natural ecosystems, including detailed protocols for assessing the likelihood of invasion in specific habitats or ecosystems. Where prediction protocols exist for landscapes comprising mosaics of ecosystems, predictions for the most vulnerable system in the landscape should dictate management decisions.
- Developing tools to factor IASs into the decision-making processes regarding land use planning and development.
- Investigating ways in which strategic and project-specific EIAs can be applied to unintentional introductions. For instance, assess large engineering projects, such as canals, tunnels, and roads that cross biogeographic zones, that might have the effect of mixing previously separated flora and fauna.

Element 7: Build Public Awareness and Engagement

Active public engagement is critical to successful IAS management. This strategy is intended to help states and organizations engage the public successfully and coordinate their efforts for greatest global benefit, leading to an informed public that supports ongoing actions to reduce the threat of IASs and key stakeholders who are actively engaged in implementation of IAS solutions. Attaining these desired outcomes will entail the following:

- Developing public awareness campaigns to support IAS management, including sharing information and coordinating information as appropriate to avoid contradiction and maximize efficiency
- Using appropriate pilot projects on IASs with high priority or visibility, or those affecting important native species, as a basis for raising public awareness, validating investment in rapid response and management systems, and building capacity through "learning by doing"

- Engaging key stakeholders, communities, and neighbors in creating solutions to the problem by linking IAS strategies wherever possible to integrated development programs, such as programs that emphasize poverty alleviation measures and other established societal priorities
- Building the capacity of local communities and groups to implement IAS management measures where they live
- Sharing experience in this strategy with other nations, states, and organizations through documentation, staff exchanges, and by other means

Element 8: Prepare National Strategies and Plans

The problems posed by IASs are not simply the responsibility of a ministry of environment or a natural resource management department. Rather, the problem extends through many economic sectors, both public and private. As with other biodiversity issues, successfully addressing IAS problems entails effective collaboration between these institutions. Drawing on experience gained in preparing National Biodiversity Strategies and Action Plans (NBSAPs), relevant agencies should collaborate, through an open consultive process, to prepare strategies and action plans for dealing with IASs or build elements for doing so into existing NBSAPs.

Elements to include in such strategies and plans include the following:

- Promoting cooperation within each country between sectors whose activities have the greatest potential to introduce IASs, including the military, economic development, forestry, agriculture, transport, health, tourism, and water supply.
- Coordinating the activities of government agencies with responsibility for human health, animal health, plant health, transport, tourism, trade, protected areas, wildlife management, water supply, and other fields relevant to IASs.
- Encouraging collaboration between different scientific disciplines and approaches that can contribute to addressing IAS problems and combining them to produce a framework for the assessment of vulnerability of systems or geographic regions to IASs. Multidisciplinary approaches should be promoted in this regard.
- Ensuring that the necessary information and policy guidance are provided to national delegations to sessions of the WTO and others responsible for setting world trade policy, with a particular focus on the Sanitary and Phytosanitary Agreement.
- Applying experience in agricultural, forestry, and human health systems to combating IASs in natural systems. For example, use quarantine facilities for agriculture to serve more broadly for all environmental pests.
- Fully involving environmental and developmental NGOs as means to address IAS issues.

At the regional and international levels, the relevant international organizations and NGOs could be more effective in addressing IAS problems by building collaboration and cooperation. This could include the following:

- Establishing close links between public health agencies (including the World Health Organization [WHO]) dealing with invasive pathogens and those dealing with other parts of the IAS issue, with a view to exchanging information about effective management approaches
- Working with the wide range of relevant international trade authorities and industry associations, with the goal of significantly reducing the risk that trade, travel, and tourism will facilitate the introduction and spread of IASs
- Encouraging, strengthening, and contributing to the development of collaborative industry standards of practice, guidelines, or codes of conduct that minimize or eliminate unintentional introductions
- Encouraging organizations such as the International Tropical Timber Organization, World Tourism Organization, FAO, Consultative Group on International Agricultural Research, United Nations Children's Fund, UNEP, and United Nations Educational, Scientific and Cultural Organization (UNESCO) to build IAS elements into their programs

Element 9: Build IAS Issues into Global Change Initiatives

Human activities are changing Earth in unprecedented ways. These changes are altering atmospheric composition (e.g., CO_2 concentrations, nitrogen deposition), changing the climate (e.g., rising temperatures, increased incidence of episodic storms), increasing the use of natural resources, changing land use (including fragmentation and altered fire regimes), and deliberately and inadvertently moving species around the globe. Global change is likely to increase opportunities for the transport and establishment of IASs. The interactions of IASs with other global change may occur in complex and unpredictable ways, acting as drivers of further change. Global change results from the cumulative impacts of local decisions, so the issues must be addressed at both international and local levels. Key actions in response to this need include the following:

- Articulating the interactions between IASs and other elements of global change (e.g., climate change, land use change)
- Quantifying the current and anticipated impacts of IASs at global and regional scales for incorporation into other global change projections
- Using scenario building as a means of incorporating uncertainty into projections of interactions between different elements of global change
- Ensuring that relevant international organizations with responsibility for global change issues (e.g., International Council of Scientific Unions, International Geosphere–Biosphere Program, WHO, UNEP, UNESCO, WWF, and FAO) include IASs as a component of global change, directly and through their member states
- Responding to global change issues without increasing the risks derived from IASs (e.g., carbon sequestration, biomass energy, mitigation of degraded lands)

Element 10: Promote International Cooperation

A wide range of approaches, strategies, models, tools, and potential partners are available for international cooperation. The most relevant approach varies for each situation.

Elements that would foster better international cooperation might include the following:

- Developing an international vocabulary, widely adopted. Note that the IPPC is promoting an initiative to encourage national agencies to use the internationally accepted phytosanitary vocabulary to facilitate communication. Wherever available, internationally adopted terminology and standards should be used in implementing legislation and regulations.
- Developing cross-sectoral collaboration between international organizations involved in trade, travel, and transport.
- Developing harmonization and links between the international institutions dealing with phytosanitary, biosafety, and biodiversity issues related to IASs and supporting them with strong links to coordinated national programs and their focal points.
- Developing joint work programs between relevant conventions, including the CBD, Ramsar Wetlands of International Importance, World Heritage, and the Convention on International Trade in Endangered Species of Wild Fauna and Flora (CITES).

Invasions often are relevant to biogeographic regions, not just jurisdictional country boundaries. Therefore, neighboring countries need to cooperate, and regional approaches to management must be encouraged, including

- Working toward regional IAS strategies.
- Identifying regional information requirements.
- Fostering regional cooperation in risk assessment, prevention, eradication, or control.
- Promoting regional cooperation in technologies and capacity building.
- Establishing at an international level a "Center for Invasive Alien Species" to provide rapid diagnosis and information on management of the spread and occurrence of new alien species threats. It would also support capacity-sharing efforts between countries in IAS prevention and management and regional quarantine capacity and systems.

Because IASs have become an issue of considerable global concern, bilateral and multilateral donor agencies should be encouraged to

- Support activities relating to sectoral and national policies on IASs
- Support better-coordinated approaches at the national level as a way to strengthen capacity for international cooperation
- Encourage intergovernmental cooperation in programs they fund
- Review planning processes with a view to ensuring that the programs they support will not include the intentional introduction of IASs and will minimize unintentional introductions

The strategy as developed via this holistic process highlights the dimensions of the problem and outlines a framework for mounting a global-scale response. Although both the problem and the scale of the solution may appear dauntingly complex, the issue presents an unparalleled opportunity to respond with actions that link biodiversity preservation with protection of the health and livelihood of the world's human populations.

Note

1. This synthesis and 10 strategic elements are derived from McNeely, J. A., H. A. Mooney, L. E. Neville, P. Schei, and J. K. Waage, eds. 2001. *A Global Strategy on Invasive Alien Species.* IUCN Gland, Switzerland, and Cambridge, UK, in collaboration with the GISP.

List of Contributors

David A. Andow
Department of Entomology
219 Hodson Hall
University of Minnesota
St. Paul, MN 55108, USA

Michael Browne
IUCN/ISSG, SEMS
University of Auckland
Private Bag 92019
Auckland, New Zealand

Françoise Burhenne-Guilmin
IUCN
Adenauerallee 214
5300 Bonn, Germany

James T. Carlton
Maritime Studies Program, Williams College
Mystic Seaport
P.O. Box 6000
Mystic, CT 06355, USA

Mick Clout
IUCN/ISSG, Biological Sciences/SEMS
University of Auckland
Private Bag 92019
Auckland, New Zealand

Matthew J. W. Cock
CABI Bioscience Centre, Switzerland
1 Rue des Grillons
CH-2800 Delémont, Switzerland

Silvana Dalmazzone
Department of Economics
University of Turin
Via Po 53
I-10124 Torino, Italy

Maj de Poorter
School of Environmental and Marine Science
University of Auckland
Private Bag 92019
Auckland, New Zealand

Eva Grotkopp
Section of Ecology and Evolution
University of California
Davis, CA 95616, USA

Steven I. Higgins
National Botanical Institute
Private Bag X7
Claremont 7735, South Africa

Richard J. Hobbs
School of Environmental Science
Murdoch University
WA 5151, Perth, Australia

Sarah Lowe
IUCN/ISSG, SEMS
University of Auckland
Private Bag 92019
Auckland, New Zealand

Richard N. Mack
Department of Biological Sciences
Washington State University
Pullman, WA 99164, USA

Jeffrey A. McNeely
IUCN Biodiversity Policy
Coordination Division
28 Rue Mauverney
CH-1196 Gland, Switzerland

Harold A. Mooney
Department of Biological Sciences
Stanford University
Stanford, CA 94305-5020, USA

Laurie E. Neville
Department of Biological Sciences
Stanford University
Stanford, CA 94305-5020, USA

Charles Perrings
Environmental Economics and
Management
University of York, Heslington
York YO1 5DD, United Kingdom

Michael J. Pitcairn
Biological Control Program
Department of Food and Agriculture,
State of California
3288 Meadowview Road
Sacramento, CA 95832, USA

Marcel Rejmánek
Section of Evolution and Ecology
University of California
Davis, CA 95616, USA

Anthony Ricciardi
Redpath Museum and McGill School
of Environment
McGill University
Montreal, Quebec H3A 2K1, Canada

David M. Richardson
Institute for Plant Conservation
Centre for Invasion Biology, Science
Faculty
University of Stellenbosch
Private Bag X1
Matieland 7602, South Africa

Gregory M. Ruiz
Smithsonian Environmental Research
Center
P.O. Box 28
Edgewater, MD 21037, USA

Peter Johan Schei
The Fridtjof Nansen Institute
P.O. Box 326
Fridtjof Nansens Vei 17
Lysaker N-1326, Norway

Clare Shine
Consultant, Environmental Law and Policy
37 Rue Erlanger
75016 Paris, France

Jeffrey K. Waage
Professor of Applied Ecology, Provost and Head of Department of Agricultural Sciences
Imperial College at Wye
Wye, Ashford Kent TN25 5AH, United Kingdom

Nattley Williams
United Nations Framework Convention on Climate Change
P.O. Box 260124
D-53153 Bonn, Germany

Mark Williamson
Department of Biology
University of York
Heslington, York YO1 5DD, United Kingdom

Rüdiger Wittenberg
CABI Bioscience Centre, Switzerland
1 Rue des Grillons
CH-2800 Delémont, Switzerland

SCOPE Series List

SCOPE 1–59 are now out of print. Selected titles from this series can be downloaded free of charge from the SCOPE Web site (http://www.icsu-scope.org).

SCOPE 1: *Global Environment Monitoring,* 1971, 68 pp
SCOPE 2: *Man-made Lakes as Modified Ecosystems,* 1972, 76 pp
SCOPE 3: *Global Environmental Monitoring Systems (GEMS): Action Plan for Phase I,* 1973, 132 pp
SCOPE 4: *Environmental Sciences in Developing Countries,* 1974, 72 pp
SCOPE 5: *Environmental Impact Assessment: Principles and Procedures,* Second Edition, 1979, 208 pp
SCOPE 6: *Environmental Pollutants: Selected Analytical Methods,* 1975, 277 pp
SCOPE 7: *Nitrogen, Phosphorus and Sulphur: Global Cycles,* 1975, 129 pp
SCOPE 8: *Risk Assessment of Environmental Hazard,* 1978, 132 pp
SCOPE 9: *Simulation Modelling of Environmental Problems,* 1978, 128 pp
SCOPE 10: *Environmental Issues,* 1977, 242 pp
SCOPE 11: *Shelter Provision in Developing Countries,* 1978, 112 pp
SCOPE 12: *Principles of Ecotoxicology,* 1978, 372 pp
SCOPE 13: *The Global Carbon Cycle,* 1979, 491 pp
SCOPE 14: *Saharan Dust: Mobilization, Transport, Deposition,* 1979, 320 pp
SCOPE 15: *Environmental Risk Assessment,* 1980, 176 pp
SCOPE 16: *Carbon Cycle Modelling,* 1981, 404 pp
SCOPE 17: *Some Perspectives of the Major Biogeochemical Cycles,* 1981, 175 pp
SCOPE 18: *The Role of Fire in Northern Circumpolar Ecosystems,* 1983, 344 pp
SCOPE 19: *The Global Biogeochemical Sulphur Cycle,* 1983, 495 pp
SCOPE 20: *Methods for Assessing the Effects of Chemicals on Reproductive Functions, SGOMSEC 1,* 1983, 568 pp
SCOPE 21: *The Major Biogeochemical Cycles and their Interactions,* 1983, 554 pp
SCOPE 22: *Effects of Pollutants at the Ecosystem Level,* 1984, 460 pp

SCOPE 47: *Long-Term Ecological Research. An International Perspective,* 1991, 312 pp
SCOPE 48: *Sulphur Cycling on the Continents: Wetlands, Terrestrial Ecosystems and Associated Water Bodies,* 1992, 345 pp
SCOPE 49: *Methods to Assess Adverse Effects of Pesticides on Non-target Organisms, SGOMSEC 7,* 1992, 264 pp
SCOPE 50: *Radioecology after Chernobyl,* 1993, 367 pp
SCOPE 51: *Biogeochemistry of Small Catchments: a Tool for Environmental Research,* 1993, 432 pp
SCOPE 52: *Methods to Assess DNA Damage and Repair: Interspecies Comparisons, SGOMSEC 8,* 1994, 257 pp
SCOPE 53: *Methods to Assess the Effects of Chemicals on Ecosystems, SGOMSEC 10,* 1995, 440 pp
SCOPE 54: *Phosphorus in the Global Environment: Transfers, Cycles and Management,* 1995, 480 pp
SCOPE 55: *Functional Roles of Biodiversity: a Global Perspective,* 1996, 496 pp
SCOPE 56: *Global Change Effects on Coniferous Forests and Grasslands,* 1996, 480 pp
SCOPE 57: *Particle Flux in the Ocean,* 1996, 396 pp
SCOPE 58: *Sustainability Indicators: a Report on the Project on Indicators of Sustainable Development,* 1997, 440 pp
SCOPE 59: *Nuclear Test Explosions: Environmental and Human Impacts,* 1999, 304 pp
SCOPE 60: *Resilience and the Behavior of Large-Scale Systems,* 2002, 287 pp
SCOPE 61: *Interactions of the Major Biogeochemical Cycles: Global Change and Human Impacts*, 2003, 384 pp
SCOPE 62: *The Global Carbon Cycle: Integrating Humans, Climate, and the Natural World,* 2004, 526 pp
SCOPE 63: *Alien Invasive Species: A New Synthesis*, 2005, 392 pp
SCOPE 64: *Sustaining Biodiversity and Ecosystem Services in Soils and Sediments*, 2003, 308 pp
SCOPE 65: *Agriculture and the Nitrogen Cycle*, 2004, 320 pp

SCOPE Executive Committee 2005–2008

President:
Prof. Osvaldo E. Sala (Argentina)

Vice-President:
Prof. Wang Rusong (China–CAST)

Past President:
Dr. Jerry M. Melillo (USA)

Treasurer:
Prof. Ian Douglas (UK)

Secretary General:
Prof. Mary C. Scholes (South Africa)

Members:
Prof. Wandera Ogana (Kenya–IGBP)
Prof. Annelies Pierrot-Bults (The Netherlands–IUBS)
Prof. Vinod P. Sharma (India)
Prof. Holm Tiessen (Germany)
Prof. Reynaldo Victoria (Brazil)

Index